Polymer Materials in Sensors, Actuators and Energy Conversion II

Polymer Materials in Sensors, Actuators and Energy Conversion II

Jung-Chang Wang

Basel • Beijing • Wuhan • Barcelona • Belgrade • Novi Sad • Cluj • Manchester

Jung-Chang Wang
Marine Engineering (D.M.E.)
National Taiwan Ocean
University (NTOU)
Keelung
Taiwan

Editorial Office
MDPI AG
Grosspeteranlage 5
4052 Basel, Switzerland

This is a reprint of articles from the Special Issue published online in the open access journal *Polymers* (ISSN 2073-4360) (available at: www.mdpi.com/journal/polymers/special_issues/ Polymer_Materials_Sensors_Actuators_Energy_Conversion_II).

For citation purposes, cite each article independently as indicated on the article page online and using the guide below:

Lastname, A.A.; Lastname, B.B. Article Title. *Journal Name* **Year**, *Volume Number*, Page Range.

ISBN 978-3-7258-2070-2 (Hbk)
ISBN 978-3-7258-2069-6 (PDF)
https://doi.org/10.3390/books978-3-7258-2069-6

© 2024 by the authors. Articles in this book are Open Access and distributed under the Creative Commons Attribution (CC BY) license. The book as a whole is distributed by MDPI under the terms and conditions of the Creative Commons Attribution-NonCommercial-NoDerivs (CC BY-NC-ND) license (https://creativecommons.org/licenses/by-nc-nd/4.0/).

Contents

About the Editor vii

Preface ix

Cancan Yan, Molin Qin, Tengxiao Guo, Lin Zhang, Junchao Yang and Yong Pan
Synthesis and Application of Polymer SXFA in the Detection of Organophosphine Agents with a SAW Sensor
Reprinted from: *Polymers* **2024**, *16*, 784, doi:10.3390/polym16060784 1

Duy Linh Vu, Quang Tan Nguyen, Pil Seung Chung and Kyoung Kwan Ahn
Flowing Liquid-Based Triboelectric Nanogenerator Performance Enhancement with Functionalized Polyvinylidene Fluoride Membrane for Self-Powered Pulsating Flow Sensing Application
Reprinted from: *Polymers* **2024**, *16*, 536, doi:10.3390/polym16040536 15

Piromya Thongkhao, Apon Numnuam, Pasarat Khongkow, Surasak Sangkhathat and Tonghathai Phairatana
Disposable Polyaniline/*m*-Phenylenediamine-Based Electrochemical Lactate Biosensor for Early Sepsis Diagnosis
Reprinted from: *Polymers* **2024**, *16*, 473, doi:10.3390/polym16040473 28

Yuhang Wang, Cancan Yan, Chenlong Liang, Ying Liu, Haoyang Li and Caihong Zhang et al.
Sensitive Materials Used in Surface Acoustic Wave Gas Sensors for Detecting Sulfur-Containing Compounds
Reprinted from: *Polymers* **2024**, *16*, 457, doi:10.3390/polym16040457 42

Youngnan Kim, Donggeun Lee, Ky Van Nguyen, Jung Hun Lee and Wi Hyoung Lee
Optimization of Gas-Sensing Properties in Poly(triarylamine) Field-Effect Transistors by Device and Interface Engineering
Reprinted from: *Polymers* **2023**, *15*, 3463, doi:10.3390/polym15163463 59

Chong Li, Liang Shen, Jiang Shao and Jiwen Fang
Simulation and Experiment of Active Vibration Control Based on Flexible Piezoelectric MFC Composed of PZT and PI Layer
Reprinted from: *Polymers* **2023**, *15*, 1819, doi:10.3390/polym15081819 70

Rong-Tsu Wang and Jung-Chang Wang
Investigations on Five PMMA Closed Types of Piezo Actuators as a Cooling Fan
Reprinted from: *Polymers* **2023**, *15*, 377, doi:10.3390/polym15020377 89

Ivan V. Portnov, Alexandra A. Larina, Rustam A. Gumerov and Igor I. Potemkin
Swelling and Collapse of Cylindrical Polyelectrolyte Microgels
Reprinted from: *Polymers* **2022**, *14*, 5031, doi:10.3390/polym14225031 109

Cheng-Ling Lee, Chao-Tsung Ma, Kuei-Chun Yeh and Yu-Ming Chen
A Dual-Cavity Fiber Fabry–Pérot Interferometer for Simultaneous Measurement of Thermo-Optic and Thermal Expansion Coefficients of a Polymer
Reprinted from: *Polymers* **2022**, *14*, 4966, doi:10.3390/polym14224966 122

Qisong Shi, Rui Xue, Yan Huang, Shifeng He, Yibo Wu and Yongri Liang
A Flexible Multifunctional PAN Piezoelectric Fiber with Hydrophobicity, Energy Storage, and Fluorescence
Reprinted from: *Polymers* **2022**, *14*, 4573, doi:10.3390/polym14214573 131

Asmaa Selim, Gábor Pál Szijjártó and András Tompos
Insights into the Influence of Different Pre-Treatments on Physicochemical Properties of Nafion XL Membrane and Fuel Cell Performance
Reprinted from: *Polymers* **2022**, *14*, 3385, doi:10.3390/polym14163385 . . . . **147**

Jia Wang, Yujian Tong, Chong Li, Zhiguang Zhang and Jiang Shao
A Novel Vibration Piezoelectric Generator Based on Flexible Piezoelectric Film Composed of PZT and PI Layer
Reprinted from: *Polymers* **2022**, *14*, 2871, doi:10.3390/polym14142871 . . . . **158**

Hongtao Dang, Yan Zhang, Yukun Qiao and Jin Li
Refractive Index and Temperature Sensing Performance of Microfiber Modified by UV Glue Distributed Nanoparticles
Reprinted from: *Polymers* **2022**, *14*, 2425, doi:10.3390/polym14122425 . . . . **172**

About the Editor

Jung-Chang Wang

Jung-Chang Wang (J.-C. Wang) is a full professor, Associate Vice President for Research and Development (R&D), and served as Dean (2020.08–2023.08) in the School of Marine Engineering (D.M.E.) at National Taiwan Ocean University (NTOU), Keelung, Taiwan. He also holds the position of Director of the Thermal-Fluid Illumination Laboratory. Prof. Wang obtained his Bachelor's and Master's degrees from National Cheng Kung University (NCKU) and his Ph.D. in mechanical engineering from National Taiwan University (NTU) in 2007. With over two decades of experience in teaching and research, his expertise lies in electronic heat transfer and renewable energy. Prof. Wang has authored more than 100 research papers in international journals and conferences, holds several patents, and has edited seven book chapters and one book. His primary research interests encompass applied and software engineering in thermal fluid science.

Preface

The first edition of the reprint titled "Polymer Materials in Sensors, Actuators, and Energy Conversion" has been successfully published. Building upon the foundation laid by the first edition, the second edition has now been completed. Polymer-based materials play an increasingly crucial role in the advancement of sensors, actuators, and energy conversion technologies, which are essential for the development of smart materials and electronic devices. These applications involve the synthesis, structural analysis, and property characterization of polymers and their composites. Significantly, these materials are employed in energy-harvesting devices and energy storage systems, which are designed for electromagnetic applications (conversion of electrical to mechanical energy) and magneto-mechanical applications (conversion of magnetic to mechanical energy). Furthermore, polymer-based materials are utilized in light-emitting devices and electrically driven sensors. The modulation of these materials and devices allows for precise control over detection, actuation, and energy management in functional devices, thereby enhancing their performance and efficiency.

The second edition expands on these topics, providing a comprehensive overview of the latest advancements and innovations in the field. It includes detailed discussions on the synthesis methods, structural properties, and functional applications of polymer-based materials, as well as their integration into various technological systems. This edition aims to serve as a valuable resource for researchers, engineers, and practitioners working in the fields of materials science, electronics, and energy conversion.

Jung-Chang Wang
Editor

Article

Synthesis and Application of Polymer SXFA in the Detection of Organophosphine Agents with a SAW Sensor

Cancan Yan, Molin Qin, Tengxiao Guo, Lin Zhang, Junchao Yang and Yong Pan *

State Key Laboratory of NBC Protection for Civilian, Beijing 102205, China; ccy805905145@163.com (C.Y.); qinmolin@139.com (M.Q.); guotengxiao@sklnbcpc.cn (T.G.); zhanglin_zju@aliyun.com (L.Z.); yangjunchao1990@163.com (J.Y.)
* Correspondence: panyong71@sina.com.cn

Abstract: The effective detection of isopropyl methylfluorophosphonate (GB, sarin), a type of organophosphine poisoning agent, is an urgent issue to address to maintain public safety. In this research, a gas-sensitive film material, poly (4-hydroxy-4,4-bis trifluoromethyl)-butyl-1-enyl)-siloxane (SXFA), with a structure of hexafluoroisopropyl (HFIP) functional group was synthesized by using methyl vinylpropyl dichlorosilane and hexafluoroacetone trihydrate as initial materials. The synthesis process products were characterized using FTIR. SXFA was prepared on a 200 MHz shear surface wave delay line using the spin-coating method for GB detection. A detection limit of <0.1 mg/m^3 was achieved through conditional experiments. Meanwhile, we also obtained a maximum response of 2.168 mV at a 0.1 mg/m^3 concentration, indicating the much lower detection limit of the SAW-SXFA sensor. Additionally, a maximum response standard deviation of 0.11 mV with a coefficient of variation of 0.01 and a maximum recovery standard deviation of 0.22 mV with a coefficient of variation of 0.02 were also obtained through five repeated experiments. The results show that the SAW-SXFA sensor has strong selectivity and reproducibility, good selectivity, positive detection ability, high sensitivity, and fast alarm performance for sarin detection.

Keywords: SXFA; SAW-SXFA sensor; organophosphorus agent; GB

Citation: Yan, C.; Qin, M.; Guo, T.; Zhang, L.; Yang, J.; Pan, Y. Synthesis and Application of Polymer SXFA in the Detection of Organophosphine Agents with a SAW Sensor. *Polymers* **2024**, *16*, 784. https://doi.org/10.3390/polym16060784

Academic Editor: Jung-Chang Wang

Received: 4 January 2024
Revised: 5 March 2024
Accepted: 6 March 2024
Published: 12 March 2024

Copyright: © 2024 by the authors. Licensee MDPI, Basel, Switzerland. This article is an open access article distributed under the terms and conditions of the Creative Commons Attribution (CC BY) license (https://creativecommons.org/licenses/by/4.0/).

1. Introduction

Sarin (GB, methylphosphonic difluoride) is a representative chemical warfare agents. It is an organophosphorus (OP) neurotoxic agent with high volatility, strong toxicity, and a short latency period. This nerve agent can be obtained easily, with characteristics of easy synthesis and difficulty in prevention and control [1,2]. Due to the high specificity and affinity of acetylcholinesterase, GB poses a great threat to human health and public safety [1–3]. Therefore, effective detection methods can qualitatively and quantitatively detect GB, improving protection capabilities.

Various gas sensing techniques have been developed for GB detection, for instance, field-effect transistors [4], fluorescence [5], flame photometry [6], ion mobility spectrometry [7], gas chromatography–mass spectrometry [8], and surface acoustic wave (SAW) [9], and each technique has special advantages and plays unique roles in GB detection. The SAW technique has been systematically and deeply studied in the detection of chemical warfare agents (CWAs), mainly due to its non-destructive nature, compact structure, ability to detect nerve agents and blister agents, and applicability to point or area detection [10–15]. A SAW sensor has a compact structure and high sensitivity and is small, inexpensive, and capable of fast responses, characteristics that are in line with the current development direction of intelligence in the field of chemical sensors, which are becoming a research hotspot in this field [13–16]. So far, SAW sensors for detecting various gases, such as H$_2$, SO$_2$, H$_2$S, and NO$_2$, have been developed and have yielded remarkable results [16,17], and sensitive film materials play a decisive role in the detection effect [16]. Among the various sensitive film materials, polymers are the most commonly used in SAW sensors [18]. One of these sensitive film materials, SXFA, is an organosilicon compound with a special structure and properties [13,14,19]. Its structural unit has a hexafluoroisopropyl (HFIP) functional group, which has a strong hydrogen-bonding effect on organophosphorus compounds [14,19–21]. Thus, it provides superior sensitivity and selectivity in the detection of organophosphorus compounds, and so far, it is one of the most widely studied and data-rich polymers [13,14,20,22–25]. However, after the synthesis of SXFA, SAW-SXFA sensors generally only exist as sensors for detecting hydrogen-bonding alkaline gases in sensor arrays, and there are few reports on the detection of organophosphorus gases individually.

Therefore, in order to individually detect GB with high sensitivity, we synthesized the polymer SXFA and constructed a SAW-SXFA sensor in this study, discussed its relevant mechanisms, analyzed its detection of organophosphorus nerve agents, and evaluated its practical performance.

2. Materials and Methods

2.1. Reagents and Instruments

Allydichloromethylsilane, 95%, Macklin, Shanghai, China; phenyltrimethylammonium, 20~25% formaldehyde solution, Tokyo Kasel Kogyo Co., LTD., Tokyo, Japan; hexafluoroacetone trihydrate (HFA·$3H_2O$), 95%, Macklin, Shanghai, China; polyepoxypropyl chloride, average Mw ~700, Macklin, Shanghai, China; ether, 99.9%, TEDIA, Fairfield, OH, USA; sulfuric acid, AR, Beijing Chemical Plant, Beijing, China; magnesium sulfate, AR, Macklin, Shanghai, China; toluene, AR, Aladdin, Seattle, WA, USA; ethanol, $\geq$99.7%, Tansoole, Shanghai, China; dry ice, Yojanbio, Beijing, China; DMMP, AR, Beijing Chemical Plant, Beijing, China; GB, 99%, State Key Laboratory of NBC Protection for Civilian, Beijing, China.

FTS-185 infrared spectrometer, Bio-Rad, Hercules, CA, USA; Q100 modulated DSC, Thermal Analysis, San Diego, CA, USA; Gel Permeation Chromatography (GPC), Waters, Mass, USA; surface acoustic wave oscillator, with central oscillation frequency of 300 MHz and a delay line surface consisting of a quartz layer, State Key Laboratory of NBC Protection for Civilian, Beijing, China; frequency counter, Proteck C3100 (Republic of Korea), Qingdao, China; equipped with RS232 interface for computer connection; Scanning Potentiometer, S4800, Fujifilm, Tokyo, Japan; Dynamic Gas Generator, State Key Laboratory of NBC Protection for Civilian, Beijing, China.

2.2. Experimental Methods

2.2.1. Mechanism of Interaction between Polymer and Gas Molecules

A SAW gas sensor utilizes sensitive film materials on piezoelectric crystals to generate characteristic responses for gas adsorption, and polymers are commonly used as the sensitive film materials on its surface [26,27]. Usually, polymers need to have the following characteristics: (1) non-volatility, which can enable polymers to remain stable on sensors for a long time; (2) viscoelasticity, which allows gas to disperse quickly within a polymer film; and (3) quick response capability, selectivity, recoverability, and the ability to be deployed on sensor surfaces [28]. Polymer sensitive film materials are mostly composed of polysiloxane as the main chain, which assumes a viscoelastic state at room temperature and has a good adsorption capacity for gas [29,30]. Before selecting polymer film materials, it is necessary to analyze the principles of interaction between polymer films and gas molecules. Generally, the main interactions between gas molecules and polymer films are van der Waals forces, polarization, and hydrogen bonding [29,30].

The process of the polymer adsorption of gas is similar to that of the dissolution of gas into liquid (Figure 1). At the interface between the polymer film surface and the gas phase, target molecules are distributed between the gas phase and the polymer phase, reaching a thermodynamic equilibrium state. Many previous studies have investigated the distribution equilibrium between adsorbate and stationary phases and proposed relevant models [24,25]. The selective adsorption of gas-phase molecules by polymers and the equilibrium between target molecules in the gas phase and the polymer phase can be expressed by the following formula:

$$Kp = Cp/Cv \qquad (1)$$

where Kp represents the equilibrium of gas entering the polymer phase from the gas phase, Cv represents the concentration of the target molecule in the gas phase, and Cp represents the concentration of the target molecule in the liquid phase. In this model, Kp quantitatively describes the equilibrium of gas entering the polymer phase from the gas phase, and a larger Kp value represents a stronger gas adsorption capacity.

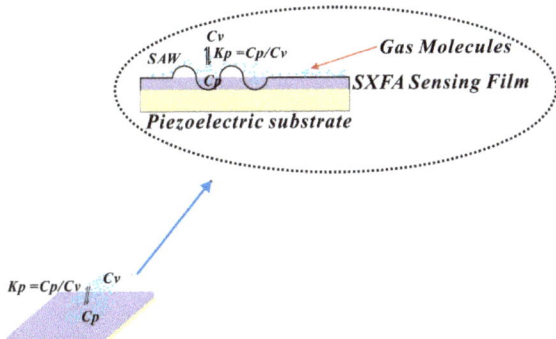

Figure 1. Equilibrium partitioning of vapor molecules between gas phase and polymer.

To determine the Kp value, Grate et al. proposed the linear solvation energy relationship (LSER) as a model for the selective adsorption of gases; this model is established by using a gas as a solute and a polymer as a solvent [31]. The solubility of a gas is expressed by a series of parameters, with LESR representing a linear combination of all forces, and the liner relationship is as follows:

$$LogK = c + r R_2 + s \pi_2^H + a \alpha_2^H + b \beta_2^H + lLogL^{16} \qquad (2)$$

Among them, the dissolution parameters R_2, π_2^H, α_2^H, β_2^H, and $LogL^{16}$ represent various solubility parameters of gases: R_2 refers to the gas hyper-molar regression parameter, and this quantitatively represents n and p electrons, which play a polarization role; π_2^H represents the gas dipole or polarization parameter; α_2^H and β_2^H represent the parameters of the gas hydrogen-bonding acid and hydrogen-bonding base, respectively; $LogL^{16}$ represents the distribution coefficient of the solute between the gas and liquid phases at 25 °C (obtained from gas–liquid chromatography) and the van der Waals force of the gas; r, s, a, b, and l are related to the properties of the polymer films; a and b, as supplements to gas hydrogen-bonding acidity and hydrogen-bonding alkalinity, represent the hydrogen-bonding acidity and hydrogen-bonding alkalinity of the polymer films; s represents the polarity and dipole effect of the polymer films; l represents the dispersion effect of the polymer films, and a larger value of l indicates a significant difference in the distribution coefficients of similar gases; r represents the polarization ability of n and π electron pairs between the polymer phase and solute molecules; and C is a constant [32].

According to the LSER equation, the interaction between a certain gas and a polymer film can be calculated [33]. Polymers should have sensitivity and selectivity towards the target gas. Additionally, LSER's polynomial coefficients, such as b/a, b/s, s/a, and l/(s + a + b), can also represent selectivity [34], and Table 1 presents the SXFA polymer's dissolution selectivity obtained based on the LSER coefficient.

Table 1. SXFA solubility selectivity examined by ratios of LESR coefficients.

Polymer	b/a	b/s	a/s	s/a	b + a + s	Dispersibility
SXFA	6.07	7.08	1.17	0.86	5.55	0.13

It is clear that, for hydrogen-bonded alkaline gases such as GB, the sought polymer should have the largest possible hydrogen-bonding acidity (b) while having the smallest possible hydrogen-bonding alkalinity (a) and dipole polarity (s), represented by b/a and b/s, respectively. As shown in Table 1, it is clear that, with a strong hydrogen-bonding acidity, the SXFA polymer's solubility values of b/a, b/s, s/a, 1/(s + a + b), and dispersibility are 6.07, 7.08, 0.86, 5.55, and 0.13, making it an ideal hydrogen-bonding acidic polymer sensitive membrane material for SAW sensors.

Additionally, SXFA is a polysiloxane film material with a low glass transition temperature [14]. The HFIP functional groups on SXFA have strong hydrogen-bonding effects on organic phosphine gases and exhibit the viscoelastic properties of polysiloxane at room temperature, enabling the selective adsorption of organic phosphine compounds [13,14], which further proves that SXFA is an ideal organic phosphine adsorption material. For instance, as shown in Figure 2, the hydrogen-bond acidity of the -OH group on the HFIP functional group was enhanced due to the influence of the neighboring -CF$_3$ group, which allowed it to selectively adsorb organophosphorus gases with alkaline hydrogen-bond interactions, achieving the selective adsorption of alkaline organophosphorus compounds; thus, enhanced SXFA is highly susceptible to forming strong hydrogen bonds with GB.

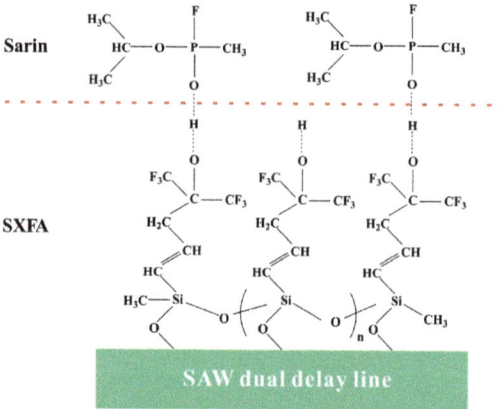

Figure 2. Schematic of sensing mechanism.

2.2.2. Synthesis Route of Hexafluoro-2-hydroxyisopropyl Polysiloxane

The synthesis of SXFA requires multiple steps and the control of reaction conditions, so its synthesis method is relatively complex. Figure 3 shows the synthesis steps of the polymer SXFA. In order to ensure the purity and quality of the final product, experimental operations need to be carried out with caution according to the designed process. On the one hand, it should be noted that, in the synthesis process, it is necessary to control the reaction temperature, reaction time, and other parameters. On the other hand, attention should also be paid to the synthesis process to avoid the generation of impurities and by-products.

The first step in the synthesis of SXFA was to prepare methylvinyl polysiloxane by adding an appropriate amount of ether to methylvinyl dichlorosilane, stirring the mixture with a magnetic stirrer, and dripping distilled water on it until complete reaction at room temperature. Then, an appropriate amount of ether was added to extract the upper liquid, and it was dried overnight with MgSO$_4$ and filtered. Then, the ether was evaporated, and siloxane was obtained. After the siloxane was prepared, it was left to sit for 3 weeks, and then MgSO$_4$ was added for drying (>24 h). It was then filtered, the ether was evaporated, and a small amount of a formaldehyde solution containing phenyl trimethylammonium hydroxide was added. Then, it was stirred at 400 K until the reaction completed and centrifuged to remove the black suspension, and polysiloxane was obtained. Finally, the polysiloxane was transferred to a high-pressure-resistant sealed tube and placed in a dry ice-cold trap. HFA·3H$_2$O was dried with H$_2$SO$_4$, then HFA gas was obtained, and the HFA gas was recovered from the sealed tube in the cold trap. After the reaction was completed, the sealed tube was heated at 380 K (in a silicone oil bath) for 48 h. After naturally cooling, the liquid was removed from the sealed tube and blown with N$_2$ overnight to remove the unreacted HFA, and, finally, the final product hexafluoro-2-hydroxyisopropyl polysiloxane (SXFA) was obtained.

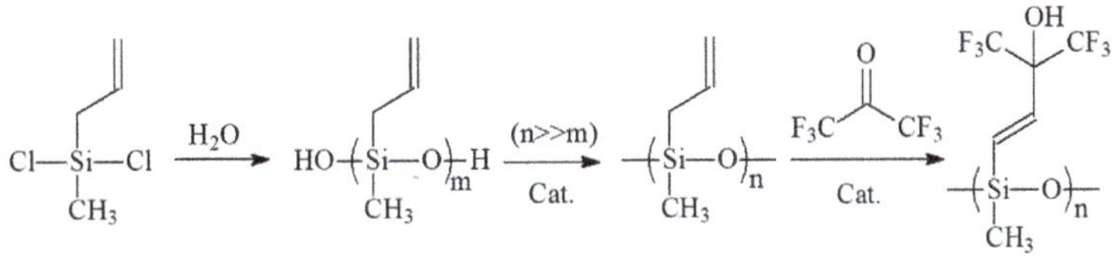

Figure 3. Synthesis route of SXFA.

2.2.3. Preparation of SAW-SXFA Sensor and Its Detection of Organophosphine Agents

The SAW-SXFA sensor consists of an interdigital transducer (IDT), a piezoelectric substrate, and an SXFA gas-sensitive thin film (Figure 4). Due to the piezoelectric effect of the piezoelectric substrate, the input interdigital transducer converts the input electrical signal into an acoustic signal, while the output interdigital transducer converts the received acoustic signal into an electrical signal output. The SXFA film can adsorb gas reversibly on the propagation path of surface acoustic waves, and the increase in its mass leads to a change in the propagation speed of the surface acoustic waves. The detection of gas is achieved by measuring its frequency or phase changes. In this research, we used a delayed linear SAW sensor device with a center frequency of 200 MHz, which was based on Y-shaped quartz cutting. To obtain low-loss and single-frequency signals, unidirectional transducers (SPUDTs) were applied, and structures were combed; the electrode widths of the SPUDTs were ~4 µm and ~2 µm. In the phase detector circuit, the electrical signal emitted by a signal source with a frequency of 200 MHz at a corresponding wavelength of ~15.8 µm and the electrical signal emitted by the SAW-SXFA sensor were output through the phase detector, and a voltage signal proportional to the phase difference of the two signals was then sent to a computer through a data transmission module.

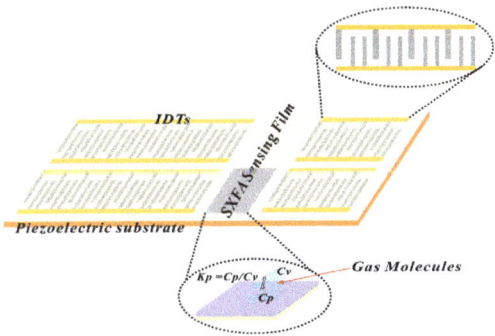

Figure 4. Schematic of SAW-SXFA sensor response mechanism.

In the preparation of the SAW-SXFA sensor, 150 nm thick aluminum was deposited on the Y-shaped quartz substrate, and a 1 mm thick photoresist was spin-coated and exposed for use in delay line patterns. Then, it was dissolved and rinsed. And, finally, a 50 nm SiO_2 film was coated on the transducer to provide good protection in the process of the gas-sensitive film coating.

The selectivity of the adsorption of a film is not only related to its structure but also to its morphology. Therefore, in order to improve the separation ability of a sensitive polymer film, the preparation method of thin films can also be controlled. In this study, SXFA polymer films were prepared on sensor components using the drop-coating method to complete the assembly of the SXFA-SAW sensor and applied to detect organophosphine agents.

3. Results and Discussions

3.1. Infrared Spectroscopy Characterization and Analysis of SXFA Material

During the synthesis process of SXFA, organosilicon compounds have a strong absorption effect on infrared. In this research, the characteristic absorption peaks of the main groups in the synthesis process are Si-Me, Si-CH=CH2, Si-O, Si-O-Si, >SiCl$_2$, and -CF$_3$ (Si-Me located at ~1260 cm^{-1} and ~765 cm^{-1}; Si-CH=CH$_2$ located at 1613 cm^{-1}, 1410~1390 cm^{-1}, 1020~1000 cm^{-1}, and 980~950 cm^{-1}; Si-OH located at 3390~3200 cm^{-1} and 910~830 cm^{-1}; Si-O located at 1100~1000 cm^{-1}; Si-O-Si located at 1080 cm^{-1}, 1025 cm^{-1}, ~1020 cm^{-1}, and ~1090 cm^{-1}; >SiCl$_2$ located at 595~535 cm^{-1}; and -CF$_3$ located at 1350~1120 cm^{-1}, 780~680 cm^{-1}, and 680~590 cm^{-1}).

Figure 5 shows the IR spectrum of methylvinyldichlorosilane; it is obvious that the peak at 1633 cm^{-1} is the stretching vibration of the vinyl double bond and that the peak at 1263 cm^{-1} is the symmetric deformation vibration of -CH$_3$. It can also be observed that there is a Si-OH peak near 3500 cm^{-1} and a Si-O peak at 1080~1025 cm^{-1}, indicating that methylvinyl dichlorosilane partially underwent spontaneous hydrolysis and condensation into siloxane. Therefore, methylvinyl dichlorosilane should be stored in a dry environment, and, during the hydrolysis reaction, to ensure the uniform polymerization of siloxane, it should first be dissolved in an organic solvent and then mixed with a suitable amount of water.

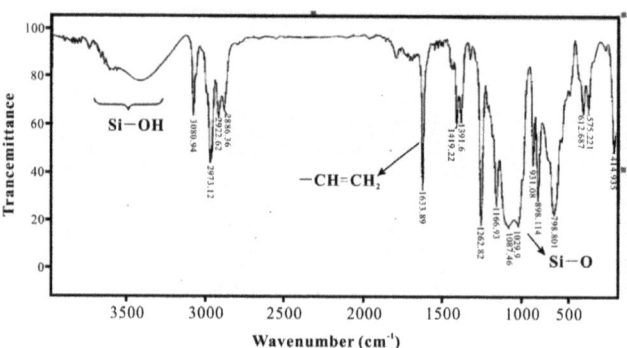

Figure 5. IR spectrum of methylallyldichlorosilane.

Figure 6 shows the IR spectrum of methylvinylpolysiloxane. Compared with Figure 4, it can be seen that the Si-OH peak near 3500 cm^{-1} disappears, indicating that the raw material was completely hydrolyzed and condensed into siloxane.

Figure 7 presents the FTIR spectrum of SXFA, which shows the appearance of an -OH peak near 3500 cm^{-1}; the original single peak at 1633 cm^{-1} for the double bond has become two adjacent double bond peaks, and there is an -CF$_3$ absorption peak between 1120 cm^{-1} and 1350 cm^{-1}. This phenomenon indicates that some vinyl double bonds underwent an addition reaction with HFA, causing the double bonds to transfer, and the appearance of hexafluoro-2-hydroxyisopropyl functional group confirms the reaction between methylvinyl polysiloxane and HFA, thus producing the final product SXFA.

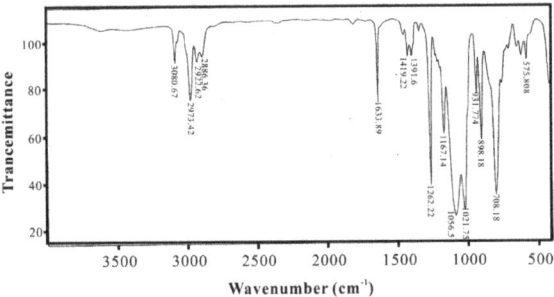

Figure 6. IR spectrum of methylvinylpolysiloxane.

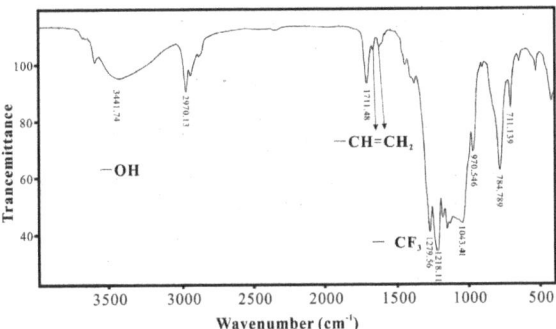

Figure 7. FTIR spectrum of SXFA.

3.2. SEM Performance

SEM performance was used to observe the morphology of the SXFA film on the surface of the SAW sensor and analyze its coverage on the sensor surface. It can be seen in Figure 8 that the SXFA film is in a porous form with a granular arrangement on the substrate, and the polymer has an irregular geometric shape. Therefore, the SXFA film is an amorphous polymer.

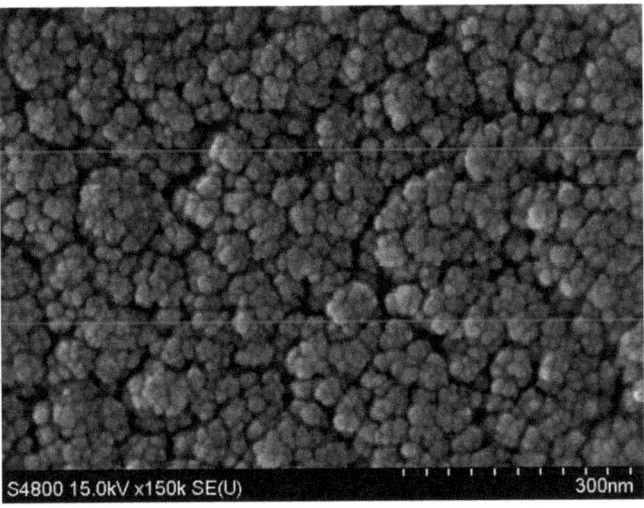

Figure 8. SEM image of SAW sensor delay line of SXFA sensitive film.

3.3. Detection of Organophosphorus Agents

3.3.1. Selective Analysis of SAW-SXFA Sensor

Superior selectivity is extremely important for SAW sensors. So, in this research, a comparative study on the adsorption effect of GB and its analog agent DMMP was conducted (Figures 9 and 10).

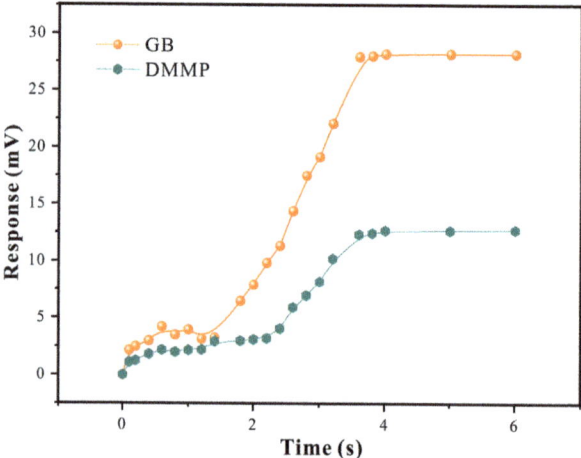

Figure 9. Response of SAW-SXFA sensor with concentration changes in GB and DMMP (19.6 °C, RH = 27%).

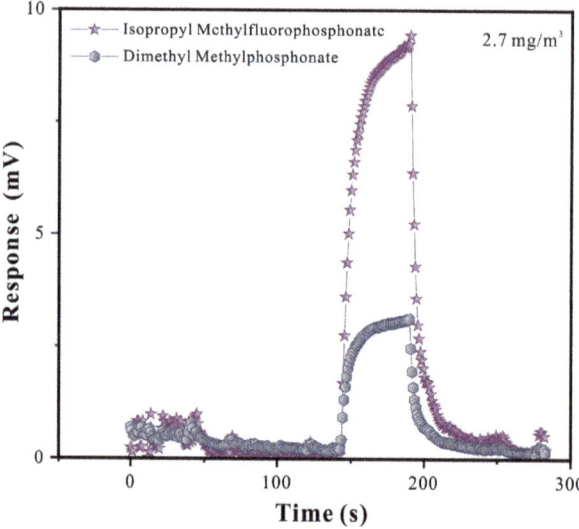

Figure 10. Comparison of GB and DMMP at 2.7 mg/m^3 (17.9 °C, RH = 28%).

In Figure 9, it can be seen that, at low concentrations, the functional group sites on the surface of SXFA were sufficient to adsorb the gas molecules in contact with them, which caused changes in the sensor mass load and thus resulted in significant changes in the sensor signal. As the concentration gradually increased, with the interaction with low concentrations, the functional group sites inside the film interacted with the gas molecules through stereo adsorption and jointly caused changes. However, due to the fact that stereo adsorption was based on the diffusion rate of the gas within the film, the time for the maximum response was prolonged. At a high concentration, both the surface sites and internal sites of the film tended to saturate, so the increase in the GB and DMMP concentrations no longer had a significant impact on the sensor response.

By comparing the responses of the adsorption equilibrium process, it could be seen that the SAW-SXFA sensor had a stronger adsorption response to GB than to DMMP. Initially, the differences in the response gradually increased with the increase in the concentration, and they gradually stabilized when the equilibrium concentration was reached (Figure 9). Through systematic research, it could be illustrated that the main reason for this was that SXFA is a linear polysiloxane-based polymer with an HFIP functional group. The hydrogen-bond acidity of the -OH group on the HFIP functional group was enhanced due to the influence of the neighboring -CF$_3$ group, which allowed it to selectively adsorb organophosphorus gases with alkaline hydrogen-bond interactions, achieving the selective adsorption of alkaline organophosphorus compounds [23]. To visually explore the selectivity of the SAW-SXFA gas sensor on GB, response–recovery curves of GB and DMMP at a concentration of 2 mg/m^3 were compared (Figure 10). As shown in Figure 8, the maximum response of GB was 9.236 mV, while the response signal of DMMP was only 3.124 mV. By comparing the responses of the two toxic gases, it was found that, at this concentration, the adsorption capacity of SXFA for GB was about three times that of DMMP. Therefore, the SAW-SXFA sensor exhibits good selectivity and detection performance for organophosphorus agents.

3.3.2. Analysis of Response of SAW-SXFA Sensor

The response of gas is crucial in the research of SAW-SXFA sensors. Therefore, relevant research was conducted in this study. In Figure 11, it can be seen that the maximum response of the SAW-SXFA sensor for GB was 6.118 mV, and the noise during the sensor equilibration process could be ignored. At the beginning of detection, due to hydrogen-bond adsorption between GB and the polymer film, the response was 2.475 mV in 10 s, accounting for 40.4% of the maximum response signal, and, furthermore, it only took 50 s to reach 80% of the maximum response. During the recovery phase, GB rapidly separated from the SXFA polymer film, and the response signal decreased rapidly by 4.283 mV within 10 s, accounting for 70% of the maximum response.

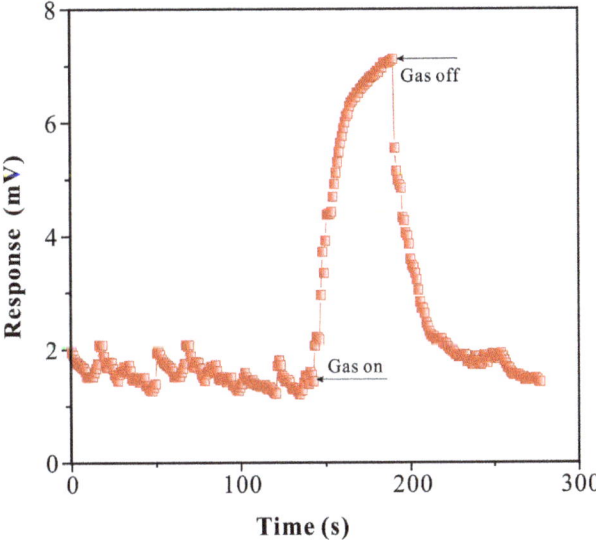

Figure 11. Detection of GB (1.6 mg/m^3) by SAW-SXFA sensor (18.6 °C, RH = 28%).

In the initial stage, the interactions between GB and the polymer mainly involved the adsorption of surface hydrogen-bonding sites, which indicated a high adsorption efficiency and therefore caused significant changes in sensor mass loading and response signals. After 1 min, some GB molecules penetrated into the interior of the polymer through steric adsorption effects, but the adsorption efficiency at the internal sites of the polymer was lower than that on the surface, so the mass loading of the sensor increased slowly, resulting in a slower response. During the recovery period, the GB molecules on the surface of the SXFA polymer quickly escaped, resulting in a significant change in the sensor signal. Immediately after, with the passage of time, the rate of sensor recovery slowed down, as the GB molecules inside the polymer film had to overcome the obstruction of the polymer chains to escape.

3.3.3. Detection Limit of SAW-SXFA Sensor

The determination of sensor detection limits plays an important role in ensuring the normal operation of sensors, improving usage efficiency, and enhancing safety. In this research, variations in the SAW-SXFA sensor response with the GB concentration were found and are shown in Table 2. When GB was at a relatively high concentration (≥ 1.0 mg/m^3), the response signals of the SAW-SXFA sensor showed an increasing trend with the increase in the GB concentration within the same time period, which conformed to the law of solid adsorption isotherms [25], and the SAW-SXFA sensor could recover more than 65% of the response signal. However, when at a relatively low concentration (<1.0 mg/m^3), the response of the SAW-SXFA sensor increased first and then decreased, and it reached a maximum response at 0.6 mg/m^3 in this period, similar to the liquid adsorption isotherm curve [25].

Table 2. Relationship between concentration of GB and response of SAW-SXFA sensor (17.3 °C, RH = 27%).

Intensity (mg/m^3)	Response (mV)	Recovery (mV)	Recovery Rate (%)
0.1	2.168	0.02	98
0.2	2.509	0.05	95
0.4	3.002	0.09	91
0.6	4.216	0.08	92
0.8	3.523	0.12	88
1.0	3.972	0.12	88
1.4	3.283	0.14	86
1.8	6.515	0.20	80
2.2	9.859	0.23	77
2.6	14.394	0.27	73
3.0	19.172	0.30	70
3.4	27.605	0.35	65

To study the sensitivity of the SAW-SXFA sensor at a low concentration, further analysis was conducted for GB at a concentration of 0.1 mg/m^3. As shown in Figure 12 and Table 2, the SAW-SXFA sensor response was 1.563 mV in 140 s, and it continued to increase with time. The initial signal change in the response within the first 10 s was 0.326 mV, which was much larger than the SAW-SXFA sensor's noise, and it could be inferred that the sensor has potential to detect lower concentrations. In addition, it was found that the SAW-SXFA sensor has an "accumulation" function when detecting low concentrations of GB. Under the conditions of a lower sample concentration and a higher alarm response, the SAW-SXFA sensor can achieve a pre-alarm purpose by accumulating adsorption over a long period of time, finally exceeding the alarm limit. Therefore, in the detection of low-concentration GB, the SAW-SXFA sensor has high sensitivity and practical performance as a rapid alarm.

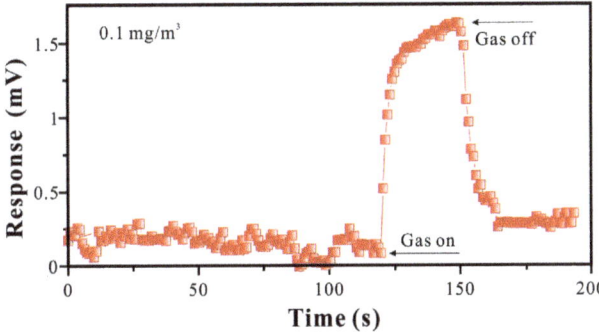

Figure 12. SAW-SXFA sensor for minimum detection concentration of GB (18.6 °C, RH = 29%).

3.3.4. Reproducibility Study of SAW-SXFA Sensor

To research the stability of the SAW-SXFA sensor, GB at the same concentration was detected continuously five times under the same conditions. The response and recovery times were both set to 120 s, and the results are shown in Figure 13 and Table 3.

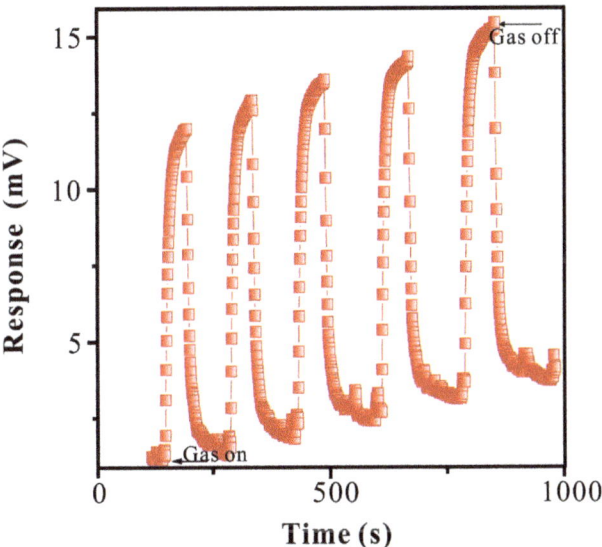

Figure 13. Reproducibility of GB by SAW-SXFA sensor (16.9 °C, RH = 28%).

Table 3. Analysis of data of reproducibility for GB by SAW-SXFA sensor (17.2 °C, RH = 27%).

Experiment No.	Response (mV)	Recovery (mV)
1	10.845	9.696
2	10.619	10.073
3	10.767	10.176
4	10.921	9.729
5	10.831	9.741
Average (mV)	10.797	9.883
Standard Deviation (mV)	0.11	0.22
Discrete Coefficient	0.01	0.022

It can be seen in Figure 13 that the changes in the SAW-SXFA sensor response and recovery times had a periodic nature of about 120 s. However, in this periodic nature of changes, due to the limited recovery time, the GB molecules could not completely desorb from the sensitive film, leading to the continuous accumulation of molecules within the sensitive film, thus resulting in a longer recovery time. Although the SAW-SXFA sensor could not fully recover to the initial value, the impact on the response within the time set in the experiment was very small (as shown in Table 3). The standard deviation of the maximum response measured 5 times was only 0.11 mV, with a coefficient of variation of 0.01, and the standard deviation of the maximum recovery signal was 0.22 mV, with a coefficient of variation of 0.02. Therefore, the SAW-SXFA sensor has good reproducibility for detecting GB, which is of great significance for the future quantitative detection of SAW-SXFA sensors.

3.3.5. Interference Gas Research

The composition of air is quite complex, and for a gas-sensitive sensor, its ability to resist interference is particularly important. Therefore, in order to verify the selectivity of the film material and evaluate sensor performance, it is important to test interference gases before detecting the target gas. The SAW-SXFA sensor was subjected to comparative experiments with various high-concentration interference gases, with each gas concentration set at 500 mg/m^3, and the results are shown in Figure 14. In Figure 14, it can be seen that the SAW-SXFA sensor had a strong response to organophosphorus gases, especially DMMP and DFP. In addition, due to the hydrogen-bond alkalinity of organophosphorus and amine gases, they could adsorb on the surface of the SXFA polymer through the hydrogen-bond interaction, resulting in a noticeable response to high concentrations of ammonia gas and N,N-dimethylacetamide. Due to the strong polarity of the HFIP group, this group adsorbed various polar gases, causing certain interference of polar gases (such as alcohols) on the SAW-SXFA sensor, and a comprehensive comparison revealed that this effect was much weaker than the hydrogen-bond adsorption effect.

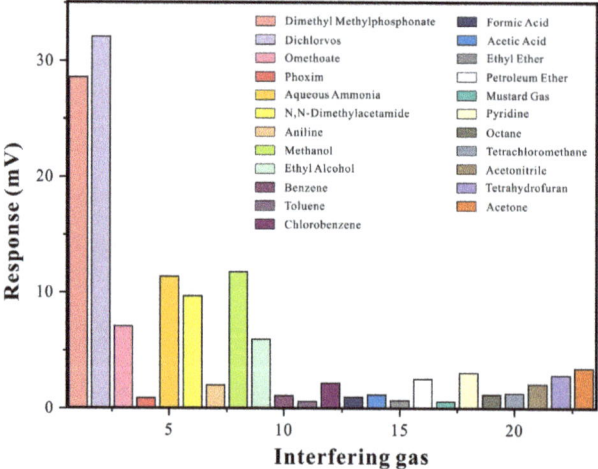

Figure 14. Column chart of interference gas response (19.3 °C, RH = 26%).

4. Conclusions

The synthesis of SXFA and the detection of organic phosphorus gas by a SAW-SXFA sensor were researched in this study, and several conclusions were drawn. Firstly, by analyzing the reaction mechanism and synthesis route of SXFA, it was found that ether plays a dispersing role in dissolving dichlorosilane, resulting in the formation of low-degree cyclic siloxane after hydrolysis and condensation. Phenyltrimethylamine hydroxide has superior catalytic activity, allowing cyclic siloxanes to undergo ring opening and generate chains like polysiloxanes with a high degree of polymerization. HFA has amphiphilic properties and undergoes an addition reaction with methylallyl dichlorosilane, generating HFIP functional groups and transferring the position of allyl double bonds. Secondly, the characterization of SXFA and its relevant materials confirmed that methylallyl polysiloxane reacts with HFA and finally generates SXFA. Then, the detection of GB and its relevant agents indicated that the SAW-SXFA sensor has strong sensitivity (detection limit < 0.1 mg/m^3), fast response and recovery speed, strong reproducibility and periodicity; additionally, due to its viscoelastic state, its response increases first and then decreases with the increase in the GB concentration at low concentrations, and its maximum response increases with the increase in the GB concentration at high concentrations. And, finally, the interfering gas experiment suggested that the SAW-SXFA sensor has good detection performance and anti-interference ability against organophosphine agents. Although high concentrations of organic phosphines, amines, and alcohol compounds may interfere with sensors, their effects are much weaker than the effects of hydrogen-bonding adsorption.

This comprehensive research suggests that the SAW-SXFA sensor has a good detection effect and interference resistance against organic phosphorus agents and has characteristics such as a short response time, good selectivity, high sensitivity, and strong reproducibility in GB detection, which proves that the SAW-SXFA sensor has superior detection performance for GB.

Author Contributions: Conceptualization, C.Y. and Y.P.; methodology, C.Y. and T.G.; software, C.Y. and Y.P.; validation, C.Y., Y.P., and L.Z.; formal analysis, C.Y. and Y.P.; investigation, C.Y.; resources, C.Y. and Y.P.; data curation, C.Y.; writing—original draft preparation, C.Y.; writing—review and editing, C.Y. and Y.P.; visualization, C.Y., Y.P., M.Q., and J.Y.; supervision, Y.P.; project administration, Y.P.; funding acquisition, Y.P. All authors have read and agreed to the published version of the manuscript.

Funding: This research received no external funding.

Institutional Review Board Statement: Not applicable.

Data Availability Statement: Data are contained within the article.

Conflicts of Interest: The authors declare no conflicts of interest.

Dual-use Research Statement:

- ∅ Explanation of Potential Risks: Our paper examines the performance of a SAW sensor using SXFA as a film material. This research is limited to providing some theoretical and experimental support for the development of impact dynamics only and does not pose a threat to public health or national security.
- ∅ Evaluation of Benefits to the General Public: Our research is limited to the academic field, and it is beneficial for the development of material science. There is no risk to the general public.
- ∅ Compliance with Laws: As an ethical responsibility, we strictly adhere to relevant national and international laws about dual-use research. And we have considered and adhered to these regulations in our paper.

References

1. Abu-Qare, A.W.; Abou-Donia, M.B. Sarin: Health effects, metabolism, and methods of analysis. *Food Chem. Toxicol.* **2002**, *40*, 1327–1333. [CrossRef]
2. Lee, E.C. Clinical manifestations of sarin nerve gas exposure. *JAMA-J. Am. Med. Assoc.* **2003**, *290*, 659–662. [CrossRef]
3. Tokuda, Y.; Kikuchi, M.; Takahashi, O.; Stein, G.H. Prehospital management of sarin nerve gas terrorism in urban settings: 10 years of progress after the Tokyo subway sarin attack. *Resuscitation* **2006**, *68*, 193–202. [CrossRef] [PubMed]
4. Yoo, J.; Kim, D.; Yang, H.; Lee, M.; Kim, S.O.; Ko, H.J.; Hong, S.; Park, T.H. Olfactory receptor-based CNT-FET sensor for the detection of DMMP as a simulant of sarin. *Sens. Actuators B Chem.* **2022**, *354*, 131188. [CrossRef]
5. Hamel, M.; Hamoniaux, J.; Rocha, L.; Normand, S. Ppb detection of Sarin surrogate in liquid solutions. *Chem. Biol. Radiol. Nucl. Explos. (CBRNE) Sens. XIV* **2013**, *8710*, 137–143.
6. O'Neill, H.J.; Brubaker, K.L.; Schneider, J.F.; Sytsma, L.F.; Kimmell, T.A. Development of an analytical methodology for sarin (GB) and soman (GD) in various military-related-wastes. *J. Chromatogr. A* **2002**, *962*, 183–195. [CrossRef] [PubMed]

7. Maziejuk, M.; Ceremuga, M.; Szyposzynska, M.; Sikora, T.; Zalewska, A. Identification of organophosphate nerve agents by the DMS detector. *Sens. Actuators B Chem.* **2015**, *213*, 368–374. [CrossRef]
8. Black, R.M.; Clarke, R.J.; Read, R.W.; Reid, M.T.J. Application of gas-chromatography mass-spectrometry and Gas-chromatography tandem mass-spectrometry to the analysis of chemical warfare samples, found to contain residues of the nerve agent sarin, sulfur mustard and their degradation products. *J. Chromatogr. A* **1994**, *662*, 301–321. [CrossRef] [PubMed]
9. William, H.K. Piezoelectric Sorption Detector. *Anal. Chem.* **1964**, *36*, 1735–1739.
10. Stevenson, A.C.; Mehta, H.M.; Sethi, R.S.; Cheran, L.E.; Thompson, M.; Davies, I.; Lowe, C.R. Gigahertz surface acoustic wave probe for chemical analysis. *Analyst* **2001**, *126*, 1619–1624. [CrossRef]
11. Fahim, F.; Mainuddin, M.; Mittal, U.; Kumar, J.; Nimal, A.T. Novel SAW CWA Detector Using Temperature Programmed Desorption. *IEEE Sens. J.* **2021**, *21*, 2915–5922. [CrossRef]
12. Raj, V.B.; Singh, H.; Nimal, A.T.; Sharma, M.U.; Gupta, V. Oxide thin films (ZnO, TeO_2, SnO_2, and TiO_2) based surface acoustic wave (SAW) E-nose for the detection of chemical warfare agents. *Sens. Actuators B Chem.* **2013**, *178*, 647. [CrossRef]
13. Pan, Y.; Zhang, G.; Guo, T.; Liu, X.; Liu, X.; Zhang, C.; Yang, J.; Cao, B.; Zhang, C.; Wang, W. Environmental characteristics of surface acoustic wave devices for sensing organophosphorus vapor. *Sens. Actuators B Chem.* **2020**, *315*, 127986. [CrossRef]
14. Pan, Y.; Qin, M.; Wang, P.; Yang, L.; Zhang, L.; Yan, C.; Zhang, C.; Wang, W. Interface and Sensitive Characteristics of the Viscoelastic Film Used in a Surface Acoustic Wave Gas Sensor. *ACS Sens.* **2022**, *7*, 612–621. [CrossRef] [PubMed]
15. Matatagui, D.; Martí, J.; Fernández, M.J.; Fontecha, J.L.; Gutiérrez, J.; Gràcia, I.; Cané, C.; Horrillo, M.C. Chemical warfare agents simulants detection with an optimized SAW sensor array. *Sens. Actuators B Chem.* **2011**, *154*, 199–205. [CrossRef]
16. Lama, S.; Kim, J.; Ramesh, S.; Lee, Y.J.; Kim, J.; Kim, J.H. Highly Sensitive Hybrid Nanostructures for Dimethyl Methyl Phosphonate Detection. *Micromachines* **2021**, *12*, 648. [CrossRef] [PubMed]
17. Lurz, F.; Ostertag, T.; Scheiner, B.; Weigel, R.; Koelpin, A. Reader Architectures for Wireless Surface Acoustic Wave Sensors. *Sensors* **2018**, *18*, 1734. [CrossRef]
18. Palla-Papavlu, A.; Voicu, S.I.; Dinescu, M. Sensitive Materials and Coating Technologies for Surface Acoustic Wave Sensors. *Chemosensors* **2021**, *9*, 105. [CrossRef]
19. Wen, W.; He, S.; Li, S.; Liu, M.; Pan, Y. Advances in SXFA-Coated SAW Chemical Sensors for Organophosphorous Compound Detection. *Sensors* **2011**, *11*, 1526–1541.
20. Liu, X.; Wang, W.; Zhang, Y.; Pan, Y.; Liang, Y.; Li, J. Enhanced Sensitivity of a Hydrogen Sulfide Sensor Based on Surface Acoustic Waves at Room Temperature. *Sensors* **2018**, *18*, 3796. [CrossRef]
21. Grate, J.W. Hydrogen Bond Acidic Polymers for Surface Acoustic Wave Vapor Sensors and Arrays. *Anal. Chem.* **1999**, *71*, 1033–1040. [CrossRef]
22. Abraham, M.H.; Andonian-Haftvan, J.; Du, C.M.; Diart, V.; Whiting, G.S.; Grate, J.W.; Andrew, M.R. Hydrogen bonding. Part 29. Characterization of 14 sorbent coatings for chemical microsensors using a new solvation equation. *J. Chem. Soc. Perkin Trans.* **1995**, *2*, 369–378. [CrossRef]
23. Freudenberg, J.; Schickfus, M.V.; Hunklinger, S. A SAW immunosensor for operation in liquid using a SiO_2 protective layer. *Sens. Actuators B Chem.* **2001**, *76*, 147–151. [CrossRef]
24. Kumar, K.V.; Gadipelli, S.; Wood, B.; Ramisetty, K.A.; Stewart, A.A.; Howard, C.A.; Brett, D.J.L.; Rodriguez-Reinoso, F. Characterization of the adsorption site energies and heterogeneous surfaces of porous materials. *J. Mater. Chem. A* **2019**, *7*, 17. [CrossRef]
25. Al-Ghouti, M.A.; Da'ana, D.A. Guidelines for the use and interpretation of adsorption isotherm models: A review. *J. Hazard. Mater.* **2020**, *393*, 122383. [CrossRef]
26. Kindlund, A.; Sundgren, H.; Lundstrm, I. Quartz crystal gas monitor with a gas concentrating stage. *Sens. Actuators* **1984**, *6*, 1–17. [CrossRef]
27. Finklea, H.O.; Phillippi, M.A.; Lompert, E.; Grate, J.W. Highly sorbent films derived from ni(scn)2(4-picoline)4 for the detection of chlorinated and aromatic hydrocarbons with quartz crystal microbalance sensors. *Anal. Chem.* **1998**, *70*, 1268–1276. [CrossRef]
28. Gregory, C.F.; Stephen, J.M. Materials characterization using surface acoustic wave devices. *Appl. Spectrosc. Rev.* **1991**, *26*, 73–149.
29. Grate, J.W.; Wenzel, S.W.; White, R.M. Flexural plate wave device for chemical analysis. *Anal. Chem.* **1991**, *63*, 1552–1561. [CrossRef]
30. Grate, J.W.; Klusty, M.; Mcgill, R.A.; Abraham, M.H.; Whiting, G.; Andonian-Haftvan, J. The predominant role of swelling-induced modulus changes of the sorbent phase in determining the responses of polymer-coated surface acoustic wave vapor sensors. *Anal. Chem.* **1992**, *64*, 610–624. [CrossRef]
31. Tascon, M.; Romero, L.M.; Acquaviva, A.; Keunchkarian, S.; Castells, C. Determinations of gas-liquid partition coefficients using capillary chromatographic columns. alkanols in squalane. *J. Chromatogr. A* **2013**, *1294*, 130–136. [CrossRef] [PubMed]
32. Abraham, M.H.; Rosés, M.; Poole, C.F.; Poole, S.K. Hydrogen bonding. 42. characterization of reversed-phase high-performance liquid chromatographic c18 stationary phases. *J. Phys. Org. Chem.* **2015**, *10*, 358–368. [CrossRef]
33. Grate, J.W.; Patrash, S.J.; Abraham, M.H. Method for estimating polymer-coated acoustic wave vapor sensor responses. *Anal. Chem.* **1995**, *67*, 2162–2169. [CrossRef]
34. Grate, J.W.; Kaganove, S.N.; Bhethanabotla, V.R. Comparisons of polymer/gas partition coefficients calculated from responses of thickness shear mode and surface acoustic wave vapor sensors. *Anal. Chem.* **1998**, *70*, 199–203. [CrossRef]

Disclaimer/Publisher's Note: The statements, opinions and data contained in all publications are solely those of the individual author(s) and contributor(s) and not of MDPI and/or the editor(s). MDPI and/or the editor(s) disclaim responsibility for any injury to people or property resulting from any ideas, methods, instructions or products referred to in the content.

Article

Flowing Liquid-Based Triboelectric Nanogenerator Performance Enhancement with Functionalized Polyvinylidene Fluoride Membrane for Self-Powered Pulsating Flow Sensing Application

Duy Linh Vu [1], Quang Tan Nguyen [2], Pil Seung Chung [1,3,*] and Kyoung Kwan Ahn [2,*]

[1] Department of Nanoscience and Engineering, Inje University, 197 Inje-ro, Gimhae-si, Gyeongsangnamdo 50834, Republic of Korea; vuduylinhbk@gmail.com
[2] School of Mechanical Engineering, University of Ulsan, 93 Daehak-ro, Nam-gu, Ulsan 44610, Republic of Korea; pax.quangtan@gmail.com
[3] Department of Energy Engineering, Inje University, 197 Inje-ro, Gimhae-si, Gyeongsangnamdo 50834, Republic of Korea
* Correspondence: pschung01@inje.ac.kr (P.S.C.); kkahn@ulsan.ac.kr (K.K.A.)

Abstract: Pulsating flow, a common term in industrial and medical contexts, necessitates precise water flow measurement for evaluating hydrodynamic system performance. Addressing challenges in measurement technologies, particularly for pulsating flow, we propose a flowing liquid-based triboelectric nanogenerator (FL-TENG). To generate sufficient energy for a self-powered device, we employed a fluorinated functionalized technique on a polyvinylidene fluoride (PVDF) membrane to enhance the performance of FL-TENG. The results attained a maximum instantaneous power density of 50.6 µW/cm^2, and the energy output proved adequate to illuminate 10 white LEDs. Regression analysis depicting the dependence of the output electrical signals on water flow revealed a strong linear relationship between the voltage and flow rate with high sensitivity. A high correlation coefficient R^2 within the range from 0.951 to 0.998 indicates precise measurement accuracy for the proposed FL-TENG. Furthermore, the measured time interval between two voltage peaks precisely corresponds to the period of pulsating flow, demonstrating that the output voltage can effectively sense pulsating flow based on voltage and the time interval between two voltage peaks. This work highlights the utility of FL-TENG as a self-powered pulsating flow rate sensor.

Keywords: fluid-based triboelectric nanogenerator; pulsating flow rate; fluorinated functionalized; polyvinylidene fluoride; self-powered sensor

Citation: Vu, D.L.; Nguyen, Q.T.; Chung, P.S.; Ahn, K.K. Flowing Liquid-Based Triboelectric Nanogenerator Performance Enhancement with Functionalized Polyvinylidene Fluoride Membrane for Self-Powered Pulsating Flow Sensing Application. *Polymers* **2024**, *16*, 536. https://doi.org/10.3390/polym16040536

Academic Editor: Jung-Chang Wang

Received: 13 January 2024
Revised: 2 February 2024
Accepted: 13 February 2024
Published: 16 February 2024

Copyright: © 2024 by the authors. Licensee MDPI, Basel, Switzerland. This article is an open access article distributed under the terms and conditions of the Creative Commons Attribution (CC BY) license (https://creativecommons.org/licenses/by/4.0/).

1. Introduction

In recent times, the rapid expansion of the Internet of Things (IoT) has led to a prevailing trend in the development of electronic technology, specifically focusing on miniaturized and portable devices. This trend underscores a growing interest in self-powered functionality, aiming to extend operational durations while reducing reliance on conventional battery usage [1–4]. Consequently, the exploration of micro/nano-technology has captivated the attention of numerous researchers, culminating in the creation of nanogenerators grounded in piezoelectric, pyroelectric, thermoelectric, and triboelectric principles [5–9]. Since it was invented in 2012, the triboelectric nanogenerator (TENG) has emerged as a prominent technology for harvesting ambient mechanical energy from various environmental sources, including vibrations, human motion, wind, and ocean waves [10–13]. The working mechanism of TENG relates to the coupling of contact electrification and electrostatic induction between diverse materials—solid, liquid, and gas [14–16]. TENG devices have undergone rapid development, showcasing the significant potential for harvesting energy from low-frequency mechanical sources [9,17–19].

Notably, TENGs have evolved into smart sensing devices, uniquely influencing input parameters and generating corresponding electrical responses. Their attributes include self-powered operation, cost-effectiveness, simple structure, easy fabrication, portability, and high reliability [20–23]. Furthermore, water-based TENGs have gained attention due to the adaptable nature of water, enabling effective contact with solid layers for enhanced triboelectric charge generation. Several studies have investigated the use of water-based TENGs in sensing devices, including those for detecting tubular flowing water, liquid level, humidity, and chemical detection [23–26]. However, limited attention has been given to the study of unsteady flow, primarily due to the intricate nature of flow dynamics. Therefore, it is necessary to conduct research on water-based TENGs for converting energy from water movement into electricity and employ them as sensing devices to measure the flow rate.

Pulsating flow is commonly used to characterize a specific type of flow, characterized by the combination of a periodically oscillating flow and a steady flow [27,28]. This concept holds significant relevance in various industries and medical fields, with applications spanning heat transfer augmentation, improved cleaning processes, fluid mixing, mass transport in porous media, and biofluid engineering [29–31]. Pulsating flow in pipes has been observed in diverse technical areas such as physiology, roller and finger pumps, transportation of blood flow, oxygen, and sanitary fluids, among others [32–37]. Scientists have shown considerable interest in studying pulsating flow, especially focusing on accurate flow rate measurement, which is crucial for evaluating system performance [38–40]. Typically, flow rate determination involves multiplying physical quantities by correction factors corresponding to measurement technologies like turbine rotational frequency [40], pressure drops through an orifice [41], electromotive force [42], ultrasonic wave transit time, or Doppler effect [43]. However, these technologies have their disadvantages; for instance, flow disturbance, high cost, complex structure, and limited application in millimeter-scale pipelines. Consequently, the flowing liquid-based TENG (FL-TENG) emerges as a promising solution to address these limitations and serve as a self-powered pulsating flow sensor. The success of FL-TENGs in practical applications relies on their ability to generate sufficient output power, usually stored in a capacitor or battery. Nevertheless, the suboptimal quality of materials used to make FL-TENGs degrades their output performance, making them unsuitable for powering electronic devices [44,45]. Therefore, there is a need to enhance the performance of FL-TENGs by increasing the transferred charge density through liquid–solid electrification. In our prior studies, we thoroughly investigated the effect of fluorinated functionalization on the output voltage and current of TENG devices utilizing a PVDF membrane. The application of functionalization was found to significantly influence the output performance of the TENG device [44,46].

This paper introduces an inventive and advanced approach to measuring pulsating flow in pipelines, utilizing a FL-TENG that exhibits heightened output performance due to the integration of a functionally enhanced triboelectric layer. To obtain a high-charge transferred density triboelectric layer, we employed a fluorinated functionalized technique on a polyvinylidene fluoride (PVDF) membrane. This involved grafting the membrane with negatively charged 1H,1H,2H,2H-Perfluorooctyltrie-thoxysilane (FOTS). The substantial negative polarizations of fluorine played a pivotal role in significantly improving the dielectric constant and the hydrophobic property of the functionalized PVDF (F-PVDF) membrane, which led to a notable increase in the performance of the TENG. The F-PVDF-based TENG reaches a maximum voltage of 10.4 V at a flow rate of 1300 mL/min, representing a 1.9-fold increase compared to the pristine PVDF-based TENG. It also attained a maximum instantaneous power density of 50.6 $\mu W/cm^2$. This energy output proved sufficient to illuminate 10 white LEDs. Furthermore, a correlation between the output electrical signals and water pulsating flow was established based on the amplitude and period of the output signals. Through regression analysis, a strong linear relationship was observed between the amplitude of the voltage and the flow rate, exhibiting high sensitivity ranging from 4.2 to 7.9 mV/mL.min. Simultaneously, an inversely proportional relationship was observed between the period of the pulsating voltage signal and the flow rate, with a constant proportionality of 74.284 s.mL/min. The high correlation coefficient

R^2, within the range [0.951, 0.998], underscores the accuracy of the proposed FL-TENG, showcasing its considerable potential as a self-powered pulsating flow sensor.

2. Experimental Section

2.1. Functionalized PVDF Membrane and FL-TENG Device Fabrication

Figure 1 illustrates the procedural steps involved in transforming the functionalized PVDF membrane into the fabrication of the FL-TENG device. Initially, a PVDF membrane (50 µm, Sigma-Aldrich, St. Louis, MO, USA) underwent treatment with an alkaline solution (7.5 M NaOH) for a duration of 3 hours at 70 °C to induce hydroxyl functionality. Subsequently, the hydroxylated PVDF membrane underwent fluorination by immersing it in a solution of 1H,1H,2H,2H-Perfluorooctyltriethoxysilane (FOTS, 98%, Sigma-Aldrich) with a concentration of 1.0 wt% for 24 h, resulting in the formation of the F-PVDF membrane.

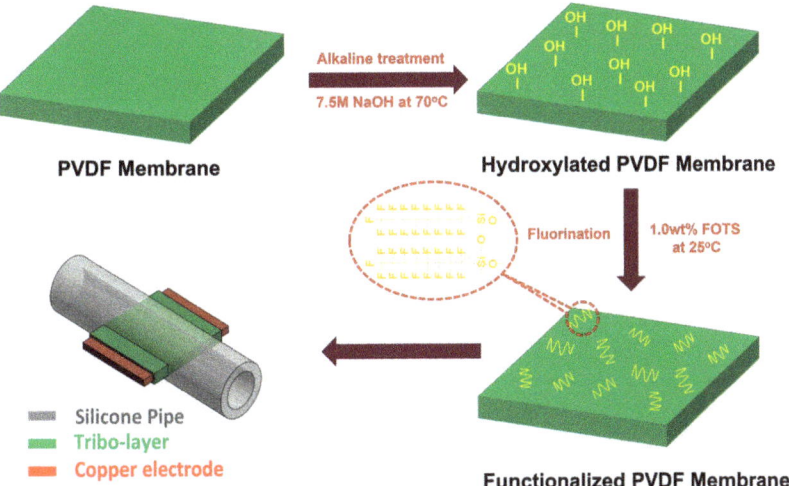

Figure 1. Schematic diagram of the procedure for functionalizing the PVDF membrane and diagram description of the FL-TENG device.

The sketch of a typical FL-TENG device, comprising a silicon pipe, a copper electrode, and a FL-PVDF membrane, is shown in Figure 1. The silicon pipes vary in inner diameters, specifically 3 mm, 5 mm, and 8 mm, denoted as 3 mm-pipe, 5 mm-pipe, and 8 mm-pipe, respectively. The width of the F-PVDF membrane corresponds to the inner diameter of the silicon pipe, determined by its size. It is important to note that the mentioned width pertains specifically to the contact area of the F-PVDF membrane inside the pipe. A copper electrode, with a thickness of 200 µm, is wrapped by the F-PVDF membrane and affixed at the center of the cross-sectional area of the pipe. For a visual representation, a real photograph of the FL-TENG device featuring an 8 mm-pipe is presented in Figure S1 (Supplementray Materials).

2.2. Characterization and Measurements

The surface structures of diverse membranes were examined using a JSM-7600 FE-SEM from JEOL Ltd., Tokyo, Japan. For the analysis of the chemical composition of these membranes, Fourier transform infrared (FTIR) analysis was conducted, using the Varian 640-IR FTIR Spectrometer, Varian Inc., Palo Alto, CA, USA. Additionally, an atomic force microscope (AFM) (MFP-3D Stand Alone AFM, Oxford Instruments, Abingdon, UK) was employed to analyze the surface roughness of the membranes. To assess hydrophobicity, the water contact angle of the membrane was measured using SmartDrop (FemtoFAB, Waltham, MA, USA). The dielectric characteristics of the membranes were determined using an impedance analyzer, specifically the 3522-50 LCR Meter from Hioki Nagano, Japan, with a frequency range spanning from 1 to 10^7 Hz.

The method of Structured Analysis and Design Technique was employed to delineate the function of the proposed FL-TENG system, as illustrated in Figure 2 and Video S1 (Supplementary Materials). For measuring the generated output, a digit graphical sampling multimeter was utilized (Keithley DMM7510, Keithley, OH, USA). To facilitate this assessment, the AC-generated voltage underwent conversion to DC using a bridge rectifier. The output power was then stored in capacitors with varying capacitance values. The typical peristaltic pump utilized in this system featured a rotor equipped with multiple rollers (specifically, a three-roller pump) attached to an external flexible tube. During rotation of the rotor, the compression of a section of the tube resulted in closure, forcing the fluid to move through the tube, thus facilitating the pulsatile flow.

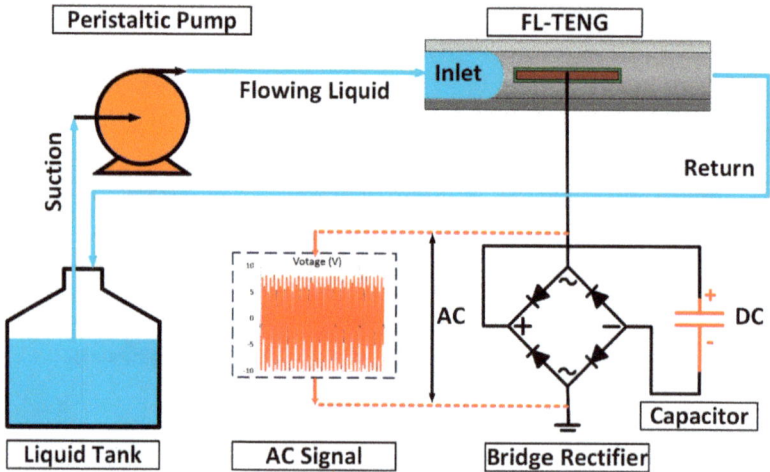

Figure 2. Schematic diagram of the experimental set up of the FL-TENG device.

3. Results and Discussions

3.1. Working Mechanisms

The working principle of the FL-TENG is based on the formation of an electric double layer (EDL) at the interface between water and the F-PVDF surface. Prior to contact, ions in water do not directly interact with those on the F-PVDF surface. However, upon contact, a bond is established with an overlap in electron clouds, creating an equilibrium distance between two atoms (Figure 3a(i)). The pressure from the water flow induces electron clouds to overlap when water ions impact the surface of the F-PVDF membrane, resulting in an interatomic distance shorter than the equilibrium distance ($X_r < d$). This facilitates electron transfer between water molecules and atoms on the F-PVDF surface (Figure 3a(ii)). Simultaneously, ionization reactions may occur on the solid surface, leading to both electron and ion transfer in the water–solid contact electrification (CE). The dominance of electron transfer in the CE is attributed to the hydrophobic of the triboelectric surface [47]. In the next stage, due to the pressure flow, water molecules adjacent to the F-PVDF surface are pushed away, increasing the interatomic distance ($X_a > d$). This diminishes the electron clouds' overlap, breaking the formed bonds. The transferred electrons then remain on the F-PVDF surface as static charge, creating a negatively charged layer on the F-PVDF surface. The charged water molecules become freely migrating ions (Figure 3a(iii)). Then, as shown in Figure 3a(iv), the loosely distributed positive ions in water are attracted to adsorb onto the F-PVDF surface through electrostatic interactions, forming an EDL. As the flow carries away adsorbed ions on the F-PVDF surface, more charges are transferred across the interface to replenish the EDL. The formation of the EDL is likely a result of contact electrification stemming from electron transfer at the water–solid interface [14]. Due to electrostatic induction in the electrode, electrons alternatively flow between the electrode and the ground through the external circuit. Figure 3b illustrates the voltage and current of the FL-TENG, along with typical signals in inset images, measured at a water flow rate of 390 mL/min.

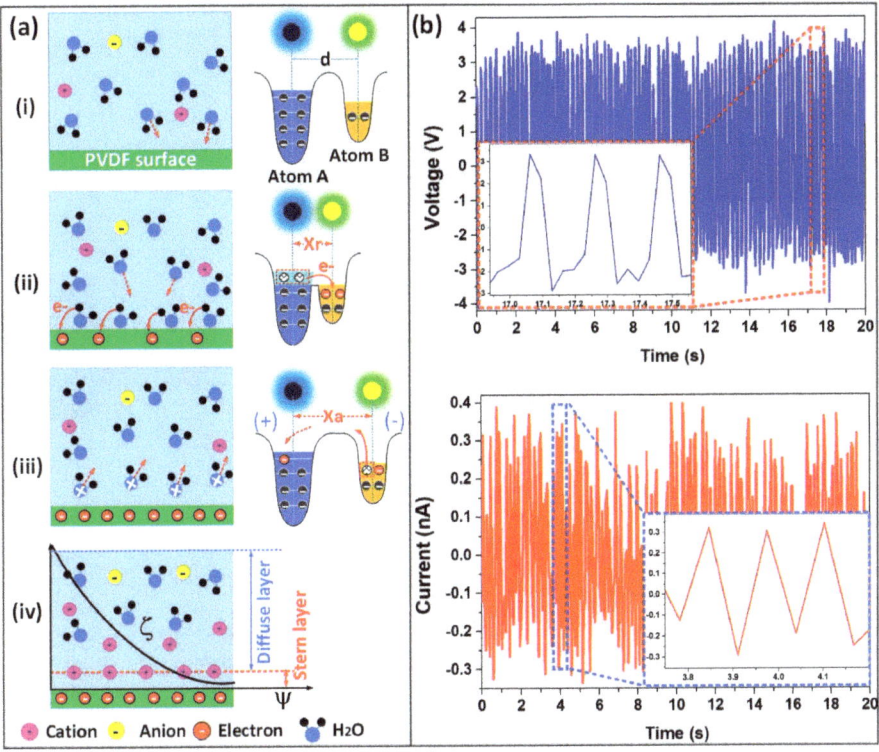

Figure 3. (a) Schematic diagram of contact electrification and the forming of EDL. (b) Output electrical signals of the SE-WTENG when water flows through the cell at a flow rate of 390 cc/min.

3.2. Electrical Output Characteristics

The analysis of the surface morphology of PVDF and F-PVDF membranes was conducted using FE-SEM and AFM, as illustrated in Figure 4a,b. The FE-SEM image reveals that the PVDF surface has a highly porous structure with evenly distributed pores ranging from 400nm to 600 nm and smooth pore walls. In contrast, the porosity of the F-PVDF membrane decreases due to the hydroxyl surface functionality process. The AFM images show a significant increase in the root mean square roughness (Rq) of the membrane, from 136 nm to 183 nm, after functionalization. This increase enhances the water contact angle of the membrane [48,49]. Moreover, the introduction of fluorine in FOTS, the most electronegative element, enhances polarizability and dipole moment, thereby manifesting superior hydrophobicity and dielectric constant properties [50,51]. The obvious increase in the water contact angle results is evident, surging from 126.3° to 145.5°. Exploring the dielectric constant of the PVDF and F-PVDF membrane, particularly in a frequency-dependent manner at room temperature, reveals a remarkable disparity (Figure 4c). Notably, at a frequency of 10^3, the dielectric constant of the F-PVDF membrane is approximately 12.3, which is 35% higher than that of the PVDF membrane. To assess the impact of functionalization on FL-TENG output performance, the output voltage of PVDF and F-PVDF-based TENG were investigated. As anticipated, the F-PVDF-based TENG exhibits a notable voltage of 10.7 V, showcasing a remarkable 1.9-fold increase compared to the PVDF-based TENG. These results highlight that the F-PVDF membrane enhances the output performance of the FL-TENG, making it advantageous for use in self-powered pulsating flow sensors.

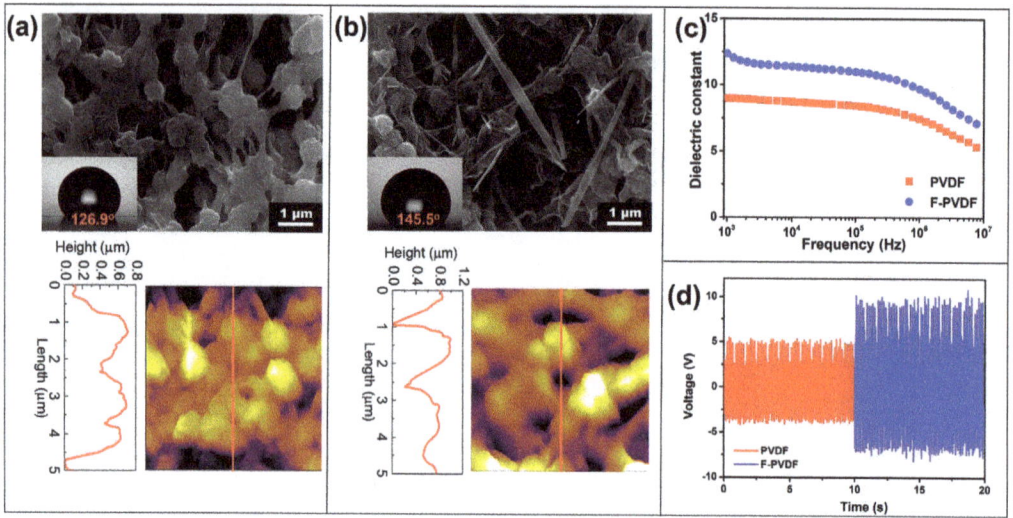

Figure 4. FE-SEM and AFM images of (**a**) PVDF membrane and (**b**) F-PVDF membrane; the inset image shows a contact angle; (**c**) frequency dependence of dielectric constant, and (**d**) output current of PVDF and F-PVDF-based TENG.

To explore the potential application of the F-PVDF-based TENG, performance assessments were conducted using DI water, and an 8 mm-pipe was utilized to evaluate the performance of the FL-TENG. Real-time measurements, depicted in Figure 5a, were taken with different water flow rates. Evidently, as the flow rate increases, the shear force exerted by the flow on the F-PVDF surface rises, leading to the generation of more charge. The result demonstrates a consistent increase in voltage with the flow rate, reaching a maximum voltage of 10.2 V at a 1170 mL/min flow rate. Furthermore, electrostatic charge transfer, as illustrated in Figure 5b, can be determined using the following equation:

$$Q_c = \frac{1}{R} \int_{t_1}^{t_2} V dt \tag{1}$$

where Q_c is the electrostatic charge transfer, R is the electric resistance, V is the voltage, and t_1 and t_2 are the related times. According to this graph, the F-PVDF-based TENG can generate a charge transfer of 5.2 nC at a 130 mL/min flow rate, reaching a maximum value of approximately 36.3 nC at 1170 mL/min. Moreover, Figure 5c illustrates the relationship between flow rate and voltage, incorporating various resistors in the external circuit (1 kΩ to 10 MΩ). Following that, instantaneous power was calculated and reached 11.5 µW at a 1170 mL/min flow rate (Figure 5d). Remarkably, this power is sufficient to directly illuminate a series of 10 white LEDs, as shown in the inset figures and Video S2 (Supplementary Materials). Correspondingly, Figure 5e presents the calculations for the power density and energy density of the F-PVDF-based TENG. At a flow rate of 1170 mL/min, the FL-TENG demonstrates a power density of 50.6 µW/cm^2 and an energy density of 0.45 µJ/cm^2. Consequently, a bridge rectifier is introduced into the external circuit to convert the output electricity into direct current (DC) for charging the capacitor, as illustrated in the inset figure of Figure 5f. Various capacitors (4.7 µF, 22 µF, and 47 µF) are utilized to measure the charging capability of the FL-TENG under a flow rate of 1170 mL/min. A 4.7 µF capacitor charges to 2.6 V in about 90 s, whereas a 47 µF capacitor takes 500 s to charge to 2.3 V. These findings indicate the significant potential of the FL-TENG for application as a self-powered device.

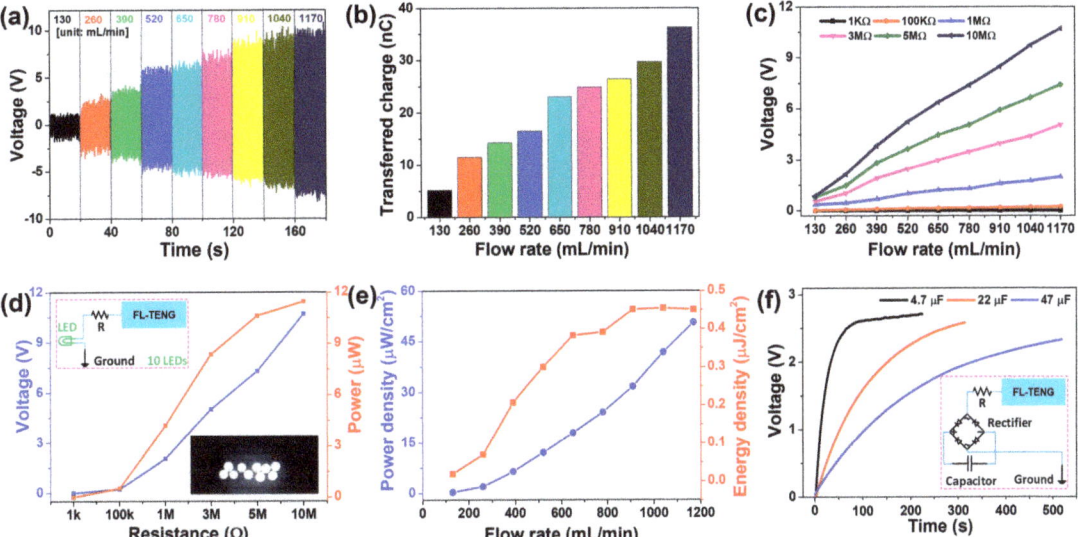

Figure 5. (**a**) Voltage and (**b**) transferred charge of the F-PVDF-based TENG depending on the flow rate (Different colors correspond to different flow rates shown as in the x-axis), (**c**) comparison of voltages measured at different resistances and flow rates, (**d**) voltage and power, and (**e**) power density and energy density of the F-PVDF-based TENG, measured at various resistances from 1 kW to 10 MW at a flow rate of 1170 mL/min, (**f**) charging of 4.7µF, 22µF, and 47µF capacitor by F-PVDF-based TENG.

3.3. Application in Self-Powered Flow Sensor

The objective of this study is to develop a self-powered sensor capable of detecting pulsating water flow within a millimeter-scale by monitoring the electrical response of the FL-TENG. Figure 6a–c shows typical output signals obtained at flow rates of 130, 390, and 780 mL/min from the experimental results of Figure 5a. In general, there are two signals that characterize the pulsating flow rate: amplitude, represented by the output voltage, and a period, represented by the time interval between two voltage peaks. As depicted in Figure 6a, at a flow rate of 130 mL/min, the F-PVDF-based TENG generates a voltage (V_1) of 1.24 V, with a time interval between two peaks (ΔT_1) measured as 0.501 s. Upon increasing the flow rate to 390 mL/min, the output voltage (V_2) also rises, reaching values of 3.72 V. Subsequently, at a flow rate of 780 mL/min, V_3 attains values of 6.21 V. In contrast, the time interval between two voltage peaks decreased with an increase in flow rate, yielding values ΔT_2 of 0.173 s and ΔT_3 of 0.105 s. It is evident that at lower flow rates, the TENG produces a relatively lower peak voltage with longer time intervals, whereas at higher flow rates, the peak voltage is higher, but the time interval is comparatively shorter.

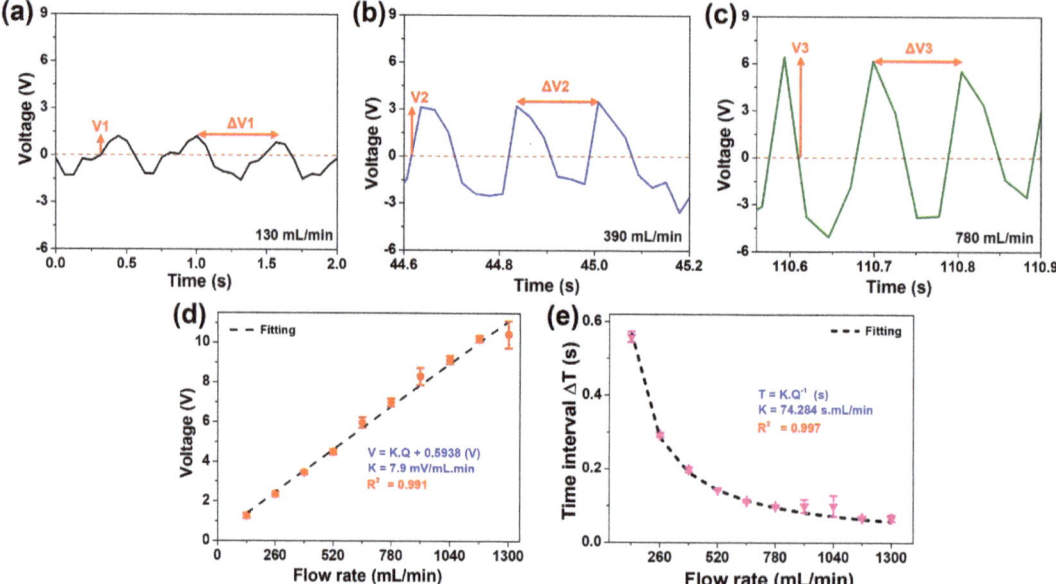

Figure 6. Electrical performance of F-PVDF-based TENG, depending on water flow conditions at a flow rate of (**a**) 130, (**b**) 390, and (**c**) 780 mL/min; regression analyses of the electrical response of FL-TENG based on (**d**) voltage and (**e**) time interval between two voltage peaks with different flow rates.

The regression analysis is carried out on the electrical signal output concerning the flow rate to uncover the relationship between the pulsed electrical response and pulsating flow. Illustrated in Figure 6d,e, it is evident that the association between the voltage and the flow rate is a robust linear relationship, whereas the time interval between two voltage peaks demonstrates an inverse proportionality to the flow rate. The regression curves for these relationships are respectively delineated by:

$$V = K_v \cdot Q + V_0 \quad (2)$$

$$T = \frac{K_t}{Q} \quad (3)$$

where V is the peak voltage (V), T is the time interval between two peaks of voltage (s), Q is the flow rate (mL/min), K_v and K_t are the constants of proportionality (or sensitivity), and V_0 is the V-interpret. Consequently, an amplitude sensitivity (K_v) of 7.9 mV/mL.min is attained for the voltage signal, accompanied by a high coefficient of determination R^2 value of 0.991. Simultaneously, the proportionality constant (K_t) for the period of time of the output signal is equal to 74.284 s.mL/min, corresponding to an R^2 value of 0.997. These elevated sensitivities and high coefficients of determination affirm an exceptional accuracy of measurement, validating the estimation of flow rate through the monitoring of the output electrical signal.

Moreover, it is interesting to draw a comparison between the correlation of flow rate with the period of the output electrical signal and the relationship between flow rate and the period of pulsating flow produced by the pump. This comparative analysis serves as evidence for the potential application of the FL-TENG. Typically, the three-roller peristaltic pump generates the pulsating flow through the interaction among the rollers and the flexible tube. A water pillow is formed between two rollers, advancing along the tube in the direction of the revolving rotor, and is subsequently propelled into the discharge outlet

(Video S3, Supplementary Materials). The operation involves alternating compression, squeezing, and release of the tube, enabling the generation of a pulsating flow of water during a single roller-step. Consequently, three pulses of flow are produced within one revolution of the rotor. Therefore, the period (T' in seconds) of the pulsating flow, referred to as the roller-step period, can be calculated by:

$$T' = \frac{60}{N \cdot n} \quad (4)$$

where N is the pump speed (rpm), and n is the number of rollers (n = 3). Moreover, the flow rate Q' in (L/min) delivered by a roller-step is given by:

$$Q' = \frac{1}{2}\pi^2 d^2 N \cdot r \quad (5)$$

where d is the inner diameter of the tube (mm), and r is the radius of the rotation(mm) that is measured from the rotor axis to the center of the rollers. The calculation of the pump is plotted in Figure S2 (Supplementary Materials). From the above equations, the relationship between the period of pulsating flow and flow rate can be determined by:

$$T' = \frac{10 \cdot \pi^2 d^2 N \cdot r}{Q'} \quad (6)$$

After applying the value of all the variables, this equation becomes:

$$T' = \frac{74.284}{Q} \quad (7)$$

Undoubtedly, there exists an inverse proportionality between the flow rate and the period of pulsating flow. Notably, this equation aligns with the regression equation established between the period of output electrical signal and the flow rate in the above discussion. By considering the analysis of both amplitude (voltage) and period (time interval) measurements of the output voltage, it is evident that the FL-TENG device has the capability to provide detailed information about the pulsating flow.

On the other hand, to demonstrate the precision of the FL-TENG in millimeter-scale measurements, various F-PVDF-based TENGs are fabricated with the 3 mm-pipe, 5 mm-pipe, and 8 mm-pipe (Figure 7a,b). The corresponding electrical responses, covering a range of flow rates from 130 to 1300 mL/min, are presented in Figure S3 (Supplementary Materials). Figure 7c–e presents the relationship between the measured time interval of two voltage peaks and the flow rate. This plot reveals that the associated time interval is prolonged at lower flow rates and shorter at higher flow rates. The regression analysis between the time interval and flow rate for all three FL-TENGs yields the same fitting curve. A constant proportionality of 74.284 s.mL/min is obtained, accompanied by a high correlation coefficient of R^2 values of 0.991, 0.998, and 0.997, respectively, for the 3 mm-pipe, 5 mm-pipe, and 8 mm-pipe. It indicates a great proportional relationship between the period of the voltage signal and the flow rate. This highlights a robust proportional connection between the period of the voltage signal and the flow rate, irrespective of the diameter of the pipe.

The regression analyses of the electrical performance of the FL-TENG based on voltage and flow rate are illustrated in Figure 7f–h. In the case of the 8 mm-pipe, a linear relationship is obtained with a sensitivity of 7.9 mV/mL.min and a correlation coefficient R^2 of 0.991, as discussed previously. Similarly, sensitivities of 7.7 and 4.2 mV/mL.min are identified for the 5 mm-pipe and 3 mm-pipe, corresponding to R^2 values of 0.974 and 0.951. Although the sensitivity of the 3 mm-pipe is significantly smaller than these two FL-TENGs, it still exhibits a high linear relationship, confirming the suitability of the FL-TENG for accurate millimeter-scale flow rate measurement. These discussions demonstrate that monitoring the voltage signals of the F-PVDF-based TENG allows for the determination of the water pulsating flow rate.

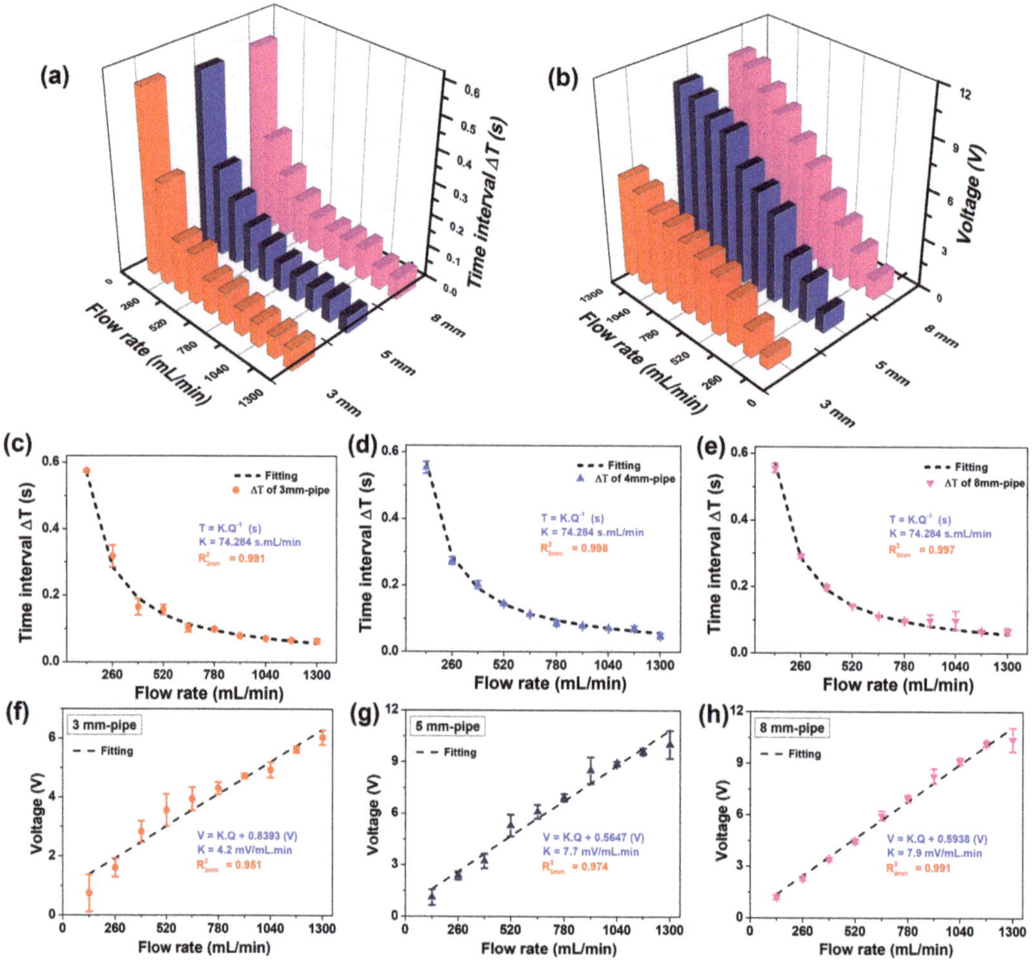

Figure 7. (a) Time interval between two voltage peaks and (b) output voltage, depending on the flow rate with different FL-TENG pipe sizes; regression analyses of the electrical response of (c) 3 mm-pipe, (d) 5 mm-pipe, (e) 8 mm-pipe-FL-TENG based on time interval between two voltage peaks and flow rates; regression analyses of the electrical response of (f) 3 mm-pipe, (g) 5 mm-pipe, and (h) 8 mm-pipe-FLTENG based on voltage and flow rate.

In practical applications, the stability and durability of a sensing device are crucial for obtaining correct and accurate measurements. To assess the stability of the F-PVDF-based TENG, experiments are conducted one month apart to validate the stability of the output signal. As illustrated in Figure S4 (Supplementary Materials), the FL-TENG exhibits a minimal change in electrical performance, indicating good working durability and stability.

4. Conclusions

In summary, a novel self-powered pulsating flow sensor has been developed using an FL-TENG, where the determination of pulsating flow involves monitoring the output electrical characteristics, specifically the amplitude and period of the pulsating voltage signal.

The performance of the FL-TENG is significantly enhanced by increasing the transferred charge density of the triboelectric layer. The F-PVDF-based TENG achieves a maximum voltage of 10.4 V at a 1300 mL/min flow rate, representing a 1.9-fold increase compared to the pristine PVDF-based TENG. Moreover, the results from regression analysis indicate a linear relationship between the output voltage and the flow rate, while the time interval between two voltage peaks is inversely proportional to the flow rate. With high sensitivity and a coefficient of determination ranging from 0.951 to 0.998, the electrical performance underscores the suitability of the FL-TENG for accurate pulsating flow measurement on a millimeter-scale. Furthermore, the device can function as a power source, contributing to the advancement of a self-powered pulsating flow sensor. The F-PVDF-based TENG exhibits notable advantages, including low cost, a straightforward structure, easy installation, reliability, and effectiveness in the millimeter-scale pipelines.

Supplementary Materials: The following supporting information can be downloaded at: https://www.mdpi.com/article/10.3390/polym16040536/s1, Figure S1: Real photograph of FL-TENG testbench; Figure S2: Characteristics of the water pump; Figure S3: Output voltage and current of the FL-TENG at a variable flow rate by different sizes of silicone pipes (a,b) 3 mm-pipe, (c,d) 5 mm-pipe, and (e,f) 8 mm-pipe; Figure S4: Stability and durability of the FL-TENG. (a,c) Output voltage measured at 390 and 650 mL/min of the FL-TENG at two moments with one month apart. (b,d) Stability test of the FL-TENG for 300 s at a flow rate of 390 and 650 mL/min; Figure S5: FTIR spectra of PVDF and F-PVDF membrane. Video S1: Experimental testbench; Video S2: Demonstration of FL-TENG performance: 10 white LEDs are directly lighted up; Video S3: Demonstration of the three-roller peristaltic pump.

Author Contributions: Conceptualization, D.L.V.; methodology, D.L.V. and Q.T.N.; validation, D.L.V. and Q.T.N.; formal analysis, D.L.V. and Q.T.N.; investigation, D.L.V. and Q.T.N.; data curation, D.L.V. and Q.T.N.; writing—original draft preparation, D.L.V.; writing—review and editing, D.L.V., P.S.C. and K.K.A.; visualization, D.L.V. and P.S.C.; supervision, P.S.C. and K.K.A.; project administration, P.S.C. and K.K.A.; funding acquisition, K.K.A. All authors have read and agreed to the published version of the manuscript.

Funding: This research was supported by "Regional Innovation Strategy (RIS)" through the National Research Foundation of Korea (NRF), funded by the Ministry of Education (MOE) (2021RIS-003), and this work was supported by the Industrial Strategic Technology Development Program (No. 20012924, Development of upcycling technology for waste CFRP with chemically degradable epoxy resins CF surface treatment and valueadded products), funded By the Ministry of Trade, Industry & Energy (MOTIE, Korea).

Institutional Review Board Statement: Not applicable.

Data Availability Statement: The authors confirm that the data supporting the findings of this study are available within the article.

Conflicts of Interest: The authors declare no conflicts of interest.

References

1. Wang, Z.L. Triboelectric Nanogenerators as New Energy Technology for Self-Powered Systems and as Active Mechanical and Chemical Sensors. *ACS Nano* **2013**, *7*, 9533–9557. [CrossRef]
2. Sharma, P.; Motte, J.F.; Fournel, F.; Cross, B.; Charlaix, E.; Picard, C. A Direct Sensor to Measure Minute Liquid Flow Rates. *Nano Lett.* **2018**, *18*, 5726–5730. [CrossRef] [PubMed]
3. Ejeian, F.; Azadi, S.; Razmjou, A.; Orooji, Y.; Kottapalli, A.; Ebrahimi Warkiani, M.; Asadnia, M. Design and Applications of MEMS Flow Sensors: A Review. *Sens. Actuators A Phys.* **2019**, *295*, 483–502. [CrossRef]
4. Yick, J.; Mukherjee, B.; Ghosal, D. Wireless Sensor Network Survey. *Comput. Netw.* **2008**, *52*, 2292–2330. [CrossRef]
5. Hanani, Z.; Izanzar, I.; Amjoud, M.; Mezzane, D.; Lahcini, M.; Uršič, H.; Prah, U.; Saadoune, I.; El Marssi, M.; Luk'yanchuk, I.A.; et al. Lead-Free Nanocomposite Piezoelectric Nanogenerator Film for Biomechanical Energy Harvesting. *Nano Energy* **2021**, *81*, 105661. [CrossRef]
6. Lee, P.C.; Hsiao, Y.L.; Dutta, J.; Wang, R.C.; Tseng, S.W.; Liu, C.P. Development of Porous ZnO Thin Films for Enhancing Piezoelectric Nanogenerators and Force Sensors. *Nano Energy* **2021**, *82*, 105702. [CrossRef]
7. Zhang, K.; Wang, Y.; Wang, Z.L.; Yang, Y. Standard and Figure-of-Merit for Quantifying the Performance of Pyroelectric Nanogenerators. *Nano Energy* **2019**, *55*, 534–540. [CrossRef]

8. Wang, N.; Feng, Y.; Zheng, Y.; Zhang, L.; Feng, M.; Li, X.; Zhou, F.; Wang, D. New Hydrogen Bonding Enhanced Polyvinyl Alcohol Based Self-Charged Medical Mask with Superior Charge Retention and Moisture Resistance Performances. *Adv. Funct. Mater.* **2021**, *31*, 2009172. [CrossRef]
9. Vu, D.L.; Ahn, K.K. Triboelectric Enhancement of Polyvinylidene Fluoride Membrane Using Magnetic Nanoparticle for Water-Based Energy Harvesting. *Polymers* **2022**, *14*, 1547. [CrossRef]
10. Luo, J.; Wang, Z.L. Recent Progress of Triboelectric Nanogenerators: From Fundamental Theory to Practical Applications. *EcoMat* **2020**, *2*, e12059. [CrossRef]
11. Kim, W.G.; Kim, D.W.; Tcho, I.W.; Kim, J.K.; Kim, M.S.; Choi, Y.K. Triboelectric Nanogenerator: Structure, Mechanism, and Applications. *ACS Nano* **2021**, *15*, 258–287. [CrossRef]
12. Bai, P.; Zhu, G.; Lin, Z.H.; Jing, Q.; Chen, J.; Zhang, G.; Ma, J.; Wang, Z.L. Integrated Multilayered Triboelectric Nanogenerator for Harvesting Biomechanical Energy from Human Motions. *ACS Nano* **2013**, *7*, 3713–3719. [CrossRef] [PubMed]
13. Huang, B.; Wang, P.; Wang, L.; Yang, S.; Wu, D. Ecent Advances in Ocean Wave Energy Harvesting by Triboelectric Nanogenerator: An Overview. *Nanotechnol. Rev.* **2020**, *9*, 716–735. [CrossRef]
14. Wang, Z.L.; Wang, A.C. On the Origin of Contact-Electrification. *Mater. Today* **2019**, *30*, 34–51. [CrossRef]
15. Chen, H.; Xing, C.; Li, Y.; Wang, J.; Xu, Y. Triboelectric Nanogenerators for a Macro-Scale Blue Energy Harvesting and Self-Powered Marine Environmental Monitoring System. *Sustain. Energy Fuels* **2020**, *4*, 1063–1077. [CrossRef]
16. Le, C.D.; Nguyen, T.H.; Vu, D.L.; Vo, C.P.; Ahn, K.K. A Rotational Switched-Mode Water-Based Triboelectric Nanogenerator for Mechanical Energy Harvesting and Vehicle Monitoring. *Mater. Today Sustain.* **2022**, *19*, 100158. [CrossRef]
17. Lai, Y.C.; Wu, H.M.; Lin, H.C.; Chang, C.L.; Chou, H.H.; Hsiao, Y.C.; Wu, Y.C. Entirely, Intrinsically, and Autonomously Self-Healable, Highly Transparent, and Superstretchable Triboelectric Nanogenerator for Personal Power Sources and Self-Powered Electronic Skins. *Adv. Funct. Mater.* **2019**, *29*, 1904626. [CrossRef]
18. Nguyen, Q.T.; Vu, D.L.; Le, C.D.; Ahn, K.K. Enhancing the Performance of Triboelectric Generator: A Novel Approach Using Solid–Liquid Interface-Treated Foam and Metal Contacts. *Polymers* **2023**, *15*, 2392. [CrossRef]
19. Ye, C.; Liu, D.; Chen, P.; Cao, L.N.Y.; Li, X.; Jiang, T.; Wang, Z.L. An Integrated Solar Panel with a Triboelectric Nanogenerator Array for Synergistic Harvesting of Raindrop and Solar Energy. *Adv. Mater.* **2023**, *35*, e2209713. [CrossRef] [PubMed]
20. Lee, J.H.; Kim, S.M.; Kim, T.Y.; Khan, U.; Kim, S.W. Water Droplet-Driven Triboelectric Nanogenerator with Superhydrophobic Surfaces. *Nano Energy* **2019**, *58*, 579–584. [CrossRef]
21. Kwak, S.S.; Lin, S.; Lee, J.H.; Ryu, H.; Kim, T.Y.; Zhong, H.; Chen, H.; Kim, S.W. Triboelectrification-Induced Large Electric Power Generation from a Single Moving Droplet on Graphene/Polytetrafluoroethylene. *ACS Nano* **2016**, *10*, 7297–7302. [CrossRef]
22. Ding, X.; Cao, H.; Zhang, X.; Li, M.; Liu, Y. Large Scale Triboelectric Nanogenerator and Self-Powered Flexible Sensor for Human Sleep Monitoring. *Sensors* **2018**, *18*, 1713. [CrossRef]
23. Nguyen, Q.T.; Vu, D.L.; Le, C.D.; Ahn, K.K. Recent Progress in Self-Powered Sensors Based on Liquid-Solid Triboelectric Nanogenerators. *Sensors* **2023**, *23*, 5888. [CrossRef] [PubMed]
24. Munirathinam, K.; Kim, D.S.; Shanmugasundaram, A.; Park, J.; Jeong, Y.J.; Lee, D.W. Flowing Water-Based Tubular Triboelectric Nanogenerators for Sustainable Green Energy Harvesting. *Nano Energy* **2022**, *102*, 107675. [CrossRef]
25. Rehman, H.M.M.U.; Prasanna, A.P.S.; Rehman, M.M.; Khan, M.; Kim, S.J.; Kim, W.Y. Edible Rice Paper-Based Multifunctional Humidity Sensor Powered by Triboelectricity. *Sustain. Mater. Technol.* **2023**, *36*, e00596. [CrossRef]
26. Wang, P.; Zhang, S.; Zhang, L.; Wang, L.; Xue, H.; Wang, Z.L. Non-Contact and Liquid–Liquid Interfacing Triboelectric Nanogenerator for Self-Powered Water/Liquid Level Sensing. *Nano Energy* **2020**, *72*, 104703. [CrossRef]
27. Nabavi, M.; Siddiqui, K. A Critical Review on Advanced Velocity Measurement Techniques in Pulsating Flows. *Meas. Sci. Technol.* **2010**, *21*, 042002. [CrossRef]
28. Dincau, B.; Dressaire, E.; Sauret, A. Pulsatile Flow in Microfluidic Systems. *Small* **2020**, *16*, 1904032. [CrossRef] [PubMed]
29. Jia, W.; Zhang, X.; Zhang, H.; Ren, Y. Turbulent Transport Dissimilarities of Particles, Momentum, and Heat. *Environ. Res.* **2022**, *211*, 113111. [CrossRef] [PubMed]
30. Jin, D.X.; Lee, Y.P.; Lee, D.Y. Effects of the Pulsating Flow Agitation on the Heat Transfer in a Triangular Grooved Channel. *Int. J. Heat Mass Transf.* **2007**, *50*, 3062–3071. [CrossRef]
31. Hassani, A.; Scaria, J.; Ghanbari, F.; Nidheesh, P.V. Sulfate Radicals-Based Advanced Oxidation Processes for the Degradation of Pharmaceuticals and Personal Care Products: A Review on Relevant Activation Mechanisms, Performance, and Perspectives. *Environ. Res.* **2023**, *217*, 114789. [CrossRef]
32. Brahma, I. Measurement and Prediction of Discharge Coefficients in Highly Compressible Pulsating Flows to Improve EGR Flow Estimation and Modeling of Engine Flows. *Front. Mech. Eng.* **2019**, *5*, 25. [CrossRef]
33. Matsusaka, S.; Fukuda, H.; Sakura, Y.; Masuda, H.; Ghadiri, M. Analysis of Pulsating Electric Signals Generated in Gas-Solids Pipe Flow. *Chem. Eng. Sci.* **2008**, *63*, 1353–1360. [CrossRef]
34. Walker, E.S.; Fedak, K.M.; Good, N.; Balmes, J.; Brook, R.D.; Clark, M.L.; Cole-Hunter, T.; Dinenno, F.; Devlin, R.B.; L'Orange, C.; et al. Acute Differences in Pulse Wave Velocity, Augmentation Index, and Central Pulse Pressure Following Controlled Exposures to Cookstove Air Pollution in the Subclinical Tests of Volunteers Exposed to Smoke (SToVES) Study. *Environ. Res.* **2020**, *180*, 108831. [CrossRef] [PubMed]
35. Tezuka, K.; Mori, M.; Suzuki, T.; Kanamine, T. Ultrasonic Pulse-Doppler Flow Meter Application for Hydraulic Power Plants. *Flow Meas. Instrum.* **2008**, *19*, 155–162. [CrossRef]

36. Tonon, D.; Willems, J.F.H.; Hirschberg, A. Self-Sustained Oscillations in Pipe Systems with Multiple Deep Side Branches: Prediction and Reduction by Detuning. *J. Sound Vib.* **2011**, *330*, 5894–5912. [CrossRef]
37. Huang, H.; Zeng, S.; Luo, C.; Long, T. Separate Effect of Turbulent Pulsation on Internal Mass Transfer in Porous Biofilms. *Environ. Res.* **2023**, *217*, 114972. [CrossRef] [PubMed]
38. Abiev, R.S.; Vasilev, M.P. Pulsating Flow Type Apparatus: Energy Dissipation Rate and Droplets Dispersion. *Chem. Eng. Res. Des.* **2016**, *108*, 101–108. [CrossRef]
39. Olczyk, A. Investigation of the Specific Mass Flow Rate Distribution in Pipes Supplied with a Pulsating Flow. *Int. J. Heat Fluid Flow* **2009**, *30*, 637–646. [CrossRef]
40. Nag, A.; Hvizdoš, P.; Dixit, A.R.; Petrů, J.; Hloch, S. Influence of the Frequency and Flow Rate of a Pulsating Water Jet on the Wear Damage of Tantalum. *Wear* **2021**, *477*, 203893. [CrossRef]
41. Ghorbani, H.; Wood, D.A.; Choubineh, A.; Tatar, A.; Abarghoyi, P.G.; Madani, M.; Mohamadian, N. Prediction of Oil Flow Rate through an Orifice Flow Meter: Artificial Intelligence Alternatives Compared. *Petroleum* **2020**, *6*, 404–414. [CrossRef]
42. Watral, Z.; Jakubowski, J.; Michalski, A. Electromagnetic Flow Meters for Open Channels: Current State and Development Prospects. *Flow Meas. Instrum.* **2015**, *42*, 16–25. [CrossRef]
43. Rothfuss, M.A.; Unadkat, J.V.; Gimbel, M.L.; Mickle, M.H.; Sejdić, E. Totally Implantable Wireless Ultrasonic Doppler Blood Flowmeters: Toward Accurate Miniaturized Chronic Monitors. *Ultrasound Med. Biol.* **2017**, *43*, 561–578. [CrossRef] [PubMed]
44. Vu, D.L.; Le, C.D.; Vo, C.P.; Ahn, K.K. Surface Polarity Tuning through Epitaxial Growth on Polyvinylidene Fluoride Membranes for Enhanced Performance of Liquid-Solid Triboelectric Nanogenerator. *Compos. Part B Eng.* **2021**, *223*, 109135. [CrossRef]
45. Lee, J.W.; Jung, S.; Lee, T.W.; Jo, J.; Chae, H.Y.; Choi, K.; Kim, J.J.; Lee, J.H.; Yang, C.; Baik, J.M. High-Output Triboelectric Nanogenerator Based on Dual Inductive and Resonance Effects-Controlled Highly Transparent Polyimide for Self-Powered Sensor Network Systems. *Adv. Energy Mater.* **2019**, *9*, 1901987. [CrossRef]
46. Vu, D.L.; Le, C.D.; Ahn, K.K. Functionalized Graphene Oxide/Polyvinylidene Fluoride Composite Membrane Acting as a Triboelectric Layer for Hydropower Energy Harvesting. *Int. J. Energy Res.* **2022**, *46*, 9549–9559. [CrossRef]
47. Lin, S.; Xu, L.; Wang, A.C.; Wang, Z.L. Quantifying Electron-Transfer in Liquid-Solid Contact Electrification and the Formation of Electric Double-Layer. *Nat. Commun.* **2020**, *11*, 399. [CrossRef]
48. Dong, Z.Q.; Ma, X.H.; Xu, Z.L.; Gu, Z.Y. Superhydrophobic Modification of PVDF-SiO2 Electrospun Nanofiber Membranes for Vacuum Membrane Distillation. *RSC Adv.* **2015**, *5*, 67962–67970. [CrossRef]
49. Kozbial, A.; Li, Z.; Conaway, C.; McGinley, R.; Dhingra, S.; Vahdat, V.; Zhou, F.; Durso, B.; Liu, H.; Li, L. Study on the Surface Energy of Graphene by Contact Angle Measurements. *Langmuir* **2014**, *30*, 8598–8606. [CrossRef]
50. Alparone, A. Dipole (Hyper)Polarizabilities of Fluorinated Benzenes: An Ab Initio Investigation. *J. Fluor. Chem.* **2012**, *144*, 94–101. [CrossRef]
51. Cappelletti, G.; Ardizzone, S.; Meroni, D.; Soliveri, G.; Ceotto, M.; Biaggi, C.; Benaglia, M.; Raimondi, L. Wettability of Bare and Fluorinated Silanes: A Combined Approach Based on Surface Free Energy Evaluations and Dipole Moment Calculations. *J. Colloid Interface Sci.* **2013**, *389*, 284–291. [CrossRef] [PubMed]

Disclaimer/Publisher's Note: The statements, opinions and data contained in all publications are solely those of the individual author(s) and contributor(s) and not of MDPI and/or the editor(s). MDPI and/or the editor(s) disclaim responsibility for any injury to people or property resulting from any ideas, methods, instructions or products referred to in the content.

Article

Disposable Polyaniline/*m*-Phenylenediamine-Based Electrochemical Lactate Biosensor for Early Sepsis Diagnosis

Piromya Thongkhao [1], Apon Numnuam [2,3], Pasarat Khongkow [1,4,5], Surasak Sangkhathat [1,4,6] and Tonghathai Phairatana [1,5,*]

1. Department of Biomedical Sciences and Biomedical Engineering, Faculty of Medicine, Prince of Songkla University, Hat Yai, Songkhla 90110, Thailand
2. Center of Excellence for Trace Analysis and Biosensor, Prince of Songkla University, Songkhla 90110, Thailand
3. Division of Physical Science, Faculty of Science, Prince of Songkla University, Songkhla 90110, Thailand
4. Translational Medicine Research Center, Faculty of Medicine, Prince of Songkla University, Songkhla 90110, Thailand
5. Institute of Biomedical Engineering, Faculty of Medicine, Prince of Songkla University, Hat Yai, Songkhla 90110, Thailand
6. Department of Surgery, Faculty of Medicine, Prince of Songkla University, Songkhla 90110, Thailand
* Correspondence: tonghathai.p@psu.ac.th

Abstract: Lactate serves as a crucial biomarker that indicates sepsis assessment in critically ill patients. A rapid, accurate, and portable analytical device for lactate detection is required. This work developed a stepwise polyurethane–polyaniline–*m*-phenylenediamine via a layer-by-layer based electrochemical biosensor, using a screen-printed gold electrode for lactate determination in blood samples. The developed lactate biosensor was electrochemically fabricated with layers of *m*-phenylenediamine, polyaniline, a crosslinking of a small amount of lactate oxidase via glutaraldehyde, and polyurethane as an outer membrane. The lactate determination using amperometry revealed the biosensor's performance with a wide linear range of 0.20–5.0 mmol L^{-1}, a sensitivity of 12.17 ± 0.02 µA·mmol^{-1}·L·cm^{-2}, and a detection limit of 7.9 µmol L^{-1}. The developed biosensor exhibited a fast response time of 5 s, high selectivity, excellent long-term storage stability over 10 weeks, and good reproducibility with 3.74% RSD. Additionally, the determination of lactate in human blood plasma using the developed lactate biosensor was examined. The results were in agreement with the enzymatic colorimetric gold standard method ($p > 0.05$). Our developed biosensor provides efficiency, reliability, and is a great potential tool for advancing lactate point-of-care testing applications in the early diagnosis of sepsis.

Keywords: biosensor; electrochemistry; lactate; layer-by-layer; *m*-phenylenediamine; polyaniline; point-of-care testing; sepsis

Citation: Thongkhao, P.; Numnuam, A.; Khongkow, P.; Sangkhathat, S.; Phairatana, T. Disposable Polyaniline/ *m*-Phenylenediamine-Based Electrochemical Lactate Biosensor for Early Sepsis Diagnosis. *Polymers* **2024**, *16*, 473. https://doi.org/10.3390/ polym16040473

Academic Editor: Jung-Chang Wang

Received: 31 December 2023
Revised: 26 January 2024
Accepted: 6 February 2024
Published: 8 February 2024

Copyright: © 2024 by the authors. Licensee MDPI, Basel, Switzerland. This article is an open access article distributed under the terms and conditions of the Creative Commons Attribution (CC BY) license (https:// creativecommons.org/licenses/by/ 4.0/).

1. Introduction

Sepsis is a time-critical medical and life-threatening condition that occurs as the body's response to an infection. It can further progress to sepsis shock, eventually leading to severe organ dysfunction and death in the emergency department [1,2]. Over 48.9 million people globally are undergoing sepsis and one-fifth of these patients have not survived [3]. Sepsis patients also have non-specific symptoms and poor prognostic pathology. According to the Surviving Sepsis Campaign, blood lactate level or blood lactate clearance is one of the most important biomarkers used for the evaluation of critically ill patients [4]. An increase in the blood lactate level is related to the high morbidity and mortality rate among ill patients. A resting blood lactate level is normally accumulated in the range of 0.5–2.2 mmol L^{-1} [5]. A lactate level of more than 4 mmol L^{-1} is considered to be a high risk for mortality. The impairment of lactate generation and clearance, particularly in critically ill patients, can cause the accumulation of blood lactate that can be found at >4.0 mmol L^{-1} [6,7]. Thus, early detection of lactate levels can reduce the mortality rate.

Techniques for lactate determination include automated devices based on enzymatic colorimetry and arterial blood gas analysis (ABG) [8]. Although these techniques are used in several clinical laboratories, they are expensive and need complicated procedures that require sample preparation by trained personnel, making them time-consuming [9]. In addition, ABG can also cause a high-risk infection when the blood sample is frequently collected from an artery [10]. To overcome these issues, electrochemical biosensors are being explored due to their high sensitivity, selectivity, simplicity, rapid response, cost-effectiveness, lack of requirement for sample preparation, and the considerable ease in developing a portable device for point-of-care (POC) diagnosis [11–13].

There are numerous research groups attempting to develop electrochemical biosensors for lactate analysis in blood samples [14,15]. However, achieving a reproducible detection range that is associated with sepsis remains a challenge. To tackle this problem, most research groups have focused on electrode modification through enzyme immobilization [16]. This approach offers a significant advantage by preserving the biosensor's stability [17], thereby enabling its application in the analysis of blood samples that demand high selectivity. Additionally, it contributes to achieving a linear range suitable for medical applications, characterized by good reproducibility [15].

An essential strategy for enzyme immobilization involves the careful selection of materials that confer unique properties to enhance sensitivity, stability, and biocompatibility. Polymers stand out as particularly attractive in this context due to their excellent suitability as supporting materials for biorecognition immobilization. Furthermore, they offer the added benefit of allowing precise control of thickness through electropolymerization [18]. Among them, polyaniline (PANI), a conducting polymer, is widely used in biosensor design [19,20]. This is because of its unique properties, i.e., tunable electrochemical characteristics, functionality-rich chemical structure, providing the amino groups (–NH$_2$), increasing specific surface area, high conductivity, biocompatibility, and great long-term stability [21–24]. Similar to PANI, *m*-phenylenediamine (*m*-PD) is another intriguing component in the construction of biosensors and various electrochemical devices. Its contribution of amino groups (–NH$_2$) to the molecular structure and enhancement of specific surface area further elevates its significance in the realm of biosensor design. Much research has also used it as a perm-selective membrane to eliminate the interfering species [15,18,25,26].

In this study, we introduced a novel electrochemical lactate oxidase-based biosensor that utilized a layer-by-layer method to modify a screen-printed gold electrode. The layers were made of three distinct polymers, including *m*-PD, PANI, and polyurethane (PU). Specifically, PU was coated as an outer layer to prevent the enzyme from leaching from the sensor, and to ensure that the detection linearity covered the lactate range associated with sepsis. The experiment parameters regarding the layer-by-layer modification were investigated and optimized, i.e., PANI thickness, PU concentration, and applied potential of lactate detection. Finally, the developed lactate biosensor was examined for human plasma analysis and compared to the results obtained from the gold standard-based enzymatic colorimetric method used in hospitals.

2. Materials and Methods
2.1. Reagents and Apparatus

Lactate oxidase (LOx) from *Aerococcus viridans* (41 units mg^{-1} lyophilized powder) was purchased from Sigma-Aldrich (St. Louis, MO, USA). Aniline (≥99.5%), *m*-Phenylenediamine (*m*-PD, Flakes, 99%), polyurethane (PU, Selectophore^{TM}), glutaraldehyde (GA, 25% in H$_2$O solution), bovine serum albumin (BSA, pH 7, ≥98%), sodium lactate, uric acid (≥99%, crystalline), D-(+)-glucose, dopamine, L-ascorbic acid (99%), and potassium hexacyanoferrate (III) (K$_3$[Fe(CN)$_6$]) were from Sigma-Aldrich (USA). Sodium dihydrogen orthophosphate (NaH$_2$PO$_4$), di-sodium hydrogen orthophosphate

(Na$_2$HPO$_4$), and potassium chloride (KCl) were from Ajax Fine chem (New South Wales, Australia). Hydrochloric acid (HCl, 37%), dimethylformamide (DMF), and tetrahydrofuran (THF) were from Merck (Darmstadt, Germany). Other chemicals were of analytical grade. All aqueous solutions were prepared with water from Milli-Q purification system (resistivity $\geq$ 18 MΩ cm^{-1}).

Electrochemical experiments were performed using an Autolab Potentiostat/Galvanostat controlled by the NOVA 2.1.4 software (Metrohm Autolab, Utrecht, The Netherlands). The measurements were carried out using the screen-printed gold electrode (Metrohm, KM Utrecht, The Netherlands) comprised of a gold working electrode (4 mm in diameter, 0.126 cm^2), a silver pseudo-reference electrode, and a gold counter electrode [27]. Scanning electron microscopic images were obtained using a scanning electron microscope (SEM, Quanta 400, FEI, Osaka, Japan). All experiments were carried out at room temperature (25 °C).

2.2. Lactate Biosensor Fabrication

Our developed lactate biosensor was constructed on a screen-printed gold electrode (SPAuE) using *m*-phenylenediamine (*m*-PD), polyaniline (PANI), and polyurethane (PU) based on layer-by-layer method. Prior to the electrode modification, the SPAuE was electrochemically pre-treated in 0.50 mol L^{-1} H$_2$SO$_4$ using cyclic voltammetry with potential scanning from 0.0 V to 1.1 V at a scan rate of 100 mV s^{-1} for 10 cycles, and was rinsed with deionized (DI) water. The step of electrode modification is illustrated in Figure 1. Initially, *m*-PD as a perm-selective membrane was deposited onto the prepared SPAuE via electropolymerization. This was accomplished using chronoamperometric method by applying a constant potential of 0.70 V for 20 min in 0.1 mol L^{-1} *m*-PD in 0.01 mol L^{-1} phosphate buffer solution (PBS, pH 7.4), followed by rinsing with DI water (*m*-PD/SPAuE). Next, a PANI layer was prepared on the *m*-PD/SPAuE by electropolymerizing 55 mmol L^{-1} aniline in 1.0 mol L^{-1} HCl, applying the potential range from −0.20 V to 1.0 V at a scan rate of 50 mV s^{-1} for 20 cycles (PANI/*m*-PD/SPAuE). Subsequently, an enzyme mixture comprising 20 µL of LOx (0.3 U), 2 µL of glutaraldehyde (GA, 2.5% *v*/*v*) as a crosslinker, and 5.0 µL of 250 mg mL^{-1} BSA as a protein stabilizer to preserve enzyme activity was prepared. A volume of 4.0 µL of this mixture was drop-casted onto the PANI/*m*-PD/SPAuE. The modified electrode was left at room temperature for an hour and then placed in a sealed system at 4 °C overnight for the crosslinking process (LOx/PANI/*m*-PD/SPAuE). Finally, the LOx/PANI/*m*-PD/SPAuE was coated with 2.0 µL of PU (2.0% *w*/*v* in THF/DMF solution) as an outer layer (PU/LOx/PANI/*m*-PD/SPAuE). The fabricated lactate biosensor was rinsed with PBS before testing and stored in a sealed system containing 0.1 mol L^{-1} PBS at 4 °C when not in use.

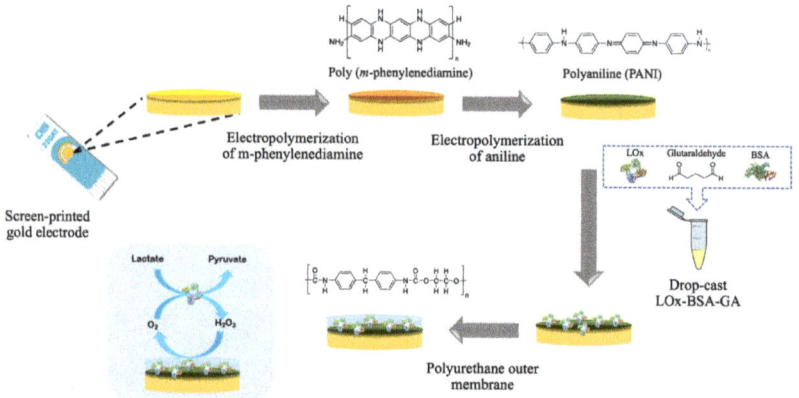

Figure 1. Schematic illustration of the lactate biosensor fabrication using layer-by-layer method.

2.3. Electrochemical Measurement

To verify that the working electrode had been successfully coated, each step of electrode modification was examined in a solution containing 5.0 mmol L^{-1} $K_3Fe(CN)_6$ and 0.10 mol L^{-1} KCl using cyclic voltammetry in the range of −0.20 V to 0.40 V at a scan rate of 50 mV s^{-1}. The lactate determination was conducted via amperometry at a constant potential of +0.70 V under stirring conditions. The amperometric response was measured as the increase in current, which was proportional to the concentration of lactate.

2.4. Optimization Studies

A set of parameters influencing the developed biosensor was studied and optimized in the range of 0.20 to 1.0 mmol L^{-1} lactate concentrations. These parameters included the strategies of polymer electrode modification, the thickness of PANI, PU concentrations, and the applied potential. Initially, four different strategies of electrode modification based on components and their order were explored, including (1) LOx/*m*-PD/SPAuE, (2) LOx/PANI/SPAuE, (3) LOx/PANI/*m*-PD/SPAuE, and (4) LOx/*m*-PD/PANI/SPAuE. The thickness of PANI was then optimized for the scan number of electropolymerization at 10, 20, 30, and 40 cycles. For PU layer, it was investigated by varying the concentration of PU (%*w*/*v*) at 1.0%, 2.0%, and 3.0%. Additionally, the effect of applying potential for lactate detection at 0.50, 0.60, 0.70, and 0.80 V was also observed. The optimum conditions were considered based on achieving the highest sensitivity (the slope of the calibration plot), a board linear range capable of covering the sepsis range, and a short response time.

2.5. Selectivity, Reproducibility, Long-Term and Storage Stability

The selectivity of the developed biosensor to lactate in the presence of potential blood-interfering substances at high physiological concentrations was examined through amperometric detection. These substances included 0.10 mmol L^{-1} ascorbic acid (AA), 0.10 mmol L^{-1} uric acid (UA), 0.10 mmol L^{-1} dopamine (DA), and 5.0 mmol L^{-1} glucose (Glu).

The reproducibility test was assessed using six different electrodes, measuring their amperometric responses to a range of lactate concentrations (0.20–5.0 mmol L^{-1}). The long-term stability of the biosensor was tested weekly by comparing the sensitivity of subsequently used electrodes to their initial values. Storage stability was also investigated using nine lactate biosensors prepared simultaneously, with their sensitivities examined through random testing every 3–5 days over a period of two months. Following each testing step, the biosensor was cleaned and stored in a sealed system containing 0.1 mol L^{-1} PBS at 4 °C when not in use.

2.6. Real Sample Analysis

Human blood plasma samples received from Songklanagarind Hospital, Hat Yai, Thailand (approved by the local ethics committee [REC.63-161-25-2]) were analyzed. Initially, the matrix effect was studied by adding a series of lactate concentrations at 4.0, 6.0, 10.0, 20.0, and 30.0 mmol L^{-1} into the tubes containing blood plasma samples. Then, the mixture (lactate and blood) was diluted with 0.1 mol L^{-1} PBS (pH 7.4) to obtain a 10-fold dilution (spiked sample). The matrix effect was determined by comparing the slope of the calibration curve (sensitivity vs. concentration of lactate) using the standard lactate solution, with that using the spiked sample through two-way ANOVA. To ascertain the lactate concentration in plasma samples using the developed lactate biosensors, eight blood plasma samples were tested, and the results were compared with values obtained from the gold standard enzymatic colorimetric method using Cobas c502 analyzer (Roche, Basel, Switzerland) used in hospitals. The comparison was conducted using the Wilcoxon signed-rank test.

3. Results and Discussion

3.1. Characterization of the Lactate Biosensor

3.1.1. Surface Morphology

The surface morphology of the modified electrode in each fabrication step was examined using SEM. Figure 2A shows a rough surface of the m-PD structure coated on the electrode surface (m-PD/SPAuE). After PANI electropolymerization, the distribution of the PANI granules with an average diameter of 390 ± 18 nm on the surface of the m-PD/SPAuE was observed (Figure 2B). Figure 2C presents the rough, gel-like layer of the enzyme matrix (LOx via the crosslinking reagent and BSA) coated on the PANI/m-PD/SPAuE. When the PU as an outer layer was coated, a smooth surface with bulky pores was observed (Figure 2D), similar to those reported earlier [28]. These SEM images could indicate that each fabrication step of the electrode modification was achieved.

3.1.2. Electrochemical Characterization

To confirm the successful fabrication of the developed lactate biosensor, the stepwise modification was electrochemically characterized using cyclic voltammetry in a redox solution of 5 mmol L^{-1} K$_3$Fe(CN)$_6$ at a scan rate of 50 mV s^{-1}, as illustrated in Figure 2E. The cyclic voltammogram (CV) of the bare SPAuE (trace a) exhibited a pair of well-defined redox peaks. When the m-PD layer was deposited on the electrode surface, the CV became a flat shape without the redox peaks (trace b), indicating the successful coating of the non-conductive m-PD layer on the SPAuE. Subsequently, after coating PANI onto the m-PD/SPAuE, the highest background current with a couple of redox peaks was observed (curve c). This indicated that PANI provided a conductive layer on the electrode surface which could be attributed to electron transfer onto the electrode surface, confirming the successful coating of PANI on the previous layer. Upon enzyme immobilization through the co-crosslink of glutaraldehyde and BSA (non-conducting materials), a decrease in the redox current signals was observed (curve d). This reduction was attributed to the non-conductive nature of the enzyme matrix, blocking electron transfer toward the electrode surface [29]. Finally, when the non-conductive PU was drop-casted, the current response slightly decreased, indicating the successful coating of the PU layer as an outer membrane onto the LOx/PANI/m-PD/SPAuE (curve e).

3.2. Optimization Studies

3.2.1. Effect of Electrode Modification Strategies

To investigate the effect of component and the sequence of PANI and m-PD on SPAuE, four strategies of electrode modification, i.e., (1) LOx/m-PD/SPAuE, (2) LOx/PANI/SPAuE, (3) LOx/PANI/m-PD/SPAuE, and (4) LOx/m-PD/PANI/SPAuE were examined based on their sensitivity using amperometry at a constant potential of 0.70 V. As seen in Figure 3A, the results revealed the significant impact of both components and the sequence of electrode modification on sensitivity. Among these modifications, the LOx/PANI/m-PD/SPAuE exhibited the highest sensitivity for lactate detection, followed by the LOx/m-PD/PANI/SPAuE, LOx/PANI/SPAuE, and LOx/m-PD/SPAuE, respectively. The use of m-PD as a supporting material for LOx immobilization (LOx/m-PD/SPAuE) resulted in lower sensitivity compared to PANI without m-PD (LOx/PANI/SPAuE). This could be attributed to the non-conductive nature of m-PD, coupled with its thin film, potentially affecting the surface area available for LOx immobilization. Moreover, the combination of m-PD and PANI in lactate biosensor fabrication was found to enhance sensitivity. PANI, known for its high conductivity, positively influenced current signal response, electrocatalysis, and the kinetics of the electron transfer process to the electrode surface [30]. The chemical structure of PANI, presenting amino group (–NH$_2$), offered a large specific surface area in granular form for enzyme immobilization. The amino groups also provided by m-PD coating further contributed to this effect. The synergistic effect of both m-PD and PANI could contribute to an increase in the specific surface area for a large amount of lactate oxidase loading, resulting in an elevated current response. Consequently, LOx/PANI/*m*-PD/SPAuE was chosen for further experiments.

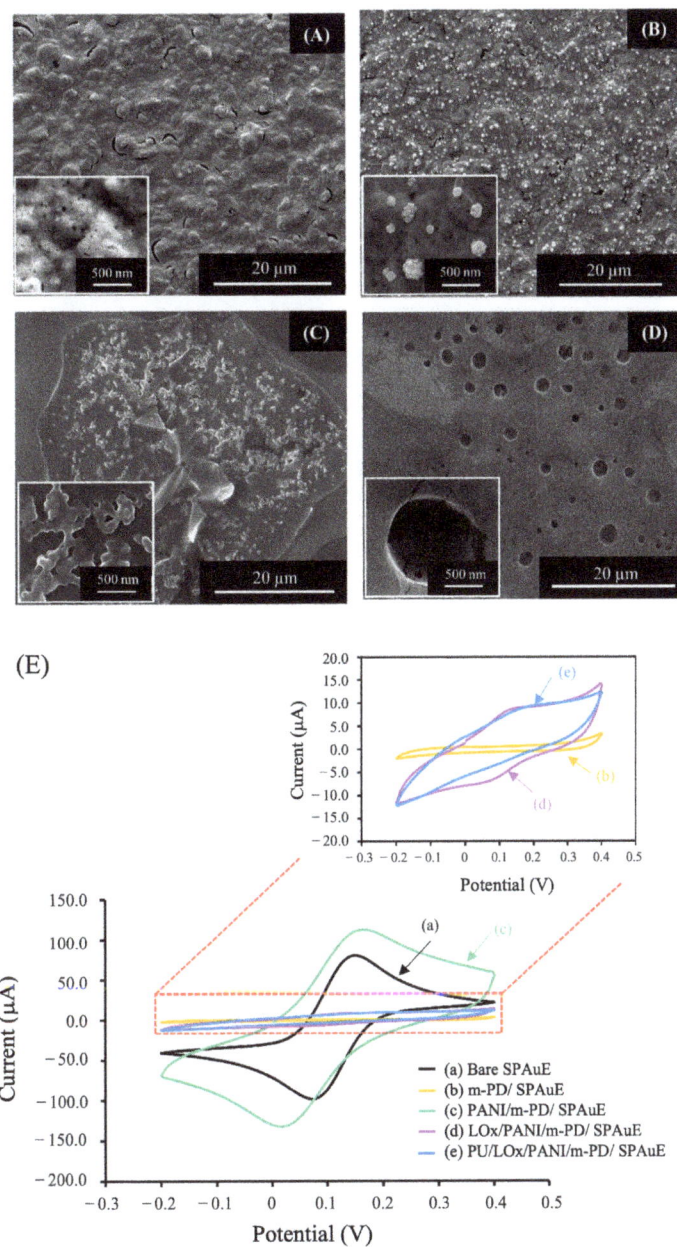

Figure 2. SEM images of the modified SPAuE: (**A**) *m*-PD/SPAuE, (**B**) PANI/*m*-PD/SPAuE, (**C**) LOx/PANI/*m*-PD/SPAuE, and (**D**) PU/LOx/PANI/*m*-PD/SPAuE, (**E**) cyclic voltammograms of the modified electrode in each fabrication step performing in 5.0 mmol L^{-1} [Fe(CN)$_6$]$^{3-/4-}$ solution containing 0.10 mol L^{-1} KCl at a scan rate of 50 mV s^{-1}.

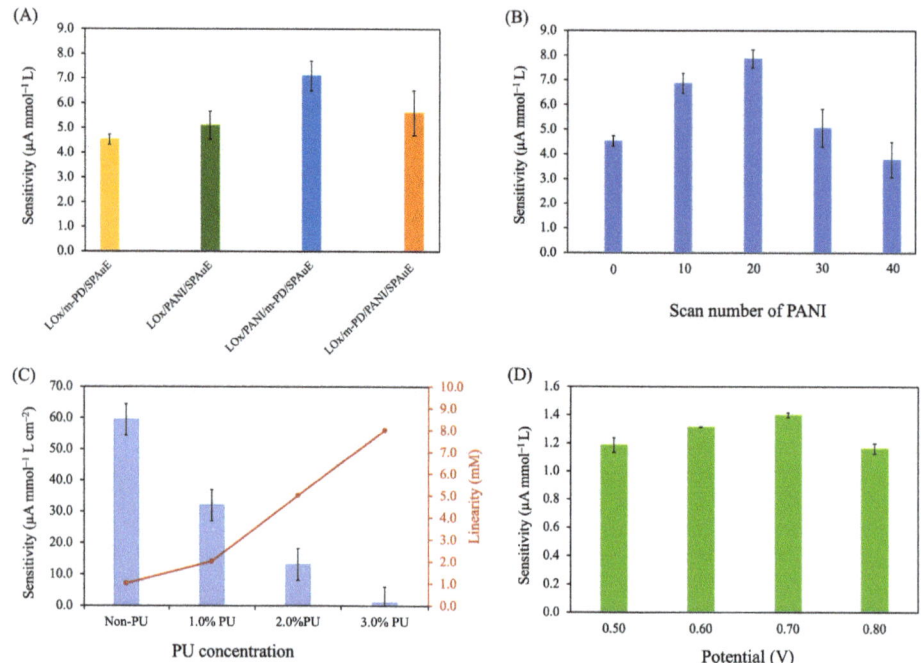

Figure 3. Optimization studies on the sensitivity for different lactate biosensor fabrication (n = 3): effects of (**A**) the components and the sequence of electrode modification, (**B**) scan cycles of PANI electropolymerization, (**C**) concentrations of PU compared with the linearity range of lactate measurements, and (**D**) constant applied potentials for lactate determination.

3.2.2. Effect of the PANI Layer

The PANI layer is another important factor that can affect the biosensor performance. PANI, as a supporting material for enzyme immobilization, can enhance the active surface area contributing to an increase in enzyme loading. However, the excessive thickness of PANI can affect the diffusion of the analyte and electron transfer kinetics. The effect of the PANI thickness on the biosensor performance was examined by varying the number of electropolymerization scans at 0, 10, 20, 30, and 40 cycles. The results revealed that 20 scan cycles exhibited the highest sensitivity, as illustrated in Figure 3B. The sensitivity increased with an increase in the number of electropolymerization scans from 0 to 20 cycles. Then, a decrease in the sensitivity was observed when the number of scans increased from 20 to 40 cycles. This could possibly be because the excessive granules of the PANI layer lead to an increase in the background current; therefore, the current response decreases. The results were confirmed by SEM images showing the surface morphology of the PANI layers at different scan cycles (Supplementary Materials Figure S1). This indicated that the higher the number of scans, the higher the density of granules distributed on the electrode surface. Thus, the PANI with the number of electropolymerization scans at 20 cycles was chosen for further studies.

3.2.3. Effect of PU Concentrations

To enhance the applicability of our developed sensor for use in human blood samples covering the sepsis range, we employed polyurethane (PU) as an outer membrane to extend the linearity of detection. The effect of PU concentration on the sensitivity and linearity was examined at 0% (non-PU), 1.0%, 2.0%, and 3.0% PU concentrations (w/v). Figure 3C illustrates the plots between the sensitivity and the linearity at the different

PU concentrations. Non-PU cases exhibited the highest sensitivity but the linear range of detection was up to 1 mmol L^{-1}, which did not cover the sepsis range. In contrast, the addition of the PU on the LOx/PANI/m-PD/SPAuE extended the linearity of detection by slowing the mass transport of the analyte. The results revealed that an increase in PU concentrations resulted in a decrease in sensitivity but an increase in linearity. This was consistent with a study reported earlier [28]. This finding could be explained that as the PU concentration increased, the PU layer became more viscous and the background current response increased, resulting in a significant reduction in its sensitivity. Regarding the developed lactate biosensors at the 1.0%, 2.0%, and 3.0% PU, they exhibited a wider linear range of lactate detection (0.2–2.0 mmol L^{-1}, 0.2–5.0 mmol L^{-1}, and 0.2–8.0 mmol L^{-1}, respectively) as compared to that of the non-PU case (0.2–1.0 mmol L^{-1}). This suggests that the addition of a PU layer can limit the diffusion of the lactate on the electrode surface. Also, the limitation of mass transfer can possibly affect the kinetics of the enzyme catalytic system, leading to a wider range of lactate detection. However, controlling the fabricated lactate biosensor at the 3.0% PU concentration posed challenges due to its high viscosity, leading to rapid solvent evaporation and difficulty in reproducing the process. Additionally, the response time of the lactate detection for each condition was also evaluated based on T$_{90}$ (the time required for the current signal to reach 90% of the final value due to an increase in lactate). The response times at non-PU, 1.0%, 2.0%, and 3.0% PU concentrations were 2.30 $\pm$ 0.47 s, 3.50 $\pm$ 0.50 s, 5.0 $\pm$ 0.82 s, and 13.50 $\pm$ 1.50 s, respectively. Taking into account sensitivity, linear range, and response time, the 2.0% PU concentration was chosen as the optimal condition for further studies. This is because it exhibited a desirable sensitivity with a short response time (5.0 s), and a linearity covering the range of 0.20–5.0 mmol L^{-1}, making it suitable for clinical applications, especially sepsis screening [31].

3.2.4. Effect of Operational Potential for Lactate Determination

A series of constant potentials within the range of 0.50 to 0.80 V was systematically investigated and optimized based on sensitivity. The operational potential plays a critical role in the oxidation of H_2O_2, a by-product of the catalytic activity of lactate oxidase. The LOx acts as a catalyst of the lactate reaction, to oxidize lactate into pyruvate and H_2O_2. In the presence of oxygen, H_2O_2 is oxidized at the electrode surface by applying a constant potential and electrons are generated as given below in Equations (1) and (2) [12]:

$$\text{L-lactate} + O_2 \xrightarrow{\text{LOx}} \text{Pyruvate} + H_2O_2 \qquad (1)$$

$$H_2O_2 \rightarrow O_2 + 2H^+ + 2e^- \qquad (2)$$

Figure 3D illustrates that the highest sensitivity for the developed biosensor in lactate detection was achieved at 0.70 V. Consequently, a constant potential of 0.70 V was chosen for lactate determination for analytical performance studies.

3.3. Analytical Performances of the Optimized PU/LOx/PANI/m-PD/SPAuE
3.3.1. Linearity and Detection Limit

Under the optimal conditions, the performance of the developed lactate biosensor (PU/LOx/PANI/m-PD/SPAuE) was evaluated at different lactate concentrations in the range of 0.2–10 mmol L^{-1}. The amperometric response obtained at a constant potential of +0.70 V is depicted in Figure 4. The result revealed that the current response exhibited a proportional increase with the concentration of lactate. As illustrated in the calibration plot inset, the biosensor exhibited a linear detection range from 0.20 to 5.0 mmol L^{-1} with a correlation coefficient of 0.9983. The sensitivity was 12.17 $\pm$ 0.02 µA mmol^{-1} L cm^{-2} with the limit of detection (LOD) of 7.9 µmol L^{-1} (3S/N) [32].

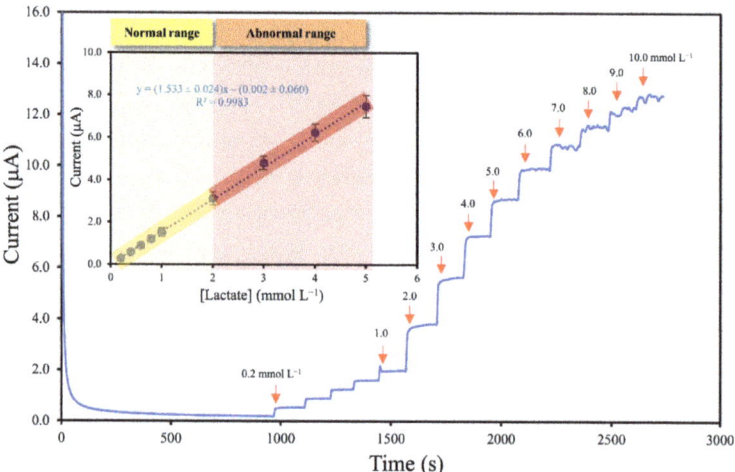

Figure 4. The current response of the PU/LOx/PANI/*m*-PD/SPAuE at different concentrations of lactate using amperometry with an inset representing the calibration plot of the developed biosensor for lactate detection covering normal range and abnormal range of lactate level (n = 3).

3.3.2. Selectivity

The selectivity of the developed lactate biosensor was evaluated by determining potential interfering species existing in blood plasma samples. These include ascorbic acid (AA), uric acid (UA), dopamine (DA), and glucose (Glu). In this study, the selectivity of the interfering species was determined at the high physiological concentration containing 0.10 mmol L^{-1} AA, 0.10 mmol L^{-1} UA, 0.10 mmol L^{-1} DA, and 5.0 mmol L^{-1} Glu. As shown in Figure 5A, the results revealed that there was no interfering effect on the developed lactate biosensor. This is possibly because *m*-PD as a perm-selective membrane is a protection layer, which helps limit access of both cations as well as anion molecules, and possible electrochemical species interference. Also, it allows H_2O_2, as a product of the lactate reaction, to pass through to the electrode surface. However, the developed lactate biosensor was sensitive to the analyte [18,26]. This indicates that the developed biosensor can be used for lactate determination.

3.3.3. Reproducibility and Long-Term and Storage Stability

The reproducibility of the developed biosensor was evaluated by preparing six biosensors and measuring the current response at different lactate concentrations (0.20–5.0 mmol L^{-1}). The sensitivities of the six developed lactate biosensors were 1.534 ± 0.050, 1.540 ± 0.038, 1.594 ± 0.019, 1.513 ± 0.005, 1.418 ± 0.007, and 1.463 ± 0.686 (Figure 5B). The average sensitivity was 1.510 ± 0.057 µA mmol^{-1} L, with a relative standard deviation (RSD) of 3.74%, indicating an acceptable reproducibility [33].

One of the developed sensors was assessed weekly to explore its long-term stability based on relative sensitivity (>90% of the initial testing is accepted value of AOAC guideline [34]). As seen in Figure 5C, the result showed that the relative sensitivity of the developed biosensor remained stable over 90% for the first four weeks. Subsequently, there was an approximately 10% increase in sensitivity from the initial value and this elevated sensitivity was sustained over the following six consecutive weeks (weeks 5–10). The initial increase in relative sensitivity during the first period could be attributed to the slight swelling of the PU membrane caused by its immersion in a PBS solution after each testing cycle. This resulted in the diffusion of the solution into the PU membrane, thereby increasing membrane permeability. However, a stable relative response was observed after the initial four weeks. The diffusion behavior can be described by the Fickian diffusion, where solution diffusion through the membrane is influenced by the thickness of the membrane [35,36]. The average relative sensitivity over the entire 10-week period was calculated to be 107.6 ± 6.5 with an RSD of 6.1%. These results indicated that the developed biosensor provided excellent long-term stability for over 10 weeks.

The storage stability was investigated by preparing nine lactate biosensors (PU/LOx/PANI/m-PD/SPAuE) at the same time and storing them in a sealed system at 4 °C in 0.10 mol L^{-1} PBS (pH 7.4) when not in use. The developed lactate biosensor was randomly examined one test at a time. Initially, the lactate measurements were conducted every 3–5 days and then every 10 days for two months. The storage stability was evaluated based on relative sensitivity. The results showed that the average relative sensitivity of 103.4 ± 4.0% with an RSD of 3.9% was obtained (Supplementary Materials Figure S2). Thus, the PU membrane layer as a protective membrane could prevent the enzyme from leaching and enhance the storage stability of the developed lactate biosensor. The results were consistent with previous studies [37,38], which reported that the characteristics of PU as the outer membrane characteristics could improve stability.

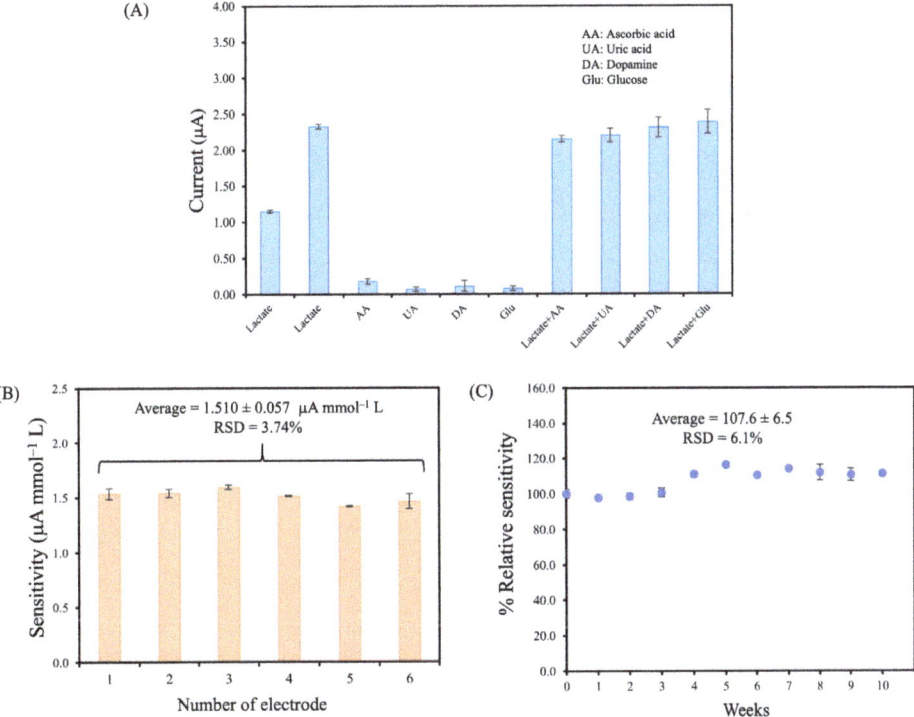

Figure 5. (**A**) Interference testing of PU/LOx/PANI/m-PD/SPAuE biosensor in 0.10 mol L^{-1} PBS containing different concentrations of the interfering substrates, including 1.0 mmol L^{-1}, and 2.0 mmol L^{-1} lactate, 0.10 mmol L^{-1} AA, 0.10 mmol L^{-1} UA, 0.10 mmol L^{-1} DA, 5.0 mmol L^{-1} Glu, and also testing in the mixture solution of 2.0 mmol L^{-1} lactate with these interferents, respectively, (**B**) reproducibility of six lactate biosensors on the sensitivity at 0.20–5.0 mmol L^{-1} lactate using amperometry and, (**C**) the stability testing for the detection of 0.40–3.0 mmol L^{-1} lactate.

3.3.4. Comparison with Other Sensors

The analytical performance of the PU/LOx/PANI/m-PD/SPAuE as a lactate biosensor was compared with other studies, as shown in Table 1. Our developed lactate sensor exhibited a wide linearity for lactate detection, distinguishing it from previously cited reports. While the linear range was not as extensive as some similar works [39], it was sufficient for early sepsis diagnosis via lactate detection. In terms of stability, this sensor outperformed other cited reports. In addition, the developed lactate sensor offers a fast response time (5.0 s) as compared with other studies [40] (30 s). This performance gives it a high potential for clinical lactate detection compared to the sensors evaluated in the cited works.

Table 1. The comparison of performance of lactate biosensors based on lactate oxidase (LOx).

Electrode Materials	Linear Range (mmol L^{-1})	Detection Limit (μmol L^{-1})	Sensitivity (μA·mmol^{-1} L cm^{-2})	Stability (Days)	Samples	Ref.
LOx/MWCNTs/CuNPs/ PANI/PEG	0.0010–2.5	0.25	NR	140	Blood plasma	[41]
LOx-Cu-MOF/CS/Pt/SPCE	0.00075–1.0 4.0–50.0	0.75	116.26 1.64	50	Sweat, saliva, wine	[42]
LOx/TiO$_2$ sol gel-Gr/ Ni foam	0.050–10.0	19	NR	8	Commercial rabbit serum	[39]
LOx-PVA-SbQ/*m*-PD/ Pt disk	0.005–1.0	5	81.6	14	Human blood serum	[40]
LOx/Pt/PANI/MXene/SPCE	0.005–5.0	5	0.62	30	Milk samples	[24]
LOx-CNDs/SPAuE	0.003–0.50	0.9	39.52	NR	Human blood serum	[43]
PU/LOx/PANI/ *m*-PD/SPAuE	0.20–5.0	7.9	12.17	>70	Human blood plasma	This work

MWCNTs: multiwalled carbon nanotubes; CuNPs: copper nanoparticles; PEG: pencil graphite electrode; Cu-MOF: copper metallic framework; CS: chitosan; SPCE: screen-printed carbon electrode; PVA-SbQ: photopolymer-containing styrylpyridine groups; CNDs: carbon nanodots; NR: not reported.

3.3.5. Analysis of Human Blood Plasma Samples

In this study, the standard curve of the developed lactate biosensor exhibited a linear range of 0.20 to 5.0 mmol L^{-1}; therefore, human blood plasma samples were diluted 10 times before testing. Initially, the matrix effect was evaluated by comparing the slope of the calibration curve of standard lactate solutions with that of the spike samples. The results showed that the sensitivity of the standard lactate solution and spike sample had no significant difference ($p > 0.05$, two-way ANOVA). Thus, the dilution at 10 times had no matrix effect. The lactate values of the diluted samples were then calculated from the calibration equation of the lactate standard curve. The lactate values of eight blood plasma samples obtained from the developed biosensor and the enzymatic colorimetric technique as a gold standard method in clinical use (from the Songklanagarind hospital) were in good agreement ($p > 0.05$), as illustrated in Figure 6. Notably, expanding the dataset in future studies is crucial for a comprehensive evaluation of false positive and false negative rates in our lactate detection method. An extended clinical study is in progress to yield valuable insights, enhancing the overall impact of our work.

The developed lactate biosensor was analyzed to verify the practicality using the recovery test based on the standard addition method. The recoveries were found to be in the range of 93.2–106.4% (Supplementary Materials Table S1), which were the accepted values according to the AOAC guidelines (80–110%) [33]. This indicates that the developed lactate biosensor exhibits high accuracy for lactate detection.

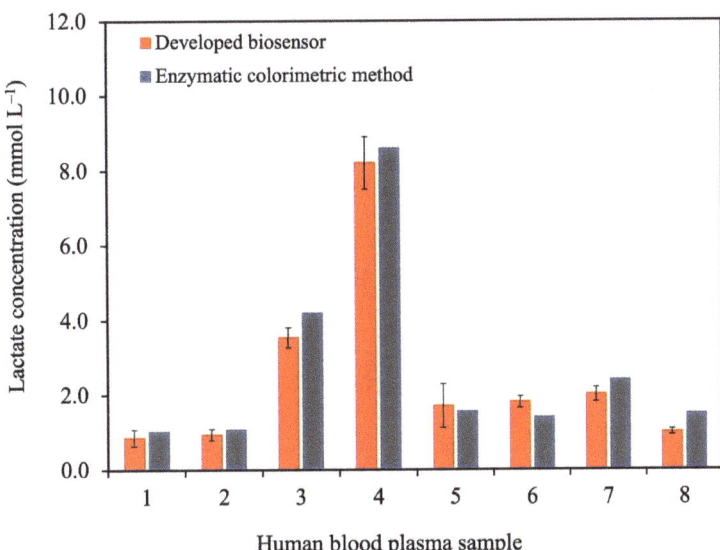

Figure 6. Comparison of the analytical results obtained from the developed biosensor and the enzymatic colorimetric method for lactate measurements in blood plasma samples.

4. Conclusions

This study presents the successful fabrication of a lactate-based electrochemical biosensor, using a polymeric layer-by-layer method with *m*-PD, PANI, and PU modifications on a screen-printed gold electrode. The developed lactate biosensors exhibited excellent sensitivity, selectivity, reproducibility, long-term stability, and high accuracy of recovery in lactate detection. In addition, the developed biosensor could effectively detect the concentration of lactate in blood plasma samples compared to the enzymatic colorimetric method. With a wide linear response (0.2–5.0 mmol L^{-1}) and a low limit of detection (7.9 µmol L^{-1}), this biosensor holds promise for early sepsis diagnosis. It stands as an efficient and reliable tool for the development of portable lactate point-of-care testing, facilitating on-site early sepsis diagnosis.

Supplementary Materials: The following supporting information can be downloaded at: https://www.mdpi.com/article/10.3390/polym16040473/s1, Figure S1: SEM images of PANI electropolymerization at different numbers of scans: (A) 10 cycles, (B) 20 cycles, (C) 30 cycles, and (D) 40 cycles, (E) cyclic voltammograms of PANI electropolymerization at four different number of scans; Figure S2: The storage stability testing for the detection of 0.40–3.0 mmol L^{-1} lactate in 0.1 mmol L^{-1} PBS (pH 7.4) using amperometry; Table S1: The recovery study was performed using the standard addition method. Figure S3: Comparison of the calibration curves between the standard lactate solution and the spiked lactate in three blood samples.

Author Contributions: Conceptualization, T.P., P.K. and S.S.; methodology, P.T., T.P. and A.N.; validation, P.T. and T.P.; formal analysis, P.T. and T.P.; investigation, P.T. and T.P.; resources, T.P.; writing—original draft preparation, P.T. and T.P.; writing—review and editing, T.P., P.K., A.N. and S.S.; visualization, T.P. and P.K.; supervision, T.P.; project administration, T.P.; funding acquisition, T.P. and S.S. All authors have read and agreed to the published version of the manuscript.

Funding: This research and innovation activity was funded by the Graduate School Funding and the Faculty of Medicine, Prince of Songkla University, and the National Research Council of Thailand (NRCT).

Institutional Review Board Statement: The study was conducted in accordance with the Declaration of Helsinki and approved by the ethics committee of the Human Research Ethics Unit, the Faculty of Medicine, Prince of Songkla University, Songkhla Thailand (REC.63-161-25-2—17 June 2020).

Data Availability Statement: Data are contained within the article and supplementary materials.

Acknowledgments: We gratefully acknowledge the student scholarship from the Faculty of Medicine and the Graduate School, Prince of Songkla University, awarded to Piromya Thongkhao. We would also like to thank the Department of Pathology at the Faculty of Medicine, the Center of Excellence for Trace Analysis and Biosensor (TAB-CoE) at Prince of Songkla University, Thailand, and the Center of Excellence on Medical Biotechnology, Mahidol University, Thailand.

Conflicts of Interest: The authors declare no conflicts of interest.

References

1. Keep, J.W.; Messmer, A.S.; Sladden, R.; Burrell, N.; Pinate, R.; Tunnicliff, M.; Glucksman, E. National Early Warning Score at Emergency Department Triage May Allow Earlier Identi Fi Cation of Patients with Severe Sepsis and Septic Shock: A Retrospective Observational Study. *Emerg. Med. J.* **2016**, *33*, 37–41. [CrossRef] [PubMed]
2. Rakpraisuthepsiri, N.; Surabenjawong, U.; Limsuwat, C.; Lertvipapath, P. Efficacy of a Modified Sepsis System on the Mortality Rate of Septic Shock Patients in the Emergency Department of Siriraj Hospital. *J. Health Sci. Med. Res.* **2022**, *40*, 543–550. [CrossRef]
3. Rudd, K.E.; Johnson, S.C.; Agesa, K.M.; Shackelford, K.A.; Tsoi, D.; Kievlan, D.R.; Colombara, D.V.; Ikuta, K.S.; Kissoon, N.; Finfer, S.; et al. Global, Regional, and National Sepsis Incidence and Mortality, 1990–2017: Analysis for the Global Burden of Disease Study. *Lancet* **2020**, *395*, 200–211. [CrossRef] [PubMed]
4. Levy, M.M.; Evans, L.E.; Rhodes, A. The Surviving Sepsis Campaign Bundle: 2018 Update. *Intensive Care Med.* **2018**, *44*, 925–928. [CrossRef] [PubMed]
5. Crapnell, R.D.; Tridente, A.; Banks, C.E.; Dempsey-Hibbert, N.C. Evaluating the Possibility of Translating Technological Advances in Non-Invasive Continuous Lactate Monitoring into Critical Care. *Sensors* **2021**, *21*, 879. [CrossRef] [PubMed]
6. Dugar, S.; Choudhary, C.; Duggal, A. Sepsis and Septic Shock: Guideline-Based Management. *Crit. Care Nephrol. Third Ed.* **2019**, *87*, 500–504.e1. [CrossRef] [PubMed]
7. Ryoo, S.M.; Lee, J.; Lee, Y.-S.; Lee, J.H.; Lim, K.S.; Huh, J.W.; Hong, S.-B.; Lim, C.-M.; Koh, Y.; Kim, W.Y. Lactate Level Versus Lactate Clearance for Predicting Mortality in Patients With Septic Shock Defined by Sepsis-3. *Crit. Care Med.* **2018**, *46*, e489–e495. [CrossRef] [PubMed]
8. Pundir, C.S.; Narwal, V.; Batra, B. Biosensors and Bioelectronics Determination of Lactic Acid with Special Emphasis on Biosensing Methods: A Review. *Biosens. Bioelectron.* **2016**, *86*, 777–790. [CrossRef]
9. Rassaei, L.; Olthuis, W.; Tsujimura, S.; Ernst, J.R.; Sudhölter, A.; van den, B. Lactate Biosensors: Current Status and Outlook. *Anal. Bioanal. Chem.* **2013**, *406*, 123–137. [CrossRef]
10. Samaraweera, S.A.; Gibbons, B.; Gour, A.; Sedgwick, P. Arterial versus Venous Lactate: A Measure of Sepsis in Children. *Eur. J. Pediatr.* **2017**, *176*, 1055–1060. [CrossRef]
11. Kiatamornrak, P.; Boobphahom, S.; Lertussavavivat, T.; Rattanawaleedirojn, P.; Chailapakul, O.; Rodthongkum, N.; Srisawat, N. A Portable Blood Lactate Sensor with a Non-Immobilized Enzyme for Early Sepsis Diagnosis. *Analyst* **2022**, *147*, 2819–2827. [CrossRef] [PubMed]
12. Rathee, K.; Dhull, V.; Dhull, R.; Singh, S. Biosensors Based on Electrochemical Lactate Detection: A Comprehensive Review. *Biochem. Biophys. Rep.* **2016**, *5*, 35–54. [CrossRef] [PubMed]
13. Bradley, Z.; Bhalla, N. Point-of-Care Diagnostics for Sepsis Using Clinical Biomarkers and Microfluidic Technology. *Biosens. Bioelectron.* **2023**, *227*, 115181. [CrossRef] [PubMed]
14. Li, S.; Zhang, H.; Zhu, M.; Kuang, Z.; Li, X.; Xu, F.; Miao, S.; Zhang, Z.; Lou, X.; Li, H.; et al. Electrochemical Biosensors for Whole Blood Analysis: Recent Progress, Challenges, and Future Perspectives. *Chem. Rev.* **2023**, *123*, 7953–8039. [CrossRef]
15. Kucherenko, I.S.; Topolnikova, Y.V.; Soldatkin, O.O. Advances in the Biosensors for Lactate and Pyruvate Detection for Medical Applications: A Review. *TrAC Trends Anal. Chem.* **2019**, *110*, 160–172. [CrossRef]
16. Kilic, N.M.; Singh, S.; Keles, G.; Cinti, S.; Kurbanoglu, S.; Odaci, D. Novel Approaches to Enzyme-Based Electrochemical Nanobiosensors. *Biosensors* **2023**, *13*, 622. [CrossRef] [PubMed]
17. Rocchitta, G.; Spanu, A.; Babudieri, S.; Latte, G.; Madeddu, G.; Galleri, G.; Nuvoli, S.; Bagella, P.; Demartis, M.I.; Fiore, V.; et al. Enzyme Biosensors for Biomedical Applications: Strategies for Safeguarding Analytical Performances in Biological Fluids. *Sensors* **2016**, *16*, 780. [CrossRef]
18. Soldatkina, O.V.; Kucherenko, I.S.; Pyeshkova, V.M.; Alekseev, S.A.; Soldatkin, O.O.; Dzyadevych, S.V. Improvement of Amperometric Transducer Selectivity Using Nanosized Phenylenediamine Films. *Nanoscale Res. Lett.* **2017**, *12*, 594. [CrossRef]
19. Ming, T.; Lan, T.; Yu, M.; Wang, H.; Deng, J.; Kong, D.; Yang, S.; Shen, Z. Platinum Black/Gold Nanoparticles/Polyaniline Modified Electrochemical Microneedle Sensors for Continuous In Vivo Monitoring of PH Value. *Polymers* **2023**, *15*, 2796. [CrossRef]

20. Banjar, M.F.; Joynal Abedin, F.N.; Fizal, A.N.S.; Muhamad Sarih, N.; Hossain, M.S.; Osman, H.; Khalil, N.A.; Ahmad Yahaya, A.N.; Zulkifli, M. Synthesis and Characterization of a Novel Nanosized Polyaniline. *Polymers* **2023**, *15*, 4565. [CrossRef]
21. Cheng, H.; Hu, C.; Ji, Z.; Ma, W.; Wang, H. A Solid Ionic Lactate Biosensor Using Doped Graphene-like Membrane of Au-EVIMC-Titania Nanotubes-Polyaniline. *Biosens. Bioelectron.* **2018**, *118*, 97–101. [CrossRef]
22. Zarrintaj, P.; Vahabi, H.; Saeb, M.R.; Mozafari, M. *Application of Polyaniline and Its Derivatives*; Elsevier: Amsterdam, The Netherlands, 2019; ISBN 9780128179154.
23. German, N.; Ramanaviciene, A.; Ramanavicius, A. Formation and Electrochemical Evaluation of Polyaniline and Polypyrrole Nanocomposites Based on Glucose Oxidase and Gold Nanostructures. *Polymers* **2020**, *12*, 3026. [CrossRef]
24. Neampet, S.; Ruecha, N.; Qin, J.; Wonsawat, W.; Chailapakul, O.; Rodthongkum, N. A Nanocomposite Prepared from Platinum Particles, Polyaniline and a Ti3C2 MXene for Amperometric Sensing of Hydrogen Peroxide and Lactate. *Microchim. Acta* **2019**, *186*, 752. [CrossRef] [PubMed]
25. Yuqing, M.; Jianrong, C.; Xiaohua, W. Using Electropolymerized Non-Conducting Polymers to Develop Enzyme Amperometric Biosensors. *Trends Biotechnol.* **2004**, *22*, 227–231. [CrossRef] [PubMed]
26. Cordeiro, C.A.; De Vries, M.G.; Cremers, T.I.F.H.; Westerink, B.H.C. The Role of Surface Availability in Membrane-Induced Selectivity for Amperometric Enzyme-Based Biosensors. *Sens. Actuators B Chem.* **2016**, *223*, 679–688. [CrossRef]
27. Phonklam, K.; Thongkhao, P.; Phairatana, T. The Stability of Gold Nanoparticles-Prussian Blue Based Sensors for Biosensor Applications in Clinical Diagnosis. *J. Health Sci. Med. Res.* **2022**, *20*, 859. [CrossRef]
28. Huang, L.; Jia, Z.; Liu, H.; Pi, X.; Zhou, J. Design of a Sandwich Hierarchically Porous Membrane with Oxygen Supplement Function for Implantable Glucose Sensor. *Appl. Sci.* **2020**, *10*, 2848. [CrossRef]
29. Sarika, C.; Rekha, K.; Narasimha Murthy, B. Studies on Enhancing Operational Stability of a Reusable Laccase-Based Biosensor Probe for Detection of Ortho-Substituted Phenolic Derivatives. *3 Biotech* **2015**, *5*, 911–924. [CrossRef]
30. Shoaie, N.; Daneshpour, M.; Azimzadeh, M.; Mahshid, S.; Khoshfetrat, S.M.; Jahanpeyma, F.; Gholaminejad, A.; Omidfar, K.; Foruzandeh, M. Electrochemical Sensors and Biosensors Based on the Use of Polyaniline and Its Nanocomposites: A Review on Recent Advances. *Microchim. Acta* **2019**, *186*, 465. [CrossRef]
31. Rattu, G.; Khansili, N.; Maurya, V.K.; Krishna, P.M. Lactate Detection Sensors for Food, Clinical and Biological Applications: A Review. *Environ. Chem. Lett.* **2020**, *19*, 1135–1152. [CrossRef]
32. Taverniers, I.; De Loose, M.; Van Bockstaele, E. Trends in Quality in the Analytical Laboratory. II. Analytical Method Validation and Quality Assurance. *TrAC—Trends Anal. Chem.* **2004**, *23*, 535–552. [CrossRef]
33. AOAC. Appendix F: Guidelines for Standard Method Performance Requirements. In *AOAC Official Methods of Analysis*; AOAC Off. Methods Anal: Rockville, MD, USA, 2016; pp. 1–17.
34. Thvenot, D.R.; Toth, K.; Durst, R.A.; Wilson, G.S. Electrochemical Biosensors: Recommended Definitions and Classification (Technical Report). *Pure Appl. Chem.* **1999**, *71*, 2333–2348. [CrossRef]
35. Wang, N.; Burugapalli, K.; Song, W.; Halls, J.; Moussy, F.; Ray, A.; Zheng, Y. Electrospun Fibro-Porous Polyurethane Coatings for Implantable Glucose Biosensors. *Biomaterials* **2013**, *34*, 888–901. [CrossRef] [PubMed]
36. Schott, H. Swelling Kinetics of Polymers. *J. Macromol. Sci. Part B* **1992**, *31*, 1–9. [CrossRef]
37. Fang, L.; Liang, B.; Yang, G.; Hu, Y.; Zhu, Q.; Ye, X. Biosensors and Bioelectronics A Needle-Type Glucose Biosensor Based on PANI Nano Fi Bers and PU/E-PU Membrane for Long-Term Invasive Continuous Monitoring. *Biosens. Bioelectron.* **2017**, *97*, 196–202. [CrossRef]
38. Ahmadi, Y.; Kim, K.-H. Functionalization and Customization of Polyurethanes for Biosensing Applications: A State-of-the-Art Review. *TrAC Trends Anal. Chem.* **2020**, *126*, 115881. [CrossRef]
39. Boobphahom, S.; Rattanawaleedirojn, P.; Boonyongmaneerat, Y.; Rengpipat, S.; Chailapakul, O.; Rodthongkum, N. TiO2 Sol/Graphene Modified 3D Porous Ni Foam: A Novel Platform for Enzymatic Electrochemical Biosensor. *J. Electroanal. Chem.* **2019**, *833*, 133–142. [CrossRef]
40. Kucherenko, I.S.; Soldatkin, O.O.; Topolnikova, Y.V.; Dzyadevych, S.V.; Soldatkin, A.P. Novel Multiplexed Biosensor System for the Determination of Lactate and Pyruvate in Blood Serum. *Electroanalysis* **2019**, *31*, 1608–1614. [CrossRef]
41. Dagar, K.; Pundir, C.S. An Improved Amperometric L-Lactate Biosensor Based on Covalent Immobilization of Microbial Lactate Oxidase onto Carboxylated Multiwalled Carbon Nanotubes/Copper Nanoparticles/Polyaniline Modified Pencil Graphite Electrode. *Enzym. Microb. Technol.* **2017**, *96*, 177–186. [CrossRef]
42. Cunha-Silva, H.; Arcos-Martinez, M.J. Dual Range Lactate Oxidase-Based Screen Printed Amperometric Biosensor for Analysis of Lactate in Diversified Samples. *Talanta* **2018**, *188*, 779–787. [CrossRef]
43. Bravo, I.; Gutiérrez-Sánchez, C.; García-Mendiola, T.; Revenga-Parra, M.; Pariente, F.; Lorenzo, E. Enhanced Performance of Reagent-Less Carbon Nanodots Based Enzyme Electrochemical Biosensors. *Sensors* **2019**, *19*, 5576. [CrossRef] [PubMed]

Disclaimer/Publisher's Note: The statements, opinions and data contained in all publications are solely those of the individual author(s) and contributor(s) and not of MDPI and/or the editor(s). MDPI and/or the editor(s) disclaim responsibility for any injury to people or property resulting from any ideas, methods, instructions or products referred to in the content.

Review

Sensitive Materials Used in Surface Acoustic Wave Gas Sensors for Detecting Sulfur-Containing Compounds

Yuhang Wang [1,†], Cancan Yan [2,†], Chenlong Liang [3,4], Ying Liu [1], Haoyang Li [1], Caihong Zhang [1,*], Xine Duan [1] and Yong Pan [2,*]

1. School of Chemistry and Chemical Engineering, Shanxi University, Taiyuan 030006, China; 202222902021@email.sxu.edu.cn (Y.W.); ly19861900620@163.com (Y.L.); vc766693@163.com (H.L.); duanxe@sxu.edu.cn (X.D.)
2. State Key Laboratory of NBC Protection for Civilian, Beijing 102205, China; ccy805905145@163.com
3. Institute of Acoustics, Chinese Academy of Sciences, Beijing 100190, China; liangchenlong22@mails.ucas.ac.cn
4. School of Electronic, Electrical and Communication Engineering, University of Chinese Academy of Sciences, Beijing 100049, China
* Correspondence: chzhang@sxu.edu.cn (C.Z.); panyong@sklnbcpc.cn (Y.P.)
† These authors contributed equally to this work.

Abstract: There have been many studies on surface acoustic wave (SAW) sensors for detecting sulfur-containing toxic or harmful gases. This paper aims to give an overview of the current state of polymer films used in SAW sensors for detecting deleterious gases. By covering most of the important polymer materials, the structures and types of polymers are summarized, and a variety of devices with different frequencies, such as delay lines and array sensors for detecting mustard gas, hydrogen sulfide, and sulfur dioxide, are introduced. The preparation method of polymer films, the sensitivity of the SAW gas sensor, the limit of detection, the influence of temperature and humidity, and the anti-interference ability are discussed in detail. The advantages and disadvantages of the films are analyzed, and the potential application of polymer films in the future is also forecasted.

Keywords: polymer; films; surface acoustic wave (SAW); gas sensor; mustard gas; hydrogen sulfide; sulfur dioxide

Citation: Wang, Y.; Yan, C.; Liang, C.; Liu, Y.; Li, H.; Zhang, C.; Duan, X.; Pan, Y. Sensitive Materials Used in Surface Acoustic Wave Gas Sensors for Detecting Sulfur-Containing Compounds. *Polymers* **2024**, *16*, 457. https://doi.org/10.3390/polym16040457

Academic Editor: Jung-Chang Wang

Received: 3 January 2024
Revised: 30 January 2024
Accepted: 1 February 2024
Published: 6 February 2024

Copyright: © 2024 by the authors. Licensee MDPI, Basel, Switzerland. This article is an open access article distributed under the terms and conditions of the Creative Commons Attribution (CC BY) license (https://creativecommons.org/licenses/by/4.0/).

1. Introduction

1.1. The Fundamental Concepts of SAW Sensors

In 1885, for the first time, British physicist Rayleigh discovered the surface acoustic wave (SAW), which was a type of elastic mechanical wave propagating along the surface of an elastic object, while he studied seismic waves [1]. Due to technological limitations at that time, SAWs were not used in practical applications. In 1965, the invention of interdigital transducers (IDTs) by White R.M. and Voltmer F.W. provided a simpler method of generating SAWs and accelerated the development of the SAW sensor [2]. Since the publication of the first paper on SAW gas sensors by Wohltjen H. and Dessy R. in 1978, SAW gas sensors have been extensively studied [3–5]; over the past 40 years, they have been used to detect many kinds of hazardous gases such as SO_2, H_2S, NO_2, NH_3, methane, hydrogen, explosives, and chemical warfare agents [6]. Because of advantages of a small size, high sensitivity, ease of integration, intelligence, and low-cost mass production, more and more researchers all over the world have paid much attention to this new field.

A SAW sensor is mainly composed of IDTs and piezoelectric materials (e.g., quartz, $LiNbO_3$, $LiTaO_3$, ZnO, AlN, $Bi_{12}GeO_{20}$, $AsGa$, piezoceramics, etc. [7–10]). When a sinusoidal wave with the same period as the IDT is applied to the input IDT, vibration will generate under the IDT, resulting in the generation of a SAW perpendicular to the IDT; the SAW propagates along the piezoelectric material in the direction away from the input IDT; when the SAW reaches the output IDT, the output IDT converts the acoustic waves to electrical signals via the piezoelectric effect [11]. The energy of SAW is primarily limited to the surface of elastic objects, and tiny variations in the surface, such as changes in temperature, pressure, and weight, will alter the acoustic wave signals received by the output IDT [12], which result in the high sensitivity of SAW gas sensors.

According to distinct structures and operating principles, SAW sensors can be categorized as a delay-line-type and resonator-type [13]. For dual-delay lines, one is coated to be used to detect harmful gases, and the other is used as a reference channel to reduce the influence of environmental factors, such as temperature or pressure. Resonator-type sensors have a high Q value, low insertion loss, and small frequency drift, which can further enhance their sensitivity and distinguishability despite their more complicated mechanism and structure [14]. In terms of functionality, SAW sensors are classified into three categories: physical sensors, chemical sensors, and biosensors. Physical sensors are primarily used to detect physical parameters such as temperature and pressure, and chemical sensors are often used to detect gases qualitatively and quantitatively, while biosensors are employed to detect substances such as deoxyribonucleic acid (DNA), proteins, etc. [15].

For a SAW chemical gas sensor, such as the delay line shown in Figure 1, the SAW might be generated from the input IDT and pass through the surface of the sensitive film covering on the piezoelectric material; when the SAW reaches the output IDT and adsorbs a certain amount of gas, a frequency shift occurs. In general, the sensitive material can adsorb target gases selectively and reversibly in two ways, physical adsorption or chemical adsorption, that is, van der Waals force, hydrogen bonds, or a chemical reaction between the target gas and film, and sometimes, the solubility of the target gas in the film is considered [11], which alters the film's physicochemical parameters, such as the mass, density, modulus of elasticity, or conductivity, and affects the wave velocity or frequency of the passing SAW [16,17]. In the detection of toxic or harmful gases with SAW sensors, the physical and chemical properties of the sensitive materials and the selection of coating conditions (e.g., film thickness, surface roughness, etc.) will have great influence on the sensitivity, selectivity, repeatability, and stability of the sensors; therefore, optimizing membrane materials is very important for preparing a SAW sensor.

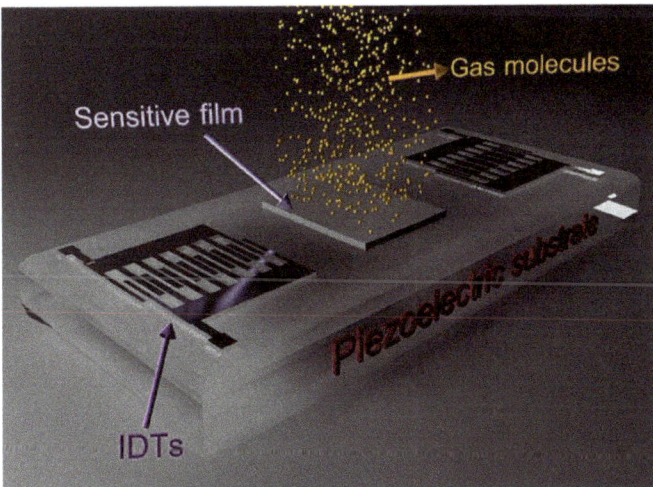

Figure 1. Scheme of SAW delay line [18]. Reproduced with permission from Mohsen Asad, Surface acoustic wave based H_2S gas sensors incorporating sensitive layers of single wall carbon nanotubes decorated with Cu nanoparticles; published by Sensors and Actuators B: Chemical, 2014.

1.2. Sulfur-Containing Hazardous Gas Species

1.2.1. Sulfur-Containing Chemical Agents

Mustard gas (HD), a vesicant chemical agent, is regarded as the king of toxic agents due to its ability to induce necrosis and tissue degeneration. As a lipophilic vesicant, it can permeate the body via the skin, eyes, and breathing system. Its alkylation reaction with proteins, DNA, and glutathione will cause cellular damage [19]. Although mustard gas has a lethality rate of only 2–5%, it has a high morbidity and psychological impact on people; this is because it has a median lethal dose (LD_{50}) of about 100 mg/kg, a median lethal concentration (LCt_{50}) of 15,000 mg min/m^3 [20], and a minimum dose of 0.2 mg to cause skin blistering [21], and the most important thing is that there is no effective antidote or treatment. Mustard gas was first synthesized by Despretz in 1822 and used in war in 1917 [22]; it caused about 1.3 million injuries and over 90,000 deaths during the First World War [23], mustard gas was also used in the Second World War and the Iran–Iraq War. According to the Chemical Weapons Convention which came into force in 1997, chemical weapons should be eliminated within ten years; however, because of the legacy of war and the fact that chemical weapons are cheap and easy to produce [24], research on detecting chemical agents are still ongoing. Table 1 lists common chemical agents along with their simulants.

Table 1. Chemical warfare simulants.

Simulant	Simulated Chemical Warfare Agent (CWA)	Median Lethal Dose (LD_{50}) Inhaled (ppm)	Ref.
Dimethyl methylphosphonate (DMMP)	Sarin (GB)	18	[25]
Dipropylene glycol monomethyl ether (DPGME)	Nitrogen mustard (HN)	180	[25]
Chloroethyl ethyl sulfide (CEES)	Distilled mustard (HD)	140	[26]
Dibutyl sulfide (DBS)	Distilled mustard (HD)	140	[27]
Chloroethyl phenyl sulfide (CEPS)	Distilled mustard (HD)	140	[27]
1,5-Dichloropentane (DCP)	Distilled mustard (HD)	140	[25]
Dimethylacetamide (DMA)	Distilled mustard (HD)	140	[25]
1,2-Dichloroethane (DCE)	Distilled mustard (HD)	140	[25]
	Soman (GD)	6	[25]
Dichloromethane (DCM)	Phosgene (CG)	800	[25]

1.2.2. Sulfur-Containing Harmful Gas

Hydrogen sulfide (H_2S), an acidic, toxic, and flammable gas, has an LD_{50} of 673 mg/kg and an odor threshold for H_2S of 11 ppb, but olfactory paralysis happens at a concentration of H_2S higher than 140 ppm [28]; people will collapse within 5 min, suffer serious eye impairment within 30 min, and face the risk of death after 30 to 60 min when the concentrations of H_2S reach 500–700 ppm [29]. H_2S naturally exists in volcanic eruptions, paper making, coal mining, chemical production, automobile exhausts, etc. Since it is one of the main causes of environmental pollution, it is essential to monitor the concentration of hydrogen sulfide gas in real time. Sulfur dioxide (SO_2), a colorless gas with an irritating odor, is one of the major pollutants in the atmosphere, it is mainly produced by natural or artificial processes such as burning fossil fuels containing sulfur (for example, coal, oil, and natural gas), volcanic eruptions, and smelting and forging sulfur-containing minerals [30]. Prolonged exposure to SO_2 can cause harm to the eyes, lungs, and throat. Additionally, SO_2 easily dissolves in water to form acid rain, which severely threatens buildings, plants, animals, and overall environmental balance [31]; therefore, monitoring SO_2 is an essential aspect of environmental protection.

2. Sensitive Functional Materials of Sulfur-Containing Agents and Their Simulants

2.1. Polymer

In 1993, Grate et al. [32] employed a 158 MHz four-channel SAW delay line sensor array to detect mustard gas and sarin. Poly(epichlorohydrin) (PECH), poly(ethylenimine)

(PEI), ethyl cellulose (ECEL), and fluoropolyol (FPOL) were utilized as the sensitive films of the sensor array, and signal processing and pattern-recognition algorithms were also employed to discriminate the target gases. Without preconcentration, mustard and sarin could be detected at concentrations as low as 2 mg/m^3 and 0.5 mg/m^3, respectively; however, when the 2 min preconcentration mode was used, the detection limits could be improved to 0.5 mg/m^3 for mustard gas and 0.01 mg/m^3 for sarin. This proved that the preconcentration mode enhanced the sensitivity of the sensors. Additionally, the study also discovered that the channel coated with PECH was more sensitive to mustard gas when humidity levels increased; however, the specific mechanism through which humidity influenced the sensor was not clearly explained and still needed to be further investigated.

In 2005, Liu et al. [33] used a 159 MHz SAW dual delay line coated with PECH as a sensitive film to detect mustard gas. To enhance the performance of sensors, the correlation between film thickness and sensitivity was investigated, the findings revealed that sensitivity increased with an increase in the film thickness. In the same year, Liu et al. [34] conducted a response test on the same sensor and found a good linear relationship between HD concentrations in the range of 2–200 mg/m^3 and the corresponding signals. The sensitivity of the sensor was determined to be 170.1 Hz/(mg/m^3). Additionally, they also found that when the temperature increased from 0 °C to 50 °C at a concentration of 2 g/m^3 of CEES, the frequency shifts decreased by 95%, and the response time and the recovery time become shorter when temperature increased. Additionally, a repeatability test showed excellent performance of the prepared sensor. In 2006, Liu et al. [35] investigated the adsorption kinetics between a PECH film and mustard gas with multimolecular layer adsorption model, they concluded that gas/liquid balance theory and van der Waals forces was very important for physical adsorption, and the related work in [36] was also summarized.

In 2007, Chen et al. [37] used an array that consisted of four 200 MHz two-port resonators with four different polymers (PECH, Silicone (SE-30), Hexafluoro-2-propanol bisphenol-substituted siloxane polymer (BSP3), fluorinated polymethyldrosiloxane (PTFP)) to detect HD, DMMP, GB, and sarin acid. Combined with a probabilistic neural network (PNN), the recognition rate could reach 90.87% successfully, and mustard gas was well recognized. In 2008, the stability, sensitivity, repeatability, consistency, and selectivity of a SAW PECH sensor were evaluated by this team [14]; the repeatability and consistency were found to have relative standard deviations of 3.27% and 2.50%, respectively, which were within the margin of error, and the detection limit was 0.3 mg/m^3.

In 2009, Matatagui et al. [38] employed a 157 MHz six-channel SAW delay line sensor array with an electrode thickness of 200 nm and a finger spacing of 5 µm to detect DMMP, DPGME, DMA, and DCE. Six polymers, including PECH, polycyanopropylmethylsiloxane (PCPMS), carbowax, polydimethylsiloxane (PDMS), PEI, and trifluoropropylmethylsiloxane–dimethylsiloxane (PMFTPMS), were prepared on the delay line, and the sensor array exhibited a rapid and significant response. The data obtained from the array were analyzed using principal component analysis (PCA) and PNN, which resulted in excellent distinguishability and a low detection limit. In their 2011 study, Matatagui et al. [25] successfully detected several substances with the same devices, including DMMP, DPGME, toluene, DCM, DCP, DMA, and DCE, they concluded that all simulants were accurately identified except DCE and DCM, as these two substances had very similar structures and could not be distinguished. In the same year, Matatagui et al. [39] developed a six-channel delay line array based on the Love wave; the sensor array was prepared with an aluminum electrode with a thickness of 200 nm and a finger spacing of 7 µm. Through spin-coating, a Novolac photoresist guide layer with a thickness of 0.8 µm was applied onto the surface of the piezoelectric material. Subsequently, the sensitive materials mentioned above were prepared on the surface of the delay line. DMMP, DPGME, DMA, DCE, DCM, and DCP were tested with the detection system, and their gas concentration and temperature were controlled, as shown in Figure 2. The sensor array exhibited excellent stability, reversibility, repeatability, and sensitivity. The CWA simulants were also accurately detected and categorized with PCA and PNN. In 2012, the same team [40] utilized a 3-micron-thick SiO$_2$ guide layer acquired through plasma-enhanced chemical vapor deposition to detect the same six target gases, and the results showed that the detection limits were 0.04 ppm, 0.25 ppm, 15 ppm, 75 ppm, 125 ppm, and 5 ppm, respectively.

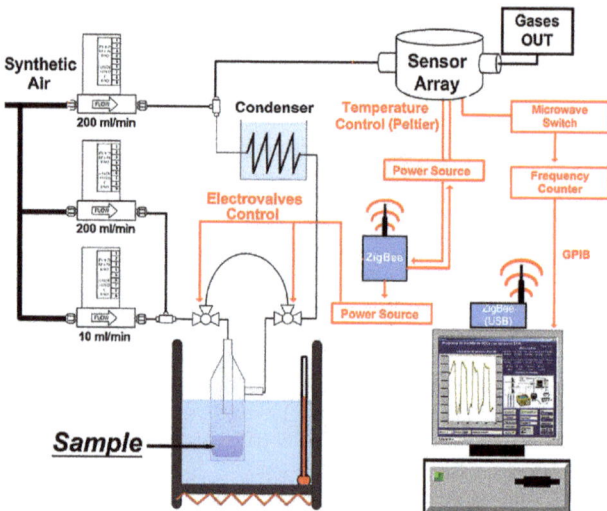

Figure 2. Scheme of the instrumentation used for data acquisition in real time [39]. Reproduced with permission from Matatagui D., Array of Love-wave sensors based on quartz/novolac to detect CWA simulants; published by Talanta, 2011.

In 2011, He et al. [41] designed a novel 300 MHz SAW dual delay line. The device was prepared with an Al/Au electrode structure and strategic phase modulation to minimize insertion loss. A PECH film was applied by solvent evaporation, and the thickness of the film was about 80 nm. This was performed under the conditions of 24 °C, RH 50%, sensor sensitivity of 25 Hz/(mg/m^3) to mustard gas, a linear range of 2–200 mg/m^3, and a repeatability error of ±10%. In 2017, Qi et al. [42] designed a 3D nanocluster resonator sensor whose surface was modified by ZnO nanoclusters to provide a larger specific surface area for the sensitive layer, thus increasing the detection sensitivity; however, it also led to an increase in the insertion loss of the sensor. When PECH, SE-30, PTFP, and BSP3 were used as sensitive films to detect a mixture of mustard gas and sarin, it obtained an identification rate of over 90%.

In 2018, Pan et al. [43] developed a SAW sensor array with a wireless communication network module and a positioning system module; in this sensor array, PECH, triethanolamine, fluoroalcoholpolysiloxane, and L-glutamic acid hydrochloride were used as sensitive films to detect CEES, H_2S, DMMP, and NH_3, respectively. Combined with pattern-recognition algorithms, target gases were successfully detected at safe concentrations outside within a range of 300 m. This study demonstrated the feasibility of using wireless sensor networks for gas detection. In light of the absence of prior research on the influence of temperature and humidity, Pan et al. [6] conducted a study in 2020 to explore the environmental adaptability of the PECH-SAW sensor in detecting CEES. The findings revealed that as the ambient temperature rose, the sensor's response value decreased, and the response time shortened. On the other hand, the detection signal exhibited an apparent increase in a higher-humidity environment; this phenomenon was attributed to the elevated environmental humidity, which amplified the solvation impact of CEES on PECH and facilitated the creation of hydrogen bond active sites. The sensor showed excellent selectivity and resistance to interfering gases in a smoke test, and its sensitivity of 233.17 Hz/(mg/m^3) along with stability over 18 months were also investigated. In 2022, Pan et al. [44] studied the physical characteristics of PECH film in detail. A viscosity of 1.969 was obtained and the glass transition temperature was found to be as low as −22.4 °C. At same time, the work of adhesion, work of immersion, and spreading coefficient were calculated, too. In general, the linear solution–energy relationship (LSER) (Equation (1)) is often used to evaluate the adsorption ability between films and target gases, and the related LSER parameters of PECH are summarized in Table 2.

$$\log K = c + rR_2 + s\pi_2^H + a\sum \alpha_2^H + b\sum \beta_2^H + l\log L^{16} \tag{1}$$

Table 2. LESR regression coefficients for PECH.

Polymer	Abbr.	Method	c	r	s	a	b	l	R	Std Error
Poly(epichlorohydrin)	PECH	SAW	−0.75	0.44	1.44	1.49	1.3	0.55	0.993	0.11

There have also been some reports on using other polymers to detect mustard gas and its simulants. In 2000, McGill et al. [45] employed an alarm system called SAWRHINO which utilized three unspecified polymer materials to detect HD, DMMP, and GD, and this is one of the few devices to use SAW technology in practice so far. In 2006, Shi et al. [11] developed liquid-phase macromolecular synthesis technology to implement molecular-level doping of poiyaniline (PANI) and phthalocyanine palladium (PdPc), which resulted in the creation of a novel organic semiconductor-sensitive material called $PdPc_{0.3}PANI_{0.7}$. It was observed that the material exhibited stability at a temperature of 300 °C by employing differential thermal analysis. The $PdPc_{0.3}PANI_{0.7}$ compound was applied onto the surface of a SAW dual delay line using vacuum-coating technology. The sensor exhibited high sensitivity of 105 kHz/(mg/m^3), and the response time was less than 5 min in detecting mustard gas. In 2014, Matatagui et al. [46] fabricated a 163 MHz six-channel sensor array; the nanofibers used in this array were prepared by electrospinning technology using polyvinyl alcohol (PVA), polyvinylpyrrolidone (PVP), polystyrene (PS), PVA+SnCl$_4$, PVA+SnCl$_4$ annealed for 4 h at 450 °C, and the copolymer PS+Poly(styrene-alt-maleic anhydride) (PS+PSMA) for detecting DMMP, DPGME, DMA, and DCE. The linear relationship between the concentration and response was found, and it was also proven that it was possible to achieve an identification rate of 100% by employing PCA. In 2015, Long et al. [47] applied a strong hydrogen-bond acidic (HBA) polymer linear fluoroalcoholic polysiloxane (PLF) as the sensitive material to detect GB, DMMP, HD, CEES, and DCP, as depicted in Figure 3. The sensor exhibited a significant response to sarin, DMMP, and CEES, while a minimal response to mustard gas and DCP was also found. The difference that existed in mustard gas, DCP, and CEES might be due to the differing polarity and electron cloud distribution of mustard gas, DCP, and CEES; this was believed to result from the chlorine atoms' strong electronegativity and the sulfur atoms' electron richness. The factors discussed above affected the formation of hydrogen bonds and diminished the detection effectiveness, so the sensor was deemed unsuitable for detecting mustard gas.

2.2. Organic Small Molecule

Katritzky et al. prepared a SAW sensor coated with organic small-molecule sensitive films to detect mustard gas and its simulants. As seen in Figure 4, pyridine 1-oxide, pyridinium salts, pyridinium betaine compounds, pyridyl ethers, and pyridinium compounds were synthesized as sensitive materials by this team in 1989 [48]. When detecting DMMP, CEES, and H$_2$O, it was found that pyridinium betaine and pyridinium sulfonate produced significant resistance changes to DMMP and CEES, respectively; however, no significant frequency shift was observed. From the point of view of the resistance response, pyridine derivatives were more easily influenced by humidity, so environmental conditions would limit their practical application. In response to this challenge, the team extended their research to acridinium betaines in 1990 [49], as shown in Figure 5, aiming to reduce humidity interference through an additional hydrocarbon mass around the ionic site. In addition to acridinium betaines, they also synthesized quaternary ammonium salts (Figure 5(**2,3**)) as sensitive materials to detect DMMP, CEES, and H_2O; they found that the compounds in Figure 5(**2,3**) had small frequency shifts and large resistance responses to CEES, but the compound in Figure 5(**3**) reacted almost as much to water vapor as CEES. For the compounds in Figure 5(**1a,1b**), CEES could be detected at frequency shifts of 9.8 kHz and 6.8 kHz, respectively, but the frequency shift resulting from the film in Figure 5(**1a**) was irreversible.

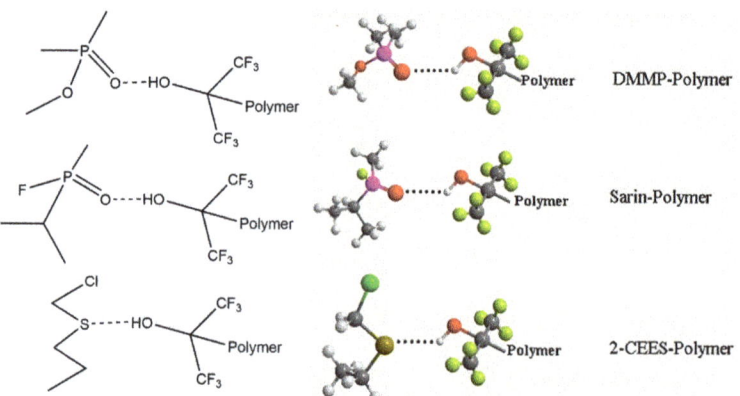

Figure 3. Hydrogen-bonding interactions between DMMP, sarin, 2-CEES, and HBA polymer PLF [47]. Reproduced with permission from Yin Long, The different sensitive behaviors of a hydrogen-bond acidic polymer-coated SAW sensor for chemical warfare agents and their simulants; published by Sensors, 2015.

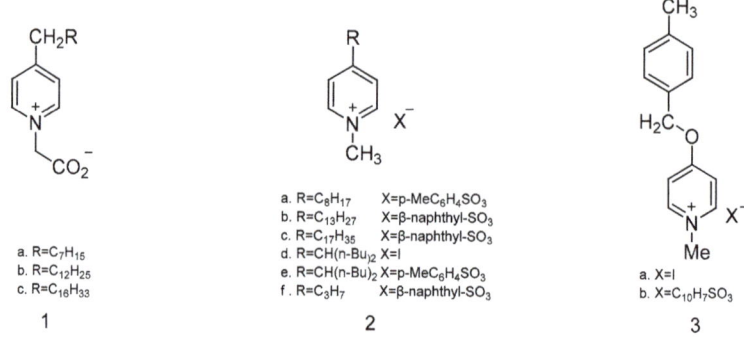

Figure 4. The structures of pyridinium betaine (**1**), pyridinium salt (**2**), and pyridine ether (**3**) [48]. Reproduced with permission from Katritzky A.R., Utilization of pyridinium salts as microsensor coatings; published by Langmuir, 1989.

Based on the speculation of the relationship between the adsorption mass and solubility of the sensitive membrane to the measured gases, in 1990, Katritzky et al. [26] sprayed phosphonic acid, phosphonate ester, and ammonium cyclohexylphosphonate, respectively, on a SAW surface, as shown in Figure 6, for the detection of DMMP, CEES, and H_2O. They expected that the sensor would achieve a better effect for DMMP than CEES and H_2O. However, the results revealed that only 4-methylbenzylphosphonic acid in Figure 6(**3b**) produced a maximum response frequency of 74.3 kHz for DMMP, while diethyl 4-dimethylaminophenylphosphonate in Figure 6(**1f**) and diethyl 2-thienylphosphonate in Figure 6(**2a**) gained frequency shifts of 77.5 kHz and 65.6 kHz for CEES, respectively, and the two compounds did not exhibit a good response for DMMP and H₂O. In 1991, they [50] employed a 52 MHz dual delay line SAW sensor and utilized several synthetic trisubstituted 1,3,5-triazines as sensitive materials to detect DMMP, CEES, and H₂O. As shown in Figure 7, 2,4-di(carboxymethylthio)-6-octanethio-1,3,5-triazine in Figure 7(**1**) and 2,4-di(carboxymethylthio)-6-dodecanethio-1,3,5-triazine in Figure 7(**2**) showed significant frequency and resistance shifts due to the interaction of carboxylic acid and phosphate functional groups. The compounds 2,4-dichloro-6-dodecylthio-1,3,5-triazine in Figure 7(**3**) and 2,4-dichloro-6-octylthio-1,3,5-triazine in Figure 7(**4**) exhibited a 37.4-fold and 34.0-fold increase in resistance to CEES, respectively. Katritzky et al. have conducted many studies on the sensitivity mechanism of sensors and the design of functional materials, and their works have great reference value for the design of sensitive materials for the detection of toxic or harmful gases.

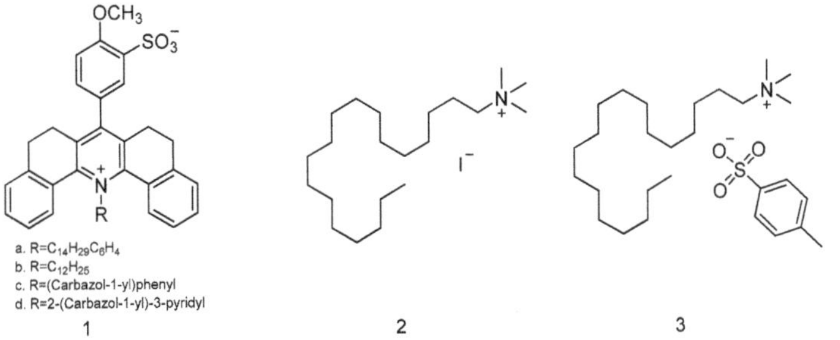

Figure 5. The structures of acridinium betaine (**1**) and two quaternary ammonium salts (**2**,**3**) [49]. Reproduced with permission from Katritzky A.R., Synthesis and response of new microsensor coatings-II Acridinium betaines and anionic surfactants; published by Talanta, 1990.

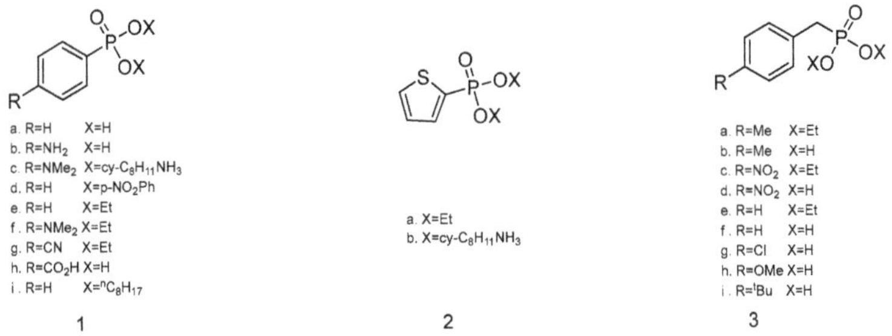

Figure 6. The structures of 4-substitutedphenylphosphonate derivatives (**1**), 2-thienylphosphonate derivatives (**2**), and 4-substitutedbenzylphosphonate derivatives (**3**) [26]. Reproduced with permission from Katritzky A.R., Synthesis of new microsensor coatings and their response to vapors-III arylphosphonic acids, salts and esters; published by Talanta, 1990.

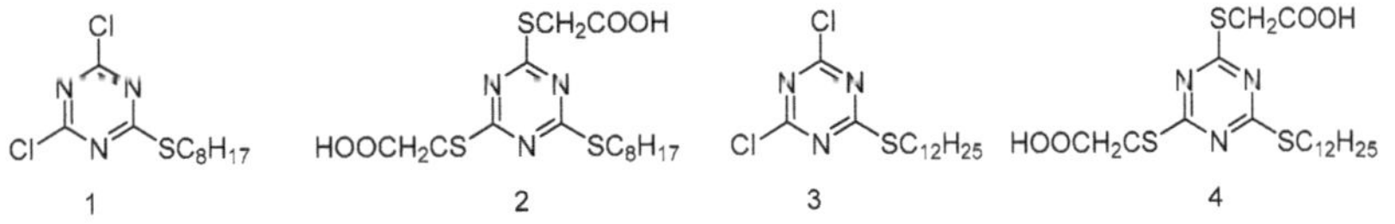

Figure 7. The structures of 2,4,6-trisubstituted-1,3,5-triazine derivatives (**1**–**4**) [50]. Reproduced with permission from Katritzky A.R., Synthesis of new microsensor coatings and their response to test vapors 2,4,6-trisubstituted-1,3,5-triazine derivatives; published by Talanta, 1991.

2.3. Other Kinds of Sensitive Materials

In 2013, to detect DMMP, diethyl cyanophosphonate, and the mustard gas simulants DBS and CEPS, Raj et al. [27] designed an electronic nose (E-nose) with four SAW sensors coated with ZnO, TeO$_2$, SnO$_2$, and TiO$_2$. The four simulants of CWA were effectively distinguished with the PCA. All simulants were clearly distinguished despite including interfering substances such as petrol, diesel, kerosene, volatile organic compounds, and water vapors in the PCA. In 2016, Sayago et al. [51] attempted to develop a Love wave sensor using graphene oxide as a sensitive film. The sensor presented good reproducibility in the detection of DMMP, DPGME, DMA, and DCE. The detection limit for DMMP was 9 ppb, and the response of graphene oxide to DMMP was much greater than that of the other gases measured, which may be due to the formation of hydrogen bonds between DMMP and graphene oxide.

There have also been some reports about the detection of sulfur-containing gases by SAW devices without a sensitive film. In 2021, Fahim et al. [52] developed an uncoated resonator SAW sensor to measure the frequency changes during programmed temperature increases to detect CEES, methyl salicylate, and DMMP. The system, combined with PCA, could identify high and sub-ppm concentrations of gases, which provided a novel method for identifying compounds. In 2022, Kumar et al. [53] investigated the impact of carrier gas on detecting sensitivity by combining gas chromatography with a SAW sensor; they used H$_2$, He, N$_2$, and air as carrier gases for the detection of CEES, DMMP, diethyl cyanophosphonate, and triethyl phosphate, as well as methanol, toluene, and xylene. The experiments revealed that higher sensitivity could be obtained with H$_2$ as the carrier gas in detecting all target gases (H$_2$ > He > air > N$_2$); therefore, it was judged that the sensitivity was affected by the density of the carrier gas.

3. Sensitive Functional Materials of Sulfur-Containing Harmful Gases

3.1. Sensitive Functional Materials for SO$_2$ Detection

Following the development of SAW sensor technology, more research has focused on its application in detecting SO$_2$. Most sensitive materials achieve an interaction with SO$_2$ through the attraction between acidic gas and alkaline sites. With a tertiary amino group as the alkaline adsorption center, N,N-dimethyl-3-aminopropyltrimethoxysilane (NND) has been a well-known material for detecting SO$_2$. In 1996, Leidl et al. [54] combined NND with hydrophobic propyltrimethoxysilane (PTMS) through co-condensation to decrease the hydrophilicity of the material. A heteropolysiloxane (NND/PTMS) consisting of 70 mol% NND and 30 mol% PTMS was obtained, and the heteropolysiloxane was then utilized as the sensitive material in a 330 MHz SAW sensor capable of detecting SO$_2$ in an RH 60% environment. In 2001, Penza et al. [55] developed resonator SAW sensors and surface transverse wave sensors with operating frequencies of 433.92 MHz and 380.0 MHz, respectively; these sensors utilized "rod-like" polymers, as shown in Figure 8, such as poly(bis(tributylphosphine)-platinum-diethynylbiphenyl) (Pt-DEBP), poly-2,5-dibutoxyethynylbenzene (DBEB), and poly-2,5-dioctyloxyethynylbenzene (DOEB), as sensitive materials. The SAW sensors generally outperformed the surface transverse wave sensors, particularly the SAW sensors with Pt-DEBP, which achieved lower detection limits of 2 ppm for SO$_2$ and 1 ppm for H$_2$S.

In 2005, Jakubik et al. [56] designed a dual delay line sensor with polyaniline as the sensitive film to detect acidic gases. However, they did not obtain a satisfactory response in detecting SO$_2$ and H$_2$S; the main reason was the thickness of the polyaniline film, which was 100 nm and was inadequate for adsorbing SO$_2$ and H$_2$S. In 2009, Wen et al. [13] utilized polyaniline as a sensitive material and opted for a film thickness of 120 μm to design a SAW dual delay line sensor consisting of three IDTs and two multistrip couplers. This design not only mitigated the impact of the environment, but also suppressed the generation of bulk acoustic waves (BAW), which ensured precise detection. The sensor also exhibited excellent linearity and sensitivity of 6.8 kHz/ppm over a measurement range from 312 ppb to 20 ppm. Reliable repeatability and long-term stability during testing SO$_2$ were displayed.

In the same year, Wen et al. [57] developed a dual delay line SAW sensor by utilizing carbon nanotube polyaniline as the sensitive material, based on their previous research. Compared to the pure polyaniline sensor, the carbon nanotube polyaniline sensor exhibited superior linearity, better sensitivity, and a lower detection limit at low concentrations. A sensitivity of 8.3 kHz/ppm and a detection limit of 0.12 ppb were obtained in the concentration range of 31.2 ppb to 20 ppm. The study determined that the application of polyaniline-coated carbon nanotubes solved the problem of pure carbon nanotubes, which tended to aggregate, so the specific surface area of polyaniline was enhanced.

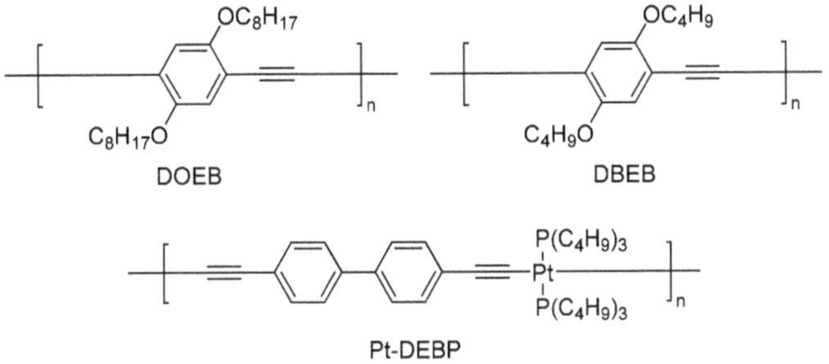

Figure 8. Three types of "rod-like" polymers [55]. Reproduced with permission from Penza M., SAW chemical sensing using poly-ynes and organometallic polymer films; published by Sensors and Actuators B: Chemical, 2001.

In 2013, Ben et al. [58] created new polyurethane imides (PUIs) with Lewis base properties by synthesizing them with N-methyldiethanolamine (MDEA), N-tert-butyldiethanolamine (tBu-DEA), N-phenyldiethanolamine (Ph-DEA), and 1,4-diethanolpiperazine-diol (Piperazine-diol) as functional monomers. Figure 9 illustrates the polymer structure. SO_2 gas was detected accurately at a concentration of 28 ppm by a three-layer Love wave sensor, the sensitivity of sensors utilizing various functional monomers could be enhanced using the following sequence: Piperazine-diol < tBu-DEA ≈ Ph-DEA << MDEA. It was found that the influence of steric hindrance on the sensitivity of the sensor was much higher than that of the alkalinity of the amino group in the functional monomer. Up to now, there have been few reports on the detection of SO_2 by SAW sensors. In addition to the polymers discussed above, many other sensitive materials have also been used to detect SO_2, such as metal oxides [59], metal sulfides [60], and small organic molecules [61,62].

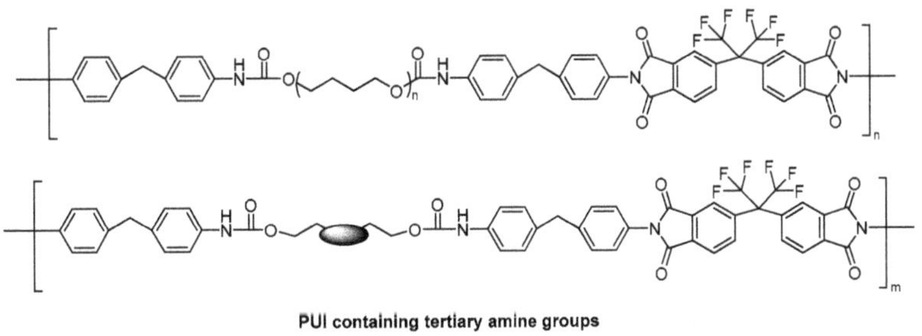

Figure 9. Synthesis of PUIs [58]. Reproduced with permission from Ismaïl Ben Youssef, Functional poly(urethane-imide)s containing Lewis bases for SO_2 detection by Love surface acoustic wave gas micro-sensors; published by Sensors and Actuators B: Chemical, 2013.

3.2. Sensitive Functional Materials for H_2S Detection

In 2001, Penza et al. [55] used a "rod-like" polymer poly(bis(tributylphosphine)-platinum-diethynylbiphenyl) (Pt-DEBP) as a sensitive material to detect H_2S, and a detection limit of 1 ppm was obtained. In 2005, Jakubik et al. [56] demonstrated that sensors utilizing polyaniline films had a suboptimal response to H_2S gas during the testing of acidic gases. The sulfur atom in the H_2S molecule exhibits distinct reactivity towards metal ions, such as Pb^{2+} or Zn^{2+}, based on this particular property. In 2020, Rabus et al. [63] synthesized a network polymer that incorporated Pb^{2+}, and used it as sensitive material in a system which could detect H_2S underground. The recognition capability was enhanced by the specific amalgamation of the lead ion with H_2S, as shown in Figure 10. Because of the irreversible reaction, its application was limited. In addition to polymers, many other sensitive films, such as metal oxides [64,65], small organic molecules [66,67], carbon nanotubes [18], and ionic liquids [68,69] have also been reported.

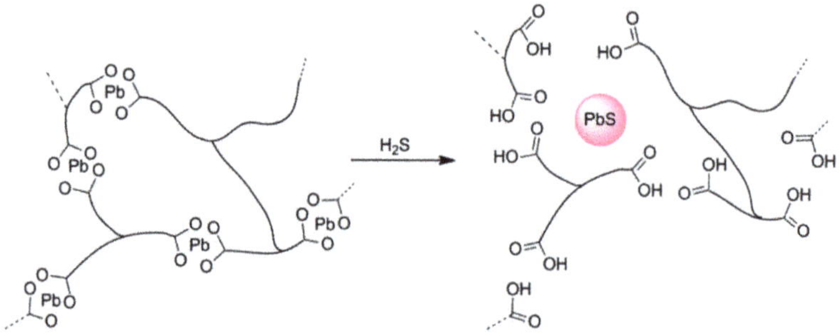

Figure 10. Response mechanism of network polymer [63]. Reproduced with permission from David Rabus, Subsurface H_2S detection by a surface acoustic wave passive wireless sensor interrogated with a ground penetrating radar; published by ACS Sensors, 2020.

4. Conclusions

This paper carried out a systematic discussion of polymer materials used in SAW sensors for detecting sulfur-containing toxic or harmful gases. The polymers discussed in this paper can be categorized into carbon-chain polymers and hetero-chain polymers based on their main chain structure, which could be modified by the insertion of functional monomers or functional groups. The sensitive materials are summarized in Table 3. Great progress has been made in the research on polymers for detecting sulfur-containing gases, and there have been many reports on the structure design, selectivity, stability and anti-interference ability of polymers, but determining how to obtain polymer materials with more selectivity for target gases is still the focus of current research. In some cases, due to the similar chemical structure of the measured gas, it is very difficult to accurately identify the target gas with a single polymer material; to solve this problem, SAW sensor arrays and pattern-recognition algorithms are always used to improve the accuracy of detection. In addition, environmental factors including temperature, humidity, and interference gases might affect the sensor during gas detection; therefore, determining how to improve the environmental adaptability of polymer materials to obtain new polymer materials is still a focal area of research.

Table 3. Summary of polymers used for detecting sulfur-containing compounds with SAW gas sensors.

Device	Polymer Types	Coating Method	Analytes	Sensitivity or Limitation of Detection	Range of Detection	Advantages	Disadvantages	Ref.
158 MHz four-channel SAW delay line sensor array	1. PECH; 2. PEI; 3. ECEL; 4.FPOL.	Spray-coating	HD	0.5 mg/m^3 in 2 min	2 mg/m^3 to 50 mg/m^3	Pattern-recognition algorithms correctly classified the analytes	Influenced by humidity	[32]
159 MHz SAW dual delay line	PECH	Solvent evaporation	HD	48.26 Hz·L·μg^{-1} and 2 mg/m^3	10 mg/m^3 to 200 mg/m^3	Response to HD was 5.6 times greater than that to GB	Not mentioned	[33]
159 MHz SAW dual delay line	PECH	Solvent evaporation	HD	170.1 Hz·m^3/mg and 2 mg/m^3	2 mg/m^3 to 200 mg/m^3	Good thermal stability, reproducibility, and linear range.	Not mentioned	[34]
159 MHz SAW dual delay line	PECH	Solvent evaporation	CEES	1.62 Hz·L·μg^{-1}	5 mg/m^3 to 100 mg/m^3	CEES was detected at low concentration.	Influenced by temperature	[35]
200 MHz four two-port SAW resonator array	1. PECH; 2. SE-30; 3. BSP3; 4. PTFP.	Spin-coating method	HD	Not mentioned	Not mentioned	Combined with PNN, the analytes were classified	Not mentioned	[37]
200 MHz two-port SAW resonator	PECH	Spin-coating method	HD	106 Hz/(mg/m^3) and 0.3 mg/m^3	1.2 mg/m^3 to 61.6 mg/m^3	Good reversibility, stability, reproducibility, and anti-interference ability	Not mentioned	[14]
157 MHz six-channel SAW delay line sensor array	1. PECH; 2. PCPMS; 3. Carbowax; 4. PDMS; 5. PEI; 6. PMFTPMS	Not mentioned	DMA DCE	Not mentioned	100 ppm to 250 ppm (DCE)	Combined with PCA, the simulants were well classified	Not mentioned	[38]
157 MHz six-channel SAW delay line sensor array	1. PECH; 2. PCPMS; 3. Carbowax; 4. PDMS; 5. PEI; 6. PMFTPMS.	Spray-coating method	DMA DCP DCE	Not mentioned	30 ppm to 150 ppm (DMA); 80 ppm to 250 ppm (DCE); 5 ppm to 100 ppm (DCP).	The array showed very good sensitivity and specificity rates	DCE and DCM cannot be classified	[25]

Table 3. Cont.

Device	Polymer Types	Coating Method	Analytes	Sensitivity or Limitation of Detection	Range of Detection	Advantages	Disadvantages	Ref.
Six-channel Love wave delay line sensor array	1. PECH; 2. PCPMS; 3. Carbowax; 4. PDMS; 5. PEI; 6. PMFTPMS.	Spin-coating method	DMA DCE DCP	25 ppm (DMA); 75 ppm (DCE); 5 ppm (DCP).	25 ppm to 250 ppm (DMA); 75 ppm to 250 ppm (DCE); 5 ppm to 25 ppm (DCP)	Good sensitivity and discrimination	Guiding layer will result in damping	[39]
Six-channel Love wave delay line sensor array	1. PECH; 2. PCPMS; 3. Carbowax; 4. PDMS; 5. PEI; 6. PMFTPMS.	Spray-coating method	DMA DCE DCP	15 ppm (DMA); 75 ppm (DCE); 5 ppm (DCP).	15 ppm to 200 ppm (DMA); 75 ppm to 300 ppm (DCE); 5 ppm to 25 ppm (DCP)	Good linearity, stability, reversibility, and accuracy; fast response; high sensitivity and selectivity	Not mentioned	[40]
300 MHz SAW dual delay line	PECH	Solvent evaporation	HD	25 Hz/(mg/m^3) and less than 2 mg/m^3	2 mg/m^3 to 200 mg/m^3	New phase-modulation methods and design resulted a great improvement in frequency stability	Not mentioned	[41]
3D nanocluster resonator sensors modified by ZnO	1.PECH; 2.SE-30; 3. PTFP; 4. BSP3.	Not mentioned	HD	Not mentioned	Not mentioned	High targeting capacity and disturbance resistance	Greater insertion loss	[42]
300 MHz five-channel two-port SAW resonator array	1. TEA; 2. PECH; 3. SXFA; 4. L-glutamic acid hydrochloride	Dipping method	CEES	14.9 Hz/ppm and less than 0.59 ppm	0.59 ppm to 14 ppm	Combined with pattern-recognition algorithms, analytes were detected within a range of 300 m	Not mentioned	[43]
150 MHz SAW dual delay line	PECH	Not mentioned	CEES	233.17 Hz/(mg/m^3) and 1.5 mg/m^3	1.2 mg/m^3 to 10 mg/m^3	High response at high humidity	Influenced by temperature	[6]
200 MHz SAW delay line	PECH	Spin-coating method	CEES	1.13 mV/(mg/m^3) and 0.85 mg/m^3	1.9 mg/m^3 to 19.6 mg/m^3	High sensitivity	Sensor poisoning at high concentration	[44]
SAW dual delay line	PdP$_{0.3}$PANI$_{0.7}$	Vacuum-coating method	HD	105 kHz/(mg/m^3)	1.5 mg/m^3 to 7.5 mg/m^3	The principle and method were feasible	The mechanism is unknown	[11]

Table 3. Cont.

Device	Polymer Types	Coating Method	Analytes	Sensitivity or Limitation of Detection	Range of Detection	Advantages	Disadvantages	Ref.
The 163 MHz six-channel SAW delay line sensor array	1. PVA; 2. PVP; 3. PS; 4. PVA+SnCl$_4$; 5. PVA+SnCl$_4$ 4-h 450 °C; 6. PS+PSMA	Electrospinning technology	DMA; DCE.	Not mentioned	50 ppm to 200 ppm (DCA) 100 ppm to 500 ppm (DCE)	The array achieved a resolution probability of 100% by PCA	Not mentioned	[46]
434-MHz two-port SAW resonator	PLF	Spray-coating	HD CEES DCP	0.01 mg/m^3 and 2.842 kHz/(mg/m^3) (CEES);	1 mg/m^3 to 20 mg/m^3 (CEES)	Significant response to CEES	Minimal response to HD and DCP	[47]
330 MHz SAW sensor	NND/PTMS	Not mentioned	SO$_2$	Not mentioned	Not mentioned	The co-condensation of NND with PTMS reduced the humidity affinity	Not mentioned	[54]
433.92 MHz SAW resonator	1. Pt-DEBP; 2. DBEB; 3. DOEB	Spin-coating method	SO$_2$; H$_2$S.	2 ppm for SO$_2$ and 1 ppm for H$_2$S (Pt-DEBP)	1 ppm to 10 ppm	High sensitivity	Not mentioned	[55]
101.764 MHz SAW dual delay line	Polyaniline	Not mentioned	SO$_2$	6.8 kHz/ppm	312 ppb to 20 ppm	New design eliminated the external perturbations and suppressed the BAW Superior linearity, better sensitivity, and lower detection limit at low concentration of SO$_2$	Not mentioned	[13]
The SAW dual delay line	Carbon nanotube polyaniline	Solvent evaporation	SO$_2$	0.12 ppb and 8.3 kHz/ppm	31.2 ppb to 20 ppm		Not mentioned	[57]
Love SAW microsensor	PUIs (MDEA, tBu-DEA, Ph-DEA, and Piperazine-diol)	Spin-coating method	SO$_2$	Not mentioned	Not mentioned	The sensitivity could be changed by changing the amino steric hindrance	The response is not completely reversible	[58]
380 MHz SAW resonator sensor	a network polymer that incorporated Pb^{2+}	Spin-coating method	H$_2$S	Not mentioned	Not mentioned	High response at high humidity	Response is irreversible	[63]

55

Author Contributions: Conceptualization, Y.P., C.Z. and X.D.; investigation, Y.W., C.Y. and C.L.; project administration, Y.P., C.Z. and X.D.; supervision, Y.P., C.Z. and X.D.; funding acquisition, C.Z.; visualization, H.L. and Y.L.; writing—original draft, Y.W., C.Y. and Y.P.; writing—review and editing, Y.W., C.L. and Y.P. All authors have read and agreed to the published version of the manuscript.

Funding: This Research Project was supported by the National Natural Science Foundation of China (No. 22177065), and Shanxi Scholarship Council of China (No. 2023-026).

Institutional Review Board Statement: Not applicable.

Data Availability Statement: Data are contained within the article.

Conflicts of Interest: The authors declare no conflict of interest.

References

1. Rayleigh, L. On waves propagated along the plane surface of an elastic solid. *Proc. Lond. Math. Soc.* **1885**, *s1-17*, 4–11. [CrossRef]
2. White, R.M.; Voltmer, F.W. Direct piezoelectric coupling to surface elastic waves. *Appl. Phys. Lett.* **1965**, *7*, 314–316. [CrossRef]
3. Wohltjen, H.; Dessy, R. Surface acoustic wave probe for chemical analysis. I. Introduction and instrument description. *Anal. Chem.* **1979**, *51*, 1458–1464. [CrossRef]
4. Wohltjen, H.; Dessy, R. Surface acoustic wave probes for chemical analysis. II. Gas chromatography detector. *Anal. Chem.* **1979**, *51*, 1465–1470. [CrossRef]
5. Wohltjen, H.; Dessy, R. Surface acoustic wave probes for chemical analysis. III. Thermomechanical polymer analyzer. *Anal. Chem.* **1979**, *51*, 1470–1475. [CrossRef]
6. Pan, Y.; Zhang, L.; Cao, B.; Xue, X.; Liu, W.; Zhang, C.; Wang, W. Effects of temperature and humidity on the performance of a PECH polymer coated SAW sensor. *RSC Adv.* **2020**, *10*, 18099–18106. [CrossRef] [PubMed]
7. Fox, C.G.; Alder, J.F. Surface acoustic wave sensors for atmospheric gas monitoring. A review. *Analyst* **1989**, *114*, 997–1004. [CrossRef]
8. Xu, G.Q.; Zhu, W.Z. Surface acoustic wave sensor and its application. *Sens. World* **1996**, *2*, 31–35.
9. Li, H.Q.; Xia, G.Q. 158MHz surface acoustic wave fixed-delay line on GaAs. *Chin. J. Semicond.* **2000**, *021*, 93–96.
10. Mandal, D.; Banerjee, S. Surface acoustic wave (SAW) sensors: Physics, materials, and applications. *Sensors* **2022**, *22*, 820. [CrossRef]
11. Shi, Y.B.; Xiang, J.J.; Feng, Q.H.; Hu, Z.P.; Zhang, H.Q.; Guo, J.Y. Binary Channel SAW Mustard Gas Sensor Based on $PdPc_{0.3}PANI_{0.7}$ Hybrid Sensitive Film. *J. Phys. Conf. Ser.* **2006**, *48*, 292. [CrossRef]
12. Liu, B.; Chen, X.; Cai, H.; Ali, M.M.; Tian, X.; Tao, L.; Yang, Y.; Ren, T. Surface acoustic wave devices for sensor applications. *J. Semicond.* **2016**, *37*, 021001. [CrossRef]
13. Wen, C.; Zhu, C.; Ju, Y.; Xu, H.; Qiu, Y. A novel dual track SAW gas sensor using three-IDT and two-MSC. *IEEE Sens. J.* **2009**, *9*, 2010–2015. [CrossRef]
14. Chen, C.Z.; Zuo, B.L.; Ma, J.Y.; Jiang, H.M. Detecting Mustard Gas Using High Q-value SAW Resonator Gas Sensors. In Proceedings of the 3rd Symposium on Piezoelectricity, Acoustic Waves and Device Applications, Nanjing, China, 5–8 December 2008.
15. Chen, Z.; Zhou, J.; Tang, H.; Liu, Y.; Shen, Y.; Yin, X.; Zheng, J.; Zhang, H.; Wu, J.; Shi, X.; et al. Ultrahigh-frequency surface acoustic wave sensors with giant mass-loading effects on electrodes. *ACS Sens.* **2020**, *5*, 1657–1664. [CrossRef]
16. Asad, M.; Sheikhi, M.H. Surface acoustic wave based H_2S gas sensors incorporating sensitive layers of single wall carbon nanotubes decorated with Cu nanoparticles. *Sens. Actuators B Chem.* **2014**, *198*, 134–141. [CrossRef]
17. Falconer, R.S. A versatile SAW-based sensor system for investigating gas-sensitive coatings. *Sens. Actuators B Chem.* **1995**, *24*, 54–57. [CrossRef]
18. Li, D.; Zu, X.; Ao, D.; Tang, Q.; Fu, Y.; Guo, Y.; Bilawal, K.; Faheem, M.B.; Li, L.; Li, S.; et al. High humidity enhanced surface acoustic wave (SAW) H_2S sensors based on sol–gel CuO films. *Sens. Actuators B Chem.* **2019**, *294*, 55–61. [CrossRef]
19. Feng, W.; Xue, M.J.; Zhang, Q.L.; Liu, S.L.; Song, Q.H. Prefluorescent probe capable of generating active sensing species in situ for detections of sulfur mustard and its simulant. *Sens. Actuators B Chem.* **2022**, *371*, 132555. [CrossRef]
20. Raber, E.; Jin, A.; Noonan, K.; McGuire, R.; Kirvel, R.D. Decontamination issues for chemical and biological warfare agents: How clean is clean enough? *Int. J. Environ. Health Res.* **2010**, *11*, 128–148. [CrossRef] [PubMed]
21. Kumar, V.; Rana, H. Selective and sensitive chromogenic and fluorogenic detection of sulfur mustard in organic, aqueous and gaseous medium. *RSC Adv.* **2015**, *5*, 91946–91950. [CrossRef]
22. Balali-Mood, M.; Hefazi, M. The pharmacology, toxicology, and medical treatment of sulphur mustard poisoning. *Fundam. Clin. Pharmacol.* **2005**, *19*, 297–315. [CrossRef]
23. Wattana, M.; Bey, T. Mustard gas or sulfur mustard: An old chemical agent as a new terrorist threat. *Prehosp. Disaster Med.* **2009**, *24*, 19–29; discussion 30–31. [CrossRef] [PubMed]
24. Chauhan, S.; D'cruz, R.; Faruqi, S.; Singh, K.K.; Varma, S.; Singh, M.; Karthik, V. Chemical warfare agents. *Environ. Toxicol. Pharmacol.* **2008**, *26*, 113–122. [CrossRef] [PubMed]
25. Matatagui, D.; Martí, J.; Fernández, M.J.; Fontecha, J.L.; Gutiérrez, J.; Gràcia, I.; Cané, C.; Horrillo, M.C. Chemical warfare agents simulants detection with an optimized SAW sensor array. *Sens. Actuators B Chem.* **2011**, *154*, 199–205. [CrossRef]

26. Katritzky, A.R.; Savage, G.P.; Offerman, R.J.; Pilarski, B. Synthesis of new microsensor coatings and their response to vapors-III arylphosphonic acids, salts and esters. *Talanta* **1990**, *37*, 921–924. [CrossRef]
27. Raj, V.B.; Singh, H.; Nimal, A.T.; Sharma, M.U.; Gupta, V. Oxide thin films (ZnO, TeO$_2$, SnO$_2$, and TiO$_2$) based surface acoustic wave (SAW) E-nose for the detection of chemical warfare agents. *Sens. Actuators B Chem.* **2013**, *178*, 636–647. [CrossRef]
28. Bhomick, P.C.; Rao, K.S. Sources and Effects of Hydrogen Sulfide. *J. Appl. Chem.* **2014**, *3*, 914–918.
29. Maldonado, C.S.; Weir, A.; Rumbeiha, W.K. A comprehensive review of treatments for hydrogen sulfide poisoning: Past, present, and future. *Toxicol. Mech. Methods* **2022**, *33*, 183–196. [CrossRef] [PubMed]
30. Zou, Y.M.; Yang, Y.W. Research progress on fluorescence detection technology of atmospheric pollutant sulfur dioxide. *Shanghai Chem. Ind.* **2019**, *44*, 39–43.
31. Khan, M.; Rao, M.; Li, Q. Recent advances in electrochemical sensors for detecting toxic gases: NO$_2$, SO$_2$ and H$_2$S. *Sensors* **2019**, *19*, 905. [CrossRef] [PubMed]
32. Grate, J.W.; Rose-Pehrsson, S.L.; Venezky, D.L.; Klusty, M.; Wohltjen, H. Smart sensor system for trace organophosphorus and organosulfur vapor detection employing a temperature-controlled array of surface acoustic wave sensors, automated sample preconcentration, and pattern recognition. *Anal. Chem.* **1993**, *65*, 1868–1881. [CrossRef]
33. Liu, W.; Yu, J.; Pan, Y.; Zhao, J.; Huang, Q. Surface acoustic wave sensor detection of mustard gas with poly(epichlorohydrin) coatings. *Chem. Sens.* **2005**, *25*, 57–60.
34. Liu, W.; Yu, J.; Pan, Y.; Zhao, J.; Huang, Q. The study of response character in the detection of HD by SAW-PECH sensor. *Chem. Sens.* **2005**, *25*, 52–54.
35. Liu, W.; Pan, Y.; Zhao, J.; Zhao, J.; Yu, J.; Wang, Y.; Wu, Z. The adsorption study of SAW-PECH sensor to organosulfur agents. *Chem. Sens.* **2006**, *26*, 64–67.
36. Liu, W.; Yu, J.; Pan, Y.; Huang, Q. Studying on the and application of SAW technology in detection of organosulfur chemical warfare agents. *Piezoelectrics Acoustooptics* **2006**, *28*, 14–16, +20.
37. Chuanzhi, C.; Jinyi, M.; Boli, Z.; Hongmin, J. A Novel Toxic Gases Detection System Based on SAW Resonator Array and Probabilistic Neural Network. In Proceedings of the 8th International Conference on Electronic Measurement and Instruments, Xi'an, China, 16–18 August 2007.
38. Matatagui, D.; Marti, J.; Fernandez, M.J.; Fontecha, J.L.; Gutierrez, J.; Gracia, I.; Cane, C.; Horrillo, M.C. Optimized design of a SAW sensor array for chemical warfare agents simulants detection. *Procedia Chem.* **2009**, *1*, 232–235. [CrossRef]
39. Matatagui, D.; Fontecha, J.; Fernández, M.J.; Aleixandre, M.; Gràcia, I.; Cané, C.; Horrillo, M.C. Array of Love-wave sensors based on quartz/novolac to detect CWA simulants. *Talanta* **2011**, *85*, 1442–1447. [CrossRef]
40. Matatagui, D.; Fernandez, M.J.; Fontecha, J.; Santos, J.P.; Gràcia, I.; Cané, C.; Horrillo, M.C. Love-wave sensor array to detect, discriminate and classify chemical warfare agent simulants. *Sens. Actuators B Chem.* **2012**, *175*, 173–178. [CrossRef]
41. He, S.; Wang, W.; Li, S.; Liu, M. Advances in Polymer-Coated Surface Acoustic Wave Gas Sensor. In Proceedings of the 2011 16th International Solid-State Sensors, Actuators and Microsystems Conference, Beijing, China, 5–9 June 2011.
42. Qi, J.; Wen, Y.M.; Li, P. Study on the detection of blister agent mustard by surface acoustic wave technology. *J. Chongqing Univ. Posts Telecommun. (Nat. Sci. Ed.)* **2017**, *29*, 494–499.
43. Pan, Y.; Mu, N.; Liu, B.; Cao, B.; Wang, W.; Yang, L. A novel surface acoustic wave sensor array based on wireless communication network. *Sensors* **2018**, *18*, 2977. [CrossRef]
44. Pan, Y.; Wang, P.; Zhang, G.; Yan, C.; Zhang, L.; Guo, T.; Wang, W.; Zhai, S. Development of a SAW poly (epichlorohydrin) gas sensor for detection of harmful chemicals. *Anal. Methods* **2022**, *14*, 1611–1622. [CrossRef] [PubMed]
45. McGill, R.A.; Nguyen, V.K.; Chung, R.; Shaffer, R.E.; DiLella, D.; Stepnowski, J.L.; Mlsna, T.E.; Venezky, D.L.; Dominguez, D. The "NRL-SAWRHINO": A nose for toxic gases. *Sens. Actuator B Chem.* **2000**, *65*, 10–13. [CrossRef]
46. Matatagui, D.; Fernández, M.J.; Fontecha, J.; Sayago, I.; Gràcia, I.; Cané, C.; Horrillo, M.C.; Santos, J.P. Characterization of an array of Love-wave gas sensors developed using electrospinning technique to deposit nanofibers as sensitive layers. *Talanta* **2014**, *120*, 408–412. [CrossRef] [PubMed]
47. Long, Y.; Wang, Y.; Du, X.; Cheng, L.; Wu, P.; Jiang, Y. The different sensitive behaviors of a hydrogen-bond acidic polymer-coated SAW sensor for chemical warfare agents and their simulants. *Sensors* **2015**, *15*, 18302–18314. [CrossRef] [PubMed]
48. Katritzky, A.R.; Offerman, R.J.; Wang, Z. Utilization of pyridinium salts as microsensor coatings. *Langmuir* **1989**, *5*, 1087–1092. [CrossRef]
49. Katritzky, A.R.; Offerman, R.J.; Aurrecoechea, J.M.; Savage, G.P. Synthesis and response of new microsensor coatings-II Acridinium betaines and anionic surfactants. *Talanta* **1990**, *37*, 911–919. [CrossRef]
50. Katritzky, A.R.; Lam, J.N.; Faid-Allah, H.M. Synthesis of new microsensor coatings and their response to test vapors 2,4,6-trisubstituted-1,3,5-triazine derivatives. *Talanta* **1991**, *38*, 535–540. [CrossRef]
51. Sayago, I.; Matatagui, D.; Fernández, M.J.; Fontecha, J.L.; Jurewicz, I.; Garriga, R.; Muñoz, E. Graphene oxide as sensitive layer in Love-wave surface acoustic wave sensors for the detection of chemical warfare agent simulants. *Talanta* **2016**, *148*, 393–400. [CrossRef]
52. Fahim, F.; Mainuddin, M.; Mittal, U.; Kumar, J.; Nimal, A.T. Novel SAW CWA detector using temperature programmed desorption. *IEEE Sens. J.* **2021**, *21*, 5914–5922. [CrossRef]
53. Kumar, J.; Nimal, A.T.; Mittal, U.; Kumar, V.; Singh, V.K. Effect of carrier gas on sensitivity of surface acoustic wave detector. *IEEE Sens. J.* **2022**, *22*, 8394–8401. [CrossRef]

54. Leidl, A.; Hartinger, R.; Roth, M.; Endres, H.E. A new SO_2 sensor system with SAW and IDC elements. *Sens. Actuators B Chem.* **1996**, *34*, 339–342. [CrossRef]
55. Penza, M.; Cassano, G.; Sergi, A.; Sterzo, C.L.; Russo, M.V. SAW chemical sensing using poly-ynes and organometallic polymer films. *Sens. Actuators B Chem.* **2001**, *81*, 88–98. [CrossRef]
56. Jakubik, W.P.; Urbanczyk, M.; Maciak, E.; Pustelny, T.; Stolarczyk, A. Polyaniline thin films as a toxic gas sensors in SAW system*. *J. De Phys. IV (Proc.)* **2005**, *129*, 121–124. [CrossRef]
57. Wen, C.B.; Zhu, C.C.; Ju, Y.F.; Qiu, Y.Z. Experimental study on SAW SO_2 Sensor Based on Carbon Nanotube-polyanilin Films. *Piezoelectrics Acoustooptics* **2009**, *31*, 157–160.
58. Youssef, I.B.; Alem, H.; Sarry, F.; Elmazria, O.; Rioboo, R.J.; Arnal-Hérault, C.; Jonquières, A. Functional poly(urethane-imide) s containing Lewis bases for SO_2 detection by Love surface acoustic wave gas micro-sensors. *Sens. Actuators B Chem.* **2013**, *185*, 309–320. [CrossRef]
59. Yang, J.; Wang, T.; Zhu, C.; Yin, X.; Dong, P.; Wu, X. AgNWs@SnO_2/CuO nanocomposites for ultra-sensitive SO_2 sensing based on surface acoustic wave with frequency-resistance dual-signal display. *Sens. Actuators B Chem.* **2023**, *375*, 132966. [CrossRef]
60. Lee, Y.J.; Kim, H.B.; Roh, Y.R.; Cho, H.M.; Baik, S. Development of a saw gas sensor for monitoring SO_2 gas. *Sens. Actuators A Phys.* **1998**, *64*, 173–178. [CrossRef]
61. Bryant, A.; Poirier, M.; Riley, G.; Lee, D.L.; Vetelino, J.F. Gas detection using surface acoustic wave delay lines. *Sens. Actuators* **1983**, *4*, 105–111. [CrossRef]
62. Liu, X.; Wang, W.; Pan, Y.; Shao, S.; Mu, N. Research on detection system of surface acoustic wave sensor array based on Internet of Things. *J. Zhengzhou Univ. (Eng. Sci.)* **2016**, *37*, 58–61.
63. Rabus, D.; Friedt, J.M.; Arapan, L.; Lamare, S.; Baqué, M.; Audouin, G.; Chérioux, F. Subsurface H_2S detection by a surface acoustic wave passive wireless sensor interrogated with a ground penetrating radar. *ACS Sens.* **2020**, *5*, 1075–1081. [CrossRef]
64. Zhao, L.; Che, J.; Cao, Q.; Shen, S.; Tang, Y. Highly sensitive surface acoustic wave H_2S gas sensor using electron-beam-evaporated CuO as sensitive layer. *Sens. Mater.* **2023**, *35*, 2293–2304. [CrossRef]
65. Wang, J.; Che, J.; Qiao, C.; Niu, B.; Zhang, W.; Han, Y.; Fu, Y.; Tang, Y. Highly porous Fe_2O_3-SiO_2 layer for acoustic wave based H_2S sensing: Mass loading or elastic loading effects? *Sens. Actuators B Chem.* **2022**, *367*, 132160. [CrossRef]
66. Liu, X.; Wang, W.; Zhang, Y.; Pan, Y.; Liang, Y.; Li, J. Enhanced sensitivity of a hydrogen sulfide sensor based on surface acoustic waves at room temperature. *Sensors* **2018**, *18*, 3796. [CrossRef] [PubMed]
67. Liu, X.; Zhang, Y.; Liang, Y.; Li, J.; Wang, W. Design of surface acoustic wave sensor for rapid detection of hydrogen sulfide. *J. Zhengzhou Univ. (Eng. Sci.)* **2019**, *40*, 43–46.
68. Murakawa, Y.; Hara, M.; Oguchi, H.; Hamate, Y.; Kuwano, H. A Hydrogen Sulfide Sensor Based on a Surface Acoustic Wave Resonator Combined with Ionic Liquid. In Proceedings of the 14th International Meeting on Chemical Sensors, Nuremberg, Germany, 20–23 May 2012.
69. Murakawa, Y.; Hara, M.; Oguchi, H.; Hamate, Y.; Kuwano, H. Surface acoustic wave based sensors employing ionic liquid for hydrogen sulfide gas detection. *Microsyst. Technol.* **2013**, *19*, 1255–1259. [CrossRef]

Disclaimer/Publisher's Note: The statements, opinions and data contained in all publications are solely those of the individual author(s) and contributor(s) and not of MDPI and/or the editor(s). MDPI and/or the editor(s) disclaim responsibility for any injury to people or property resulting from any ideas, methods, instructions or products referred to in the content.

Article

Optimization of Gas-Sensing Properties in Poly(triarylamine) Field-Effect Transistors by Device and Interface Engineering

Youngnan Kim [1], Donggeun Lee [1], Ky Van Nguyen [1], Jung Hun Lee [2] and Wi Hyoung Lee [1,*]

1 Department of Organic and Nano System Engineering, School of Chemical Engineering, Konkuk University, Seoul 05029, Republic of Korea
2 Department of Materials Science and Engineering, Northwestern University, Evanston, IL 60208, USA
* Correspondence: whlee78@konkuk.ac.kr

Abstract: In this study, we investigated the gas-sensing mechanism in bottom-gate organic field-effect transistors (OFETs) using poly(triarylamine) (PTAA). A comparison of different device architectures revealed that the top-contact structure exhibited superior gas-sensing performance in terms of field-effect mobility and sensitivity. The thickness of the active layer played a critical role in enhancing these parameters in the top-contact structure. Moreover, the distance and pathway for charge carriers to reach the active channel were found to significantly influence the gas response. Additionally, the surface treatment of the SiO_2 dielectric with hydrophobic self-assembled mono-layers led to further improvement in the performance of the OFETs and gas sensors by effectively passivating the silanol groups. Under optimal conditions, our PTAA-based gas sensors achieved an exceptionally high response (>200%/ppm) towards NO_2. These findings highlight the importance of device and interface engineering for optimizing gas-sensing properties in amorphous polymer semiconductors, offering valuable insights for the design of advanced gas sensors.

Keywords: poly(triarylamine); gas sensor; organic field-effect transistor; device structure; surface treatment; sensitivity

Citation: Kim, Y.; Lee, D.; Nguyen, K.V.; Lee, J.H.; Lee, W.H. Optimization of Gas-Sensing Properties in Poly(triarylamine) Field-Effect Transistors by Device and Interface Engineering. *Polymers* **2023**, *15*, 3463. https://doi.org/10.3390/polym15163463

Academic Editors: Jung-Chang Wang and Tao Chen

Received: 14 July 2023
Revised: 4 August 2023
Accepted: 15 August 2023
Published: 18 August 2023

Copyright: © 2023 by the authors. Licensee MDPI, Basel, Switzerland. This article is an open access article distributed under the terms and conditions of the Creative Commons Attribution (CC BY) license (https://creativecommons.org/licenses/by/4.0/).

1. Introduction

Nitrogen dioxide (NO_2) is a toxic volatile organic compound that originates from industrial sources such as automobiles. Therefore, it is essential to monitor NO_2 concentrations down to the part-per-billion (ppb) level. While several NO_2-sensing techniques, such as combustion-type sensors, are available, there is a need for instant detection methods like amperometric sensing. Metal oxide-based gas sensors offer precise sensing platforms, but they require high operating temperatures [1]. In this regard, the development of a room temperature NO_2 sensor compatible with a plastic substrate is necessary. We have previously identified that a polymer semiconductor can serve as an alternative gas-sensing element for a room-temperature-operating NO_2 gas sensor [2–4]. This approach offers advantages such as low processing temperature and low production cost [5]. It is worth noting that polymer semiconductors can be easily deposited using solution processing techniques such as spin-coating, inkjet printing, and roll-to-roll printing [6,7].

Since the conductivity of polymer semiconductors is typically low, it is necessary to amplify the charge carrier density of the polymer semiconductor. Therefore, the structure of an organic field-effect transistor (OFET), including the semiconductor, dielectric, and source/drain/gate electrodes, is preferred. With this FET structure, the application of a gate bias can enhance the source–drain current through the field-effect phenomenon [8,9]. Gas sensors based on OFETs require several key performance parameters; namely, sensitivity, selectivity, and stability [10]. Specifically, sensitivity is strongly influenced by OFET characteristics such as field-effect mobility and subthreshold voltage. As gas molecule detection relies on the modulation of the source–drain current in the active channel region, high field-effect mobility facilitates the fast detection of target gas molecules. Extensive research has shown that both the microstructure and molecular structure of the polymer semiconductor affect the performance of OFETs [11]. Additionally, a few studies have explored the structure–property relationship in OFET-based gas sensors. Our group recently demonstrated that the presence of a glycol side chain in a diketopyrrolopyrrole-based polymer offers advantages for selective NO_2 detection below the ppb level [12,13]. Although the glycol side chain degrades the field-effect mobility, it enables efficient gas diffusion for gas absorption and desorption. Consequently, the NO_2-gas-sensing performance is inversely proportional to the crystallinity of the polymer semiconductor used. It is important to note that the simple logic of increasing crystallinity for high-performance OFETs does not apply to OFET-based gas sensors.

From the literature, it has been found that the amorphous polymer poly(triarylamine) (PTAA) can serve as an excellent active layer for OFET-based NO_2 sensors [14,15]. PTAA possesses the highest occupied molecular orbital (HOMO) level of -5.14 eV [16], enabling stable operation in ambient conditions. Although the field-effect mobility of PTAA FETs is approximately 10^{-5} cm^2/Vs, PTAA sensors have demonstrated the ability to detect NO_2 concentrations as low as 10 ppb [15]. In PTAA FETs, the PTAA film functions as the active sensing layer, while the source, drain, and gate electrodes are employed for electrical measurements. By adjusting the gate voltage, the current flow through the PTAA film can be controlled, allowing for the measurement of the response to NO_2 exposure. The mechanism of NO_2 sensing relies on the adsorption of NO_2 on the PTAA surface, which induces changes in the electrical properties of the PTAA FETs. This modulation is typically observed as variations in the charge carrier density and field-effect mobility. However, the precise mechanism of gas detection requires further study, including investigations into gas dynamics (such as diffusion) and device physics. Continued research efforts are aimed at optimizing PTAA-based sensors, which involve the development of novel device structures, surface functionalization techniques, and integration with other materials or technologies to enhance overall sensor performance [11,14,17,18].

In this report, we investigated the NO_2-sensing performance of PTAA FETs with varying device structures; namely, top-contact and bottom-contact, as well as different thicknesses of the PTAA active layer. This study marks the first attempt to compare the gas-sensing properties of bottom-contact FET sensors with the top-contact structure. We used a common SiO_2 gate dielectric because of the ease in surface functionalization with a silane coupling agent. Additionally, we examined the impact of surface treatment on the SiO_2 dielectric layer and its influence on FET performance. Surface treatment plays a crucial role in enhancing the device performance of PTAA FETs and also affects their gas-sensing properties. Note that charge carrier transport occurs mainly at the interface between the semiconductor and dielectric layer. We analyzed the NO_2-sensing performance by evaluating response and recovery rates. Furthermore, we compared the measurement of transfer characteristics before and after NO_2 injection. Finally, we proposed a mechanism for gas sensing that takes into consideration different device architectures and surface treatments; this is a novel aspect not previously explored in existing reports.

2. Materials and Methods

2.1. Fabrication of PTAA FET Sensor

PTAA (Ossila, molecular weight 6312 g/mol) was dissolved in chloroform (Sigma Aldrich Co., St. Louis, MO, USA) at concentrations of 2.5 mg/mL and 5.0 mg/mL to control the thickness of the PTAA film. The thickness of the PTAA film from a 2.5 mg/mL solution was measured to be 18.7 nm, whereas the thickness of the PTAA film from a 5.0 mg/mL solution was 38.7 nm. A silicon wafer with a thermally grown SiO_2 layer (300 nm thickness) was obtained from Fine Science. The wafer was cut into 1.8 cm by 1.8 cm pieces and cleaned for 20 min using acetone and isopropyl alcohol by ultrasonication. The silicon substrates were dried with N_2 and then treated with UV-ozone for 30 min to make the SiO_2/Si surface hydrophilic. Optionally, the SiO_2/Si substrate was treated with octadecyl-trichlorosilane (ODTS) using a dipping method. This treatment resulted in the formation of ODTS self-assembled monolayers (SAMs) on the SiO$_2$ surface through a chemical reaction with silanol groups in SiO$_2$. However, the surface treatment of the SiO$_2$ dielectric with ODTS SAMs did not affect the thickness of the PTAA film. To fabricate bottom-gate top-contact PTAA FETs, the PTAA solution (2.5 mg/mL and 5.0 mg/mL) was spin-cast onto the substrate at 1500 rpm for 60 s. Thin-film gold (Au) source/drain electrodes with a thickness of 60 nm were thermally deposited onto the PTAA film using a shadow mask. The resulting FETs had a channel length of 70 µm and a width of 2000 µm. For the bottom-gate bottom-contact PTAA FETs, a 60 nm thick Au film was deposited onto the cleaned SiO$_2$/Si substrate through a shadow mask to create the source/drain electrodes with a channel length of 70 µm and a width of 2000 µm, respectively. Subsequently, the PTAA solution (2.5 mg/mL and 5.0 mg/mL) was spin-cast at 1500 rpm for 60 s to form a PTAA thin-film. PTAA FET gas sensors were fabricated by connecting the gate, source, and drain electrodes to a gas sensor module using silver wire and silver paste.

2.2. Characterization

The morphologies of the PTAA thin-films were characterized using atomic force microscopy (AFM, Park Scientific Instrument, Suwon, Republic of Korea) to investigate their surface structures. The electrical characteristics of the PTAA FETs were measured using a Keithley 2612A semiconductor parametric analyzer connected to a gas chamber (GASENTEST, Precision Sensor System Inc., Daejeon, Republic of Korea). The transfer curves were obtained under the following conditions: the gate voltage was swept from V_{GS} = 40 to −80 V with a source/drain voltage of V_{DS} = −80 V. To evaluate the gas-sensing properties of the PTAA FETs, a gas chamber (GASENTEST, Precision Sensor System Inc.) was employed. The source/drain voltage was fixed at −20 V, while the applied gate voltage varied depending on the FETs. The applied gate voltage was determined by subtracting 10 V from each threshold voltage of the FETs to accurately compare the gas sensitivity. Target NO$_2$ gas was used, and the concentrations of the target gas were adjusted using N$_2$ as a carrier gas. Dynamic gas-sensing properties were measured by periodically introducing the target gases at specific time intervals. To assess the response and recovery rates of the NO$_2$ sensors based on PTAA FETs, the time-dependent source–drain current at the given gate voltage was monitored. The response to the target NO$_2$ gas was calculated by dividing the change in the current flow by the initial value, using the following equation:

$$Response = (I_G - I_B)/I_B \times 100 \; (\%)$$

Here, I_B represents the initial base current at the initial measurement, while I_G indicates the source–drain current at the given condition. Similarly, the recovery was expressed as the ratio of returning to the initial current value (I_R) compared to the total amount of current change caused by the NO$_2$ target gas.

$$Recovery = (I_G - I_R)/(I_G - I_0) \times 100 \; (\%)$$

In this equation, I_R represents the current at the end of the recovery, and I_G represents the current immediately after the target gas injection. Additionally, to analyze gas-sensing characteristics based on the exposure time to the NO$_2$ gas, the response rate and recovery rate were determined by dividing the response and recovery values by the corresponding exposure time, respectively.

3. Results and Discussion

Figure 1a presents the chemical structure of PTAA. In contrast to semicrystalline conjugated polymers like poly(3-hexylthiophene), PTAA incorporates two methyl groups into the aromatic phenyl group. These short aliphatic side chains do not contribute to structural ordering. An AFM image of the PTAA thin-film obtained after spin-casting the PTAA solution is shown in Figure 1b. The image displays a featureless morphology, and the Root Mean Square (RMS) roughness of the PTAA thin-film was measured to be

0.33 nm. These characteristics provide evidence for the amorphous nature of the PTAA thin-film. Figure 2 shows the fabrication steps for gas sensors based on top-contact versus bottom-contact FETs. Detailed fabrication steps are illustrated in the Materials and Methods section. As shown in Figure 3, we fabricated PTAA FETs with different device architectures, while highly doped Si and SiO_2 (thickness of 300 nm) layers serve as the gate electrode and gate dielectric, respectively. The pathways for charge carriers are indicated by the yellow arrows. In the top-contact structure, injected charge carriers move toward the semiconductor–dielectric interface, and adsorption of the target gas can affect both the carrier injection and carrier transport [19]. In the bottom-contact structure, charge injection mainly occurs at the edge of the source/drain electrodes. Although the area of adsorption in the bottom-contact structure is larger than that in the top-contact structure, the effect of charge injection after gas adsorption is limited, possibly due to the shorter injection path.

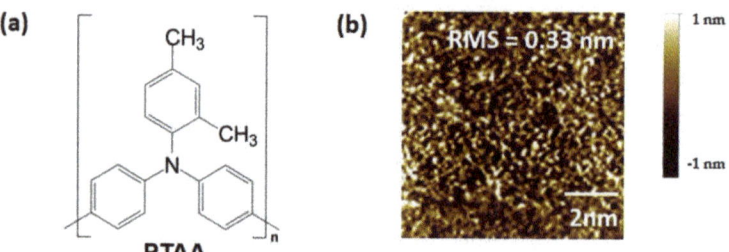

Figure 1. (**a**) Molecular structure of the PTAA. (**b**) Atomic force microscopy (AFM) image of PTAA thin-film spin-cast from a 5.0 mg/mL PTAA solution.

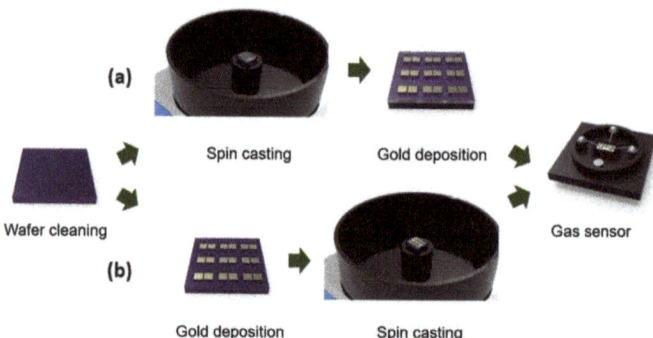

Figure 2. Fabrication steps for gas sensors based on (**a**) bottom-gate/top-contact FETs, and (**b**) bottom-gate/bottom-contact FETs.

In Figure 4a, the transfer characteristics of bottom-gate OFETs based on different concentrations of PTAA and device structures are presented. The corresponding device parameters extracted from these transfer characteristics are summarized in Table 1. The term "Top" represents the top-contact structure, while the term "Bottom" represents the bottom-contact structure of the PTAA FETs. The term "Bare" represents the SiO_2 dielectric without modification, while the term "ODTS" represents the SiO_2 dielectric with ODTS SAMs. For the top-contact structure, an increase in PTAA concentration from 2.5 mg/mL to 5.0 mg/mL (thickness from 18.7 nm to 38.7 nm) resulted in a decrease in both the on-current and threshold voltage, while the off-current remained unchanged. In this structure, the field-effect charge carriers injected from the source electrode are expected to move towards the active channel region near the gate–dielectric layer and subsequently reach the drain electrode, as shown in Figure 3a. To ensure a shorter injection pathway, a thin semiconducting layer is preferred. It was observed that the field-effect mobility degraded five times when the concentration was doubled (from 2.5 mg/mL to 5.0 mg/mL). On the other hand, in the bottom-contact structure, the on-current and field-effect mobility remained unchanged with varying concentrations. The pathway for charge transport did not significantly change with an increase in the semiconducting layer, as depicted in Figure 3b. It is important to note that field-effect charge carriers are mostly located in the active channel region near the gate–dielectric layer and the source–drain electrodes [20]. Therefore, the thick overlayer of amorphous PTAA film in the bottom-contact structure does not play a significant role in the current modulation.

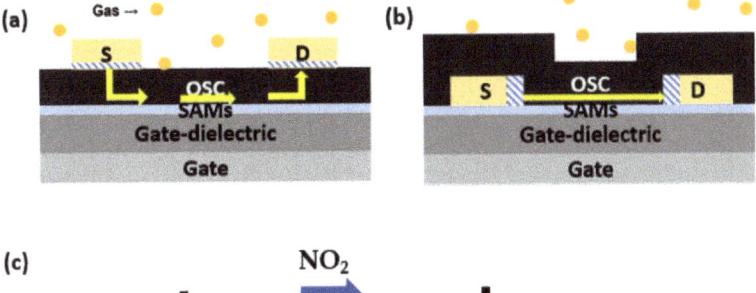

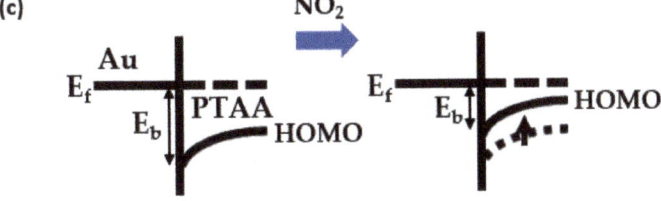

Figure 3. A schematic diagram showing the operating mechanism of PTAA gas sensors based on (**a**) bottom-gate/top-contact FETs, and (**b**) bottom-gate/bottom-contact FETs. (**c**) Change in band diagram at the interface between Au and PTAA under NO_2 exposure.

The surface treatment of the gate–dielectric is crucial in controlling the trap density at the interface between the active channel and the gate–dielectric [20,21]. Previous studies have reported that hydrophobic self-assembled monolayers (SAMs) can protect silanol groups in SiO_2/Si interfaces [22]. In Figure 4b, the transfer characteristics of bottom-gate OFETs based on PTAA with different surface treatments and device structures are depicted. A PTAA concentration of 5.0 mg/mL was used because using 2.5 mg/mL resulted in thin-film dewetting on hydrophobic ODTS SAMs. The surface treatment of the SiO_2 with ODTS SAMs improved the on-current and field-effect mobility, irrespective of the device geometry (top-contact or bottom-contact). The significant increases in field-effect mobility can be directly attributed to the coverage of silanol groups with hydrophobic ODTS SAMs [23,24]. Treatment of the SiO_2 surface with ODTS SAMs reduces the number of silanol groups, thereby decreasing the trapping of hole carriers. As PTAA is an amorphous polymer semiconductor, the structural effect of PTAA due to surface treatment is minimal compared to the dominant trap-covering effect. Additionally, the subthreshold slope, which indicates the switching capability, is an important factor. After ODTS surface treatment, the subthreshold slope decreases significantly, indicating an improved switching performance. Simple surface treatment with ODTS SAMs proved advantageous for enhancing the device's performance in both top-contact and bottom-contact PTAA FETs. Specifically, top-contact FETs based on PTAA thin-films prepared from 5.0 mg/mL and featuring an ODTS interfacial layer represent the optimal conditions for achieving the best switching performance.

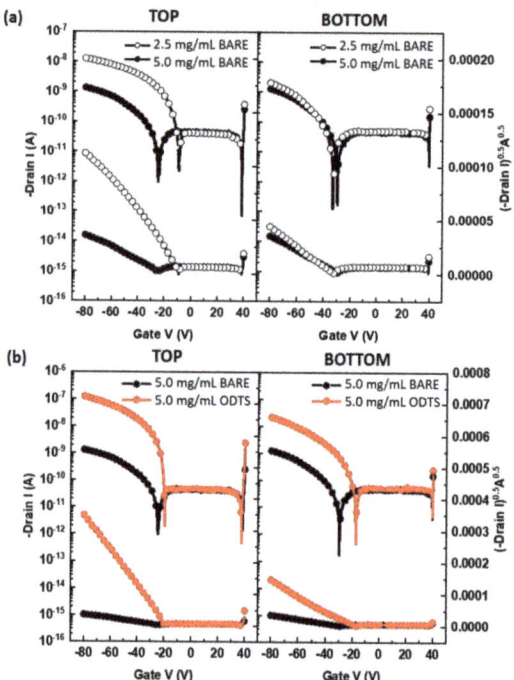

Figure 4. Transfer characteristics of (**a**) FETs based on different concentrations of PTAA (2.5 mg/mL, 5 mg/mL) and device structures (top-contact, bottom-contact), and (**b**) FETs based on different surface treatments (bare, ODTS) and device structures (top-contact, bottom-contact). V_{DS} was fixed at −80 V at all measurements while gate voltage was swept from 40 V to −80 V.

Table 1. Electrical characteristics of PTAA FETs based on different concentrations of PTAA, surface treatments, and device structures. V_{TH}: threshold voltage, SS: subthreshold slope.

	Mobility, μ [10^{-5} cm^2/(V·s)]	V_{TH} [V]	Turn On V [V]	SS [V/dec]
2.5 mg/mL Top Bare	3.28	−8.21	−9.30	3.56
2.5 mg/mL Bottom Bare	0.535	−31.7	−32.9	5.72
5 mg/mL Top Bare	0.603	−23.1	−24.4	5.33
5 mg/mL Bottom Bare	0.598	−30.0	−29.6	5.63
5 mg/mL Top ODTS	44.3	−19.4	−19.6	3.33
5 mg/mL Bottom ODTS	4.62	−16.9	−17.5	4.75

To measure the gas-sensing response of the PTAA FETs, NO$_2$ gas was periodically injected into the gas chamber, which was connected to the current monitoring setup. Since NO$_2$ is an oxidizing gas and PTAA is a p-type semiconductor, the adsorption of NO$_2$ to PTAA results in an increase in the accumulation of hole carriers. Figure 5 illustrates the gas-sensing characteristics of the PTAA FETs based on different concentrations of PTAA and device structures after exposure to 50 ppm NO$_2$. From these curves, the response rate and recovery rate were extracted and are presented in Table 2. The response rate is highly dependent on the sensor type, while the recovery rate remains nearly the same regardless of the sensor type. During the 50 s injection of NO$_2$ (indicated in the grey region), the current in the PTAA FETs increases abruptly, while the current recovers to its initial state during the 1000 s N$_2$ purging. The sluggish recovery observed after the initial fast recovery may be attributed to the interaction between PTAA and NO$_2$ [12]. To enhance the recovery behavior, thermal annealing can be applied. In the top-contact structure, a decrease in the PTAA concentration (from 5.0 mg/mL to 2.5 mg/mL) results in an increase in the response. The electrical properties of the PTAA FETs, such as field-effect mobility and subthreshold slope, are superior in the 2.5 mg/mL device. The fast-switching speed in this device is advantageous for the rapid detection of target NO_2 molecules. In the bottom-contact structure, on the other hand, the change in current upon NO_2 exposure is not significantly affected by the thickness of the PTAA film, which corresponds to the concentration of the PTAA solution. This finding suggests that the adsorption and diffusion of NO_2 onto the PTAA film plays a crucial role in modulating the current in top-contact FETs.

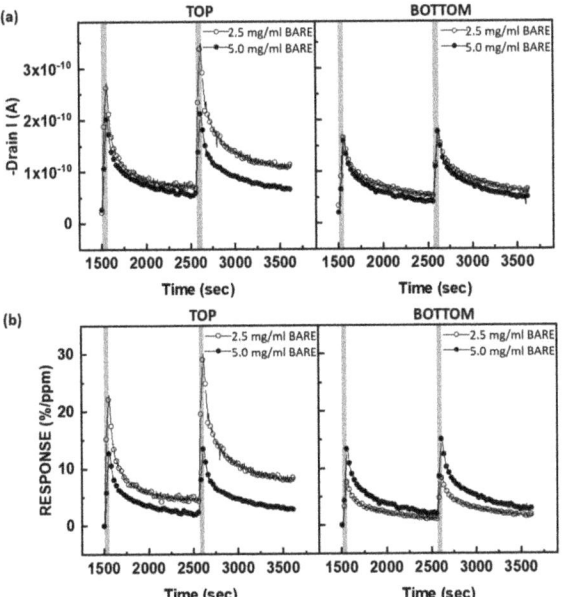

Figure 5. Gas-sensing characteristics of PTAA FETs based on different concentrations of PTAA and device structures by exposure to 50 ppm of NO_2: (**a**) raw data of the source–drain current, (**b**) normalized gas response (%/ppm). V_{DS} was fixed at -20 V, and V_{GS} was fixed at a voltage lower than 10 V for each Turn-On Voltage.

Table 2. Summary of gas sensor performance.

	Response Rate (s^{-1})	Recovery Rate (s^{-1})
2.5 mg/mL Top Bare	0.198	0.000816
2.5 mg/mL Bottom Bare	0.0666	0.000878
5 mg/mL Top Bare	0.106	0.000861
5 mg/mL Bottom Bare	0.118	0.000870
5 mg/mL Top ODTS	1.77	0.000979
5 mg/mL Bottom ODTS	0.521	0.000927

To evaluate the impact of surface treatment and device structure, the NO_2-sensing performance was compared in Figure 6. The surface treatment of SiO_2 with hydrophobic ODTS SAMs improved the gas response in both the top-contact and bottom-contact devices, which correlates with the enhanced device performance observed in the FETs (as shown in Table 1). Because the adsorption of NO_2 occurs at the PTAA surface, the hydrophobic character of ODTS did not decrease the gas adsorption and diffusion behaviors. In particular, the top-contact PTAA FETs with ODTS SAMs exhibited an exceptionally high response (>200%/ppm) towards NO_2. This outcome suggests that amorphous PTAA film is well-suited for detecting NO_2 at levels as low as parts per million (ppm).

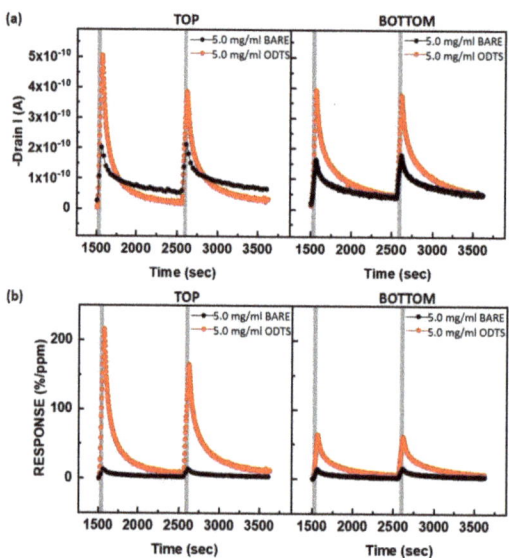

Figure 6. Gas-sensing characteristics of PTAA FETs based on different surface treatments and device structures by exposure to 50 ppm of NO_2: (**a**) raw data of the source–drain current, (**b**) normalized gas response (%/ppm). V_{DS} was fixed at −20 V, and V_{GS} was fixed at a voltage lower than 10 V for each Turn-On Voltage.

Several reasons can be proposed to explain the excellent NO_2-sensing performance in top-contact PTAA FETs. Firstly, gas diffusion is facilitated within the amorphous PTAA film. It is speculated that partially positive NO_2 molecules can migrate towards the semiconductor–dielectric interface, particularly at the interface between PTAA and SiO_2. This migration is driven by the electric field generated by the gate bias, and it is applicable regardless of the device structure, whether it is top-contact or bottom-contact. Secondly, the presence of NO_2 molecules within the PTAA film can induce the generation of additional hole carriers, enhancing charge injection along the electrode-active channel pathway. It is worth noting that the HOMO level of PTAA has been reported to be approximately −5.14 eV [16]. This HOMO level creates a barrier relative to the Fermi level of the Au electrode. It has been observed that evaporated Au electrodes typically have a work function of approximately 4.8 eV [25,26]. Due to the generation of hole carriers from the NO_2 molecules, the Fermi level of PTAA could shift upward, facilitating charge injection (Figure 3c). Indeed, it is noticeable that the top-contact structure has a longer injection path, which can contribute to the enhanced sensing performance in this configuration. On the other hand, in the bottom-contact structure, the injection pathway is short and is weakly affected by the charge carriers generated from the adsorbed NO_2 molecules. As a result, the response in the bottom-contact structure tends to be lower compared to the response observed in the top-contact structure. This is because the distance and pathway for charge carriers to reach the active channel are less favorable for sensing NO_2 gas in the bottom-contact structure.

To further understand the gas-sensing mechanism, transfer characteristics before and after gas injection were compared in Figure 7. All the curves exhibited significant increases in on-currents, while the off-currents remained unchanged. The minor shift in threshold voltage may be attributed to the combined effects of the NO_2-induced generation of hole carriers and gate bias instability resulting from hole trapping. Notably, there was a substantial increase in field-effect mobility after gas injection, which is calculated and summarized in Table 3. The increase in field-effect mobility can be attributed to the trap-filling effect in the PTAA film. As mentioned earlier, NO_2 molecules induce the generation of extra hole carriers. These hole carriers fill the trap sites within the PTAA film, leading to an enhancement in the field-effect mobility. It was observed that thicker PTAA films exhibited a higher rate of increase in field-effect mobility, suggesting a higher amount of adsorbed NO_2 molecules in thicker films. However, in the top-contact structure, the increase in mobility (from 3.28×10^{-5} cm^2/(V·s) to 21.0×10^{-5} cm^2/(V·s)) was greater in a thinner PTAA film (2.5 mg/mL Top Bare) compared to in a thicker PTAA film (5.0 mg/mL Top Bare), which exhibited an increase from 0.603×10^{-5} cm^2/(V·s) to 6.65×10^{-5} cm^2/(V·s). This higher increase in mobility in the thin PTAA film correlates with the higher gas response observed in the 2.5 mg/mL device. In contrast, the bottom-contact structure exhibited a lower increase in mobility and a relatively weaker thickness effect compared to the top-contact structure. This finding supports the assumption that the distance and pathway for charge carriers to reach the active channel play a critical role in the gas response. The top-contact structure with ODTS SAMs exhibited the highest increase in mobility, further supporting the significant effect of surface treatment. It can be proposed that the trap-filling effect is more dominant in the top-contact structure; therefore, top-contact PTAA FETs with ODTS SAMs demonstrate the best sensing performance.

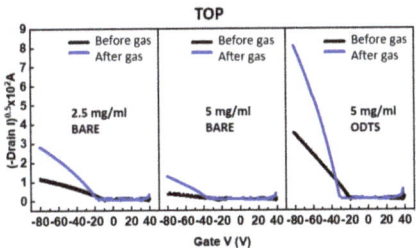

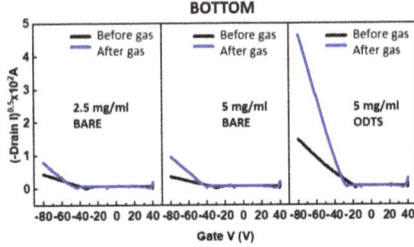

Figure 7. Transfer characteristics of PTAA FETs before and after NO_2 gas injection. V_{DS} was fixed at -80 V.

Table 3. The field-effect mobility of PTAA FETs based on different concentrations of PTAA, surface treatments, and device structures before and after NO_2 gas injection.

	Before Mobility, μ_{Before} [10^{-5} cm^2/(V·s)]	After Mobility, μ_{After} [10^{-5} cm^2/(V·s)]	μ_{After}/μ_{Before}
2.5 mg/mL Top Bare	3.28	21.0	6.40
2.5 mg/mL Bottom Bare	0.535	1.24	2.32
5 mg/mL Top Bare	0.603	6.65	11.0
5 mg/mL Bottom Bare	0.598	2.28	3.81
5 mg/mL Top ODTS	44.3	309	6.98
5 mg/mL Bottom ODTS	4.62	29.6	6.41

4. Conclusions

In summary, our study investigated the gas-sensing properties of PTAA FETs based on different concentrations of PTAA, surface treatments, and device structures. The amorphous nature of the PTAA film was confirmed through AFM analysis, and the device structures were fabricated as top-contact and bottom-contact configurations. The transfer characteristics of the PTAA FETs revealed that an increase in PTAA concentration in the top-contact structure resulted in a decrease in the on-current and threshold voltage, while the off-current remained unchanged. In contrast, the bottom-contact structure showed no significant changes in the on-current and field-effect mobility with varying PTAA concentrations. Surface treatment of SiO_2 with hydrophobic ODTS SAMs was found to enhance the gas response and field-effect mobility in both top-contact and bottom-contact PTAA FETs. The improved performance was attributed to the coverage of silanol groups by the ODTS SAMs, leading to a decrease in subthreshold slope and enhanced switching capabilities.

Gas-sensing experiments with NO_2 gas demonstrated that the current in PTAA FETs increased abruptly upon exposure to NO_2 and recovered during N_2 purging. The decrease in PTAA concentration led to an increase in response, particularly in the top-contact structure. The adsorption of NO_2 onto the PTAA film played a critical role in the current change. Comparative analysis of the NO_2-sensing performance in top-contact and bottom-contact PTAA FETs revealed that the top-contact structure exhibited a higher response. The longer injection path in the top-contact structure facilitated interaction between NO_2 molecules and the PTAA film, resulting in a higher sensitivity to NO_2 gas. Further investigation of transfer characteristics before and after gas injection showed significant increases in on-currents and field-effect mobility. The trap-filling effect in the PTAA film was identified as the reason for the increase in field-effect mobility. The top-contact structure demonstrated a higher increase in mobility, supporting the notion that the distance and pathway for charge carriers play a critical role in the gas response.

Based on the findings, it can be concluded that the gas-sensing performance of top-contact PTAA FETs with ODTS SAMs was superior. The amorphous PTAA film was found to be well-suited for detecting NO_2 at low concentrations, highlighting its potential for gas-sensing applications.

Author Contributions: Conceptualization, Y.K. and W.H.L.; methodology, Y.K. and D.L.; software, Y.K. and K.V.N.; validation, Y.K., D.L. and W.H.L.; formal analysis, Y.K.; investigation, Y.K.; resources, Y.K. and J.H.L.; data curation, Y.K.; writing—original draft preparation, Y.K.; writing—review and editing, K.V.N. and W.H.L.; visualization, Y.K., J.H.L. and D.L.; supervision, W.H.L.; project administration, W.H.L.; funding acquisition, K.V.N. and W.H.L. All authors have read and agreed to the published version of the manuscript.

Funding: This research was supported by grants from the Brain Pool program (2022H1D3A2A02063358) and the Basic Science Research Program (2023-00208902) of the National Research Foundation of Korea (NRF) funded by the Ministry of Science and ICT.

Institutional Review Board Statement: Not applicable.

Data Availability Statement: The data presented in this study are available on request from the corresponding author.

Conflicts of Interest: The authors declare no conflict of interest.

References

1. Kim, H.-J.; Lee, J.-H. Highly sensitive and selective gas sensors using p-type oxide semiconductors: Overview. *Sens. Actuators B Chem.* **2014**, *192*, 607–627. [CrossRef]
2. Bai, H.; Shi, G. Gas sensors based on conducting polymers. *Sensors* **2007**, *7*, 267–307. [CrossRef]
3. Zhang, C.; Chen, P.; Hu, W. Organic field-effect transistor-based gas sensors. *Chem. Soc. Rev.* **2015**, *44*, 2087–2107. [CrossRef] [PubMed]
4. Wang, S.; Kang, Y.; Wang, L.; Zhang, H.; Wang, Y.; Wang, Y. Organic/inorganic hybrid sensors: A review. *Sens. Actuators B Chem.* **2013**, *182*, 467–481. [CrossRef]
5. Forrest, S.R. The path to ubiquitous and low-cost organic electronic appliances on plastic. *Nature* **2004**, *428*, 911–918. [CrossRef]

6. Sirringhaus, H.; Kawase, T.; Friend, R.; Shimoda, T.; Inbasekaran, M.; Wu, W.; Woo, E.P. High-resolution inkjet printing of all-polymer transistor circuits. *Science* 2000, *290*, 2123–2126. [CrossRef] [PubMed]
7. Søndergaard, R.R.; Hösel, M.; Krebs, F.C. Roll-to-Roll fabrication of large area functional organic materials. *J. Polym. Sci. B Polym.* 2013, *51*, 16–34. [CrossRef]
8. Sirringhaus, H. 25th anniversary article: Organic field-effect transistors: The path beyond amorphous silicon. *Adv. Mater.* 2014, *26*, 1319–1335. [CrossRef]
9. Lee, W.H.; Park, Y.D. Organic semiconductor/insulator polymer blends for high-performance organic transistors. *Polymers* 2014, *6*, 1057–1073. [CrossRef]
10. Liu, X.; Zheng, W.; Kumar, R.; Kumar, M.; Zhang, J. Conducting polymer-based nanostructures for gas sensors. *Coord. Chem. Rev.* 2022, *462*, 214517. [CrossRef]
11. Lee, J.H.; Chun, J.H.; Chung, H.-J.; Lee, W.H. Microstructural Control of Soluble Acene Crystals for Field-Effect Transistor Gas Sensors. *Nanomaterials* 2022, *12*, 2564. [CrossRef] [PubMed]
12. Kang, Y.; Kwak, D.H.; Kwon, J.E.; Kim, B.-G.; Lee, W.H. NO_2-Affinitive Conjugated Polymer for Selective Sub-Parts-Per-Billion NO_2 Detection in a Field-Effect Transistor Sensor. *J. Am. Chem. Soc.* 2021, *13*, 31910–31918. [CrossRef] [PubMed]
13. Ahn, Y.; Hwang, S.; Kye, H.; Kim, M.S.; Lee, W.H.; Kim, B.-G. Side-Chain-Assisted Transition of Conjugated Polymers from a Semiconductor to Conductor and Comparison of Their NO_2 Sensing Characteristics. *Materials* 2023, *16*, 2877. [CrossRef]
14. Lee, J.H.; Lee, S.; Lee, H.; Choi, H.H.; Chae, H.; Kim, Y.; Yang, S.J.; Anthony, J.E.; Jang, H.W.; Won, S.M. Marangoni Flow Driven via Hole Structure of Soluble Acene–Polymer Blends for Selective Nitrogen Dioxide Sensing. *Adv. Funct. Mater.* 2023, *33*, 2215215. [CrossRef]
15. Das, A.; Dost, R.; Richardson, T.; Grell, M.; Morrison, J.J.; Turner, M.L. A nitrogen dioxide sensor based on an organic transistor constructed from amorphous semiconducting polymers. *Adv. Mater.* 2007, *19*, 4018–4023. [CrossRef]
16. Zhang, W.; Smith, J.; Hamilton, R.; Heeney, M.; Kirkpatrick, J.; Song, K.; Watkins, S.E.; Anthopoulos, T.; McCulloch, I. Systematic improvement in charge carrier mobility of air stable triarylamine copolymers. *J. Am. Chem. Soc.* 2009, *131*, 10814–10815. [CrossRef]
17. Smith, J.; Hamilton, R.; Qi, Y.; Kahn, A.; Bradley, D.D.; Heeney, M.; McCulloch, I.; Anthopoulos, T.D. The Influence of Film Morphology in High-Mobility Small-Molecule: Polymer Blend Organic Transistors. *Adv. Funct. Mater.* 2010, *20*, 2330–2337. [CrossRef]
18. Wedge, D.C.; Das, A.; Dost, R.; Kettle, J.; Madec, M.-B.; Morrison, J.J.; Grell, M.; Kell, D.B.; Richardson, T.H.; Yeates, S. Real-time vapour sensing using an OFET-based electronic nose and genetic programming. *Sens. Actuators B Chem.* 2009, *143*, 365–372. [CrossRef]
19. Oh, S.; Khan, M.R.R.; Choi, G.; Seo, J.; Park, E.; An, T.K.; Park, Y.D.; Lee, H.S. Advanced Organic Transistor-Based Sensors Utilizing a Solvatochromic Medium with Twisted Intramolecular Charge-Transfer Behavior and Its Application to Ammonia Gas Detection. *ACS Appl. Mater. Interfaces* 2021, *13*, 56385–56393. [CrossRef]
20. Park, Y.D.; Lim, J.A.; Lee, H.S.; Cho, K. Interface engineering in organic transistors. *Mater. Today* 2007, *10*, 46–54. [CrossRef]
21. Yoon, M.-H.; Kim, C.; Facchetti, A.; Marks, T.J. Gate dielectric chemical structure−organic field-effect transistor performance correlations for electron, hole, and ambipolar organic semiconductors. *J. Am. Chem. Soc.* 2006, *128*, 12851–12869. [CrossRef] [PubMed]
22. Kim, S.; Yoo, H. Self-assembled monolayers: Versatile uses in electronic devices from gate dielectrics, dopants, and biosensing linkers. *Micromachines* 2021, *12*, 565. [CrossRef] [PubMed]
23. Lee, H.S.; Kim, D.H.; Cho, J.H.; Hwang, M.; Jang, Y.; Cho, K. Effect of the phase states of self-assembled monolayers on pentacene growth and thin-film transistor characteristics. *J. Am. Chem. Soc.* 2008, *130*, 10556–10564. [CrossRef] [PubMed]
24. Ito, Y.; Virkar, A.A.; Mannsfeld, S.; Oh, J.H.; Toney, M.; Locklin, J.; Bao, Z. Crystalline ultrasmooth self-assembled monolayers of alkylsilanes for organic field-effect transistors. *J. Am. Chem. Soc.* 2009, *131*, 9396–9404. [CrossRef] [PubMed]
25. Schultz, T.; Lenz, T.; Kotadiya, N.; Heimel, G.; Glasser, G.; Berger, R.; Blom, P.W.; Amsalem, P.; de Leeuw, D.M.; Koch, N. Reliable work function determination of multicomponent surfaces and interfaces: The role of electrostatic potentials in ultraviolet photoelectron spectroscopy. *Adv. Mater. Interfaces* 2017, *4*, 1700324. [CrossRef]
26. Lee, W.H.; Park, J.; Sim, S.H.; Jo, S.B.; Kim, K.S.; Hong, B.H.; Cho, K. Transparent flexible organic transistors based on monolayer graphene electrodes on plastic. *Adv. Mater.* 2011, *23*, 1752–1756. [CrossRef]

Disclaimer/Publisher's Note: The statements, opinions and data contained in all publications are solely those of the individual author(s) and contributor(s) and not of MDPI and/or the editor(s). MDPI and/or the editor(s) disclaim responsibility for any injury to people or property resulting from any ideas, methods, instructions or products referred to in the content.

Article

Simulation and Experiment of Active Vibration Control Based on Flexible Piezoelectric MFC Composed of PZT and PI Layer

Chong Li *, Liang Shen, Jiang Shao and Jiwen Fang

School of Mechanical Engineering, Jiangsu University of Science and Technology, Zhenjiang 212100, China
* Correspondence: lichong@just.edu.cn; Tel.: +86-511-8444-5385

Abstract: In order to improve the vibration suppression effect of the flexible beam system, active control based on soft piezoelectric macro-fiber composites (MFCs) consisting of polyimide (PI) sheet and lead zirconate titanate (PZT) is used to reduce the vibration. The vibration control system is composed of a flexible beam, a sensing piezoelectric MFC plate, and an actuated piezoelectric MFC plate. The dynamic coupling model of the flexible beam system is established according to the theory of structural mechanics and the piezoelectric stress equation. A linear quadratic optimal controller (LQR) is designed based on the optimal control theory. An optimization method, designed based on a differential evolution algorithm, is utilized for the selection of weighted matrix Q. Additionally, according to theoretical research, an experimental platform is built, and vibration active control experiments are carried out on piezoelectric flexible beams under conditions of instantaneous disturbance and continuous disturbance. The results show that the vibration of flexible beams is effectively suppressed under different disturbances. The amplitudes of the piezoelectric flexible beams are reduced by 94.4% and 65.4% under the conditions of instantaneous and continuous disturbances with LQR control.

Keywords: piezoelectric polymer; flexible beam; LQR; differential evolution algorithm; vibration control

Citation: Li, C.; Shen, L.; Shao, J.; Fang, J. Simulation and Experiment of Active Vibration Control Based on Flexible Piezoelectric MFC Composed of PZT and PI Layer. *Polymers* **2023**, *15*, 1819. https://doi.org/10.3390/polym15081819

Academic Editors: Jung-Chang Wang and Roman A. Surmenev

Received: 17 February 2023
Revised: 3 April 2023
Accepted: 5 April 2023
Published: 7 April 2023

Copyright: © 2023 by the authors. Licensee MDPI, Basel, Switzerland. This article is an open access article distributed under the terms and conditions of the Creative Commons Attribution (CC BY) license (https://creativecommons.org/licenses/by/4.0/).

1. Introduction

Piezoelectric materials have been widely used in the vibration control of structures due to their characteristics of light weight, flexible size, fast response speed, and wide frequency response range [1–3]. By using the positive and inverse piezoelectric effects of piezoelectric materials, sensors and actuators can be made embedded into the structure's surface in order to control the active vibration of the system.

The piezoelectric materials commonly used for vibration control include single-layer piezoelectric ceramics, multilayer piezoelectric ceramics, and piezoelectric fiber composite materials. Single-layer piezoelectric ceramics are ferroelectric ceramics which possess a piezoelectric effect after being polarized by a high-voltage DC electric field. They have the advantages of low density, good response accuracy, high-frequency response, and high output [4–6]. Multilayer piezoelectric ceramics can generate more significant displacement than single-layer piezoelectric ceramics under the existing driving conditions in order to meet the need for vibration reductions in large vibration amplitude [7–9]. Although piezoelectric ceramics have achieved good results in vibration control, their application scope is limited due to low toughness and high brittleness. With the complexity and diversification of mechanical structures, flexibility has become one of the essential indicators of piezoelectric materials. Piezoelectric fiber composites can be bonded in various types of systems or embedded into composite structures, bending and twisting forms under the action of the applied voltage in order to generate or counteract vibration. They can be used as strain gauges to sense deformation, noise, and vibration when there is no applied voltage [10–12]. Piezoelectric fiber composites can also be used in precision sensor technology in microscopic mass sensing and measurements of viscosity, stiffness, and many other physical quantities using methods of influencing the electrical substitution model of piezoresonators [13,14].

In the field of active control with piezoelectric materials, researchers have carried out a series of studies and achieved numerous achievements. Hosseini et al. [15] proposed an active vibration control system for monitoring and suppressing human forearm fibrillation. It operates in the ways that are outlined here. Firstly, a dynamic model of the forearm is established. The forearm is simplified into a uniform flexible continuous beam model. The upper surface of the beam is covered with a layer of piezoelectric sensors, and the lower surface is covered with a layer of piezoelectric actuators, forming a vibration control system. The closed-loop active control of forearm vibration is realized by this control system. In addition, the effects of control gain, piezoelectric coefficient, and dielectric constant on the vibration response are investigated. The experimental results showed that the proposed active vibration control system can effectively suppress forearm flutter. Huang et al. [16] studied the active vibration control of piezoelectric sandwich plates. The structure used consists of two layers of piezoelectric sensors and a plate. The method used is based on the constitutive equation of piezoelectric material, whereby the active vibration control dynamic equation of the sandwich structure is established using the hypothesis mode method and Hamiltonian principle. The results show that the natural frequency of the structure is greatly affected by the different boundary conditions and the position of the piezoelectric plate. It can be said that the effect of active control is proportional to the velocity feedback coefficient. Additionally, Sohn et al. [17] used the piezoelectric actuators to control the active vibration of the ship shell structure. Watanabe et al. [18] analyzed and tested the active vibration control of the piezoelectric plate on the swing of the rear edge of the aircraft wing. National Aeronautics and Space Administration (NASA) researchers applied piezoelectric active control to the vibration of the rotating blade [19]. Furthermore, Callipari et al. [20] studied the active control of elastic vibration of large space structures by using bias piezoelectric actuators and tested the control model using the cantilever plate of solar panels. Hashemi et al. [21] designed an intelligent active vibration control system composed of a piezoelectric actuator and linear quadratic regulator in order to control the transverse deflection of wind turbine blades. To apply the control rules to fan blades, an advanced semi-analytical solution is proposed for the transverse displacement of fan blades under external loads.

Although piezoelectric ceramics have been widely used in piezoelectric vibration control, they can be used for the vibration reduction of some specific structures because of their poor elasticity. However, piezoelectric fiber composites have good elasticity and flexibility and they can adapt to different structural surfaces. This means that they have more applications in the field of vibration control. A novel composite sandwich beam with an adaptive active control system was proposed by Lu et al. [22]. The macro-fiber composite (MFC) piezoelectric patches were employed to construct the adaptive closed-loop control system. The resulted show that the vibrations induced by complex multi-frequency excitation with different amplitudes were effectively reduced with MFC piezoelectric patches. Wang et al. [23] investigated vibration suppression for a high-speed macro–micro manipulator with structural flexibility and parameter perturbation. In the study, the macro–micro manipulator contained an air-floating macro-motion platform and an MFC micromanipulator. The electromechanical dynamic model was thus derived. The results showed the proposed control strategy improved manipulation stability, robustness, and accuracy. Furthemore, the topology optimization of piezoelectric macro-fiber composite patches on laminated plates for vibration suppression was analyzed by Padoin et al. [24]. The linear–quadratic regulator (LQR) optimal control technique was used to find the optimum localization of the MFC piezoelectric patch for vibration control. The proposed MFC structure interaction model agreed well with experiments and numerical simulations of models. Dubey et al. [25] studied the shear-based attenuation of the vibration of an annular sandwich plate using shear actuated fiber composite (SAFC) and balanced laminate of PFC (BL-PFC) methods. The results showed that the BL-PFC is the best for shear-based attenuation of vibration of the annular sandwich plate.

For piezoelectric-based flexible beam systems, many scholars have performed numerous studies and obtained valuable results. Mohammed et al. [26] investigated the active vibration control of a cantilever beam. The optimal LQR controller was designed to reduce the vibration of the smart beam. The effect of piezoelectric vibration suppression was studied by changing the beam length. The results indicated that the maximum reduction percentage for settling time related to the free vibration of the smart beam reaches 80%. Lu et al. [27] studied the active vibration control of thin-plate structures with partial smart constrained layer damping (SCLD) treatment. In their research, the emphasis was placed on the feedback control system in order to attenuate the vibration of plates with SCLD treatments. When the external incentive was obtained from a single-frequency signal, vibration response amplitude attenuated by up to nearly 60%. Ezzraimi et al. [28] compared the control effects of different control algorithms and optimized the parameters of the controller by using genetic algorithms. The active vibration control using two types of LQR and PID controllers with different control parameters was tested and compared for the two recovery configurations of the piezoelectric elements. Furthermore, the active vibration control of composite cantilever beams was investigated by Huang et al. [29]. The linear quadratic regulator (LQR) feedback gain was optimized based on the particle swarm optimization (PSO) algorithm. The study showed that the optimal feedback gain of the controller effectively balanced the control effect and the control cost. The vibration of the cantilever was reduced by more than 50%. A control system was designed by Grzybek for the multi-input and simple-output piezoelectric beam actuators based on macro-fiber composite [30]. Additionally, the LQR control algorithm was used to generate control voltages. Furthermore, multiple PZT actuators were employed to suppress the vibration of the composite laminate plate by Her [31].

Piezoelectric polymers allow researchers to overcome the drawbacks of ceramic materials. The popular piezoelectric polymers can be divided into polyvinylidene fluoride (PVDF), polylactic acids (PLA), polyurethanes (PU) and PI [32–35]. Piezoelectric polymers have demonstrated better dielectric behavior and field strength, as well as an ability to tolerate high driving voltage [36,37]. It is because of the above characteristics that piezoelectric polymers can be used for the active vibration control of structures. At present, the efficiency of active control methods with piezoelectric polymers has been greatly improved. However, the issue of how to improve the vibration reduction effect of piezoelectric active control has been the focus of scholars' research.

Therefore, the purpose of this paper is to improve the vibration reduction effect of the flexible beam based on soft piezoelectric film composed of PZT and PI Layer. To achieve the goal, a differential evolution algorithm is used to realize the optimal control effect of a vibration reduction on piezoelectric flexible beams. Compared with other algorithms, the differential evolution algorithm has the advantages of few setting parameters and high search efficiency; thus, it can complete the optimization of the weighted matrix in a short time. This paper will carry out research on the following aspects of this issue. First of all, a dynamic model of piezoelectric flexible beam technology is established. Then, a linear quadratic optimal controller (LQR) based on the differential evolution algorithm is designed. Finally, the effectiveness of active vibration control with piezoelectric polymers is verified by experimental tests.

2. Dynamic Model of Piezoelectric-Based Flexible Beam

A piezoelectric-based flexible cantilever beam structure is a simplified model commonly found in engineering mechanics. In this paper, the Euler–Bernoulli beam is selected as the research object. The piezoelectric flexible beam system consists of three parts: flexible beam, sensing piezoelectric MFC plate, and actuating piezoelectric MFC plate, as shown in Figure 1.

The MFC-5628 was selected in this paper. It can be divided into P1 and P2 two types, and both of them can be used as piezoelectric actuators. The maximum displacement and output of P1 type are larger than that of P2 type. However, the driving voltage of P1 type ranges from −500 V to 1500 V, while that of P2 type only ranges from −60 V to 360 V. Therefore, under the same working conditions, the assessment of P2 type is simpler. Therefore, a P2 piezoelectric plate is selected in this paper. The piezoelectric MFC film is composed of a piezoelectric material that has two sides and one side is attached to an adhesive backing sheet. The slicing of the piezoelectric material constitutes a plurality of piezoelectric fibers in juxtaposition. A conductive film is then adhesively bonded to the other side of the piezoelectric material, and the adhesive backing sheet is removed. The conductive film has first and second conductive patterns formed thereon. These are electrically isolated from one another and are in electrical contact with the piezoelectric material. The first and second conductive patterns of the conductive film each have a plurality of electrodes with which to form a pattern of interdigitated electrodes. A second film is then bonded to the other side of the piezoelectric material. The second film may have a pair of conductive patterns similar to the conductive patterns of the first film [37,38].

The piezoelectric sensing and actuating MFC plates are pasted symmetrically into the upper and lower surfaces of the cantilever beam, and their parameter information is shown in Table 1. The length of the cantilever is L_b, the width is b, and the thickness is t_b. Furthermore, the length of the sensing/actuating piezoelectric plate is L_p, the width is the same as that of the flexible beam, and the thickness is t_p. The distance between the left and right ends of the beam's fixed end is x_1 and x_2, respectively.

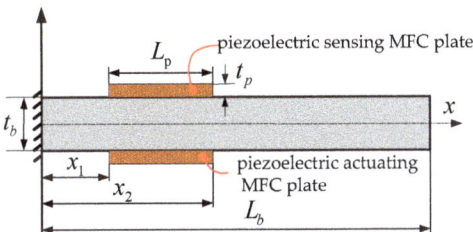

Figure 1. Piezoelectric flexible beam system.

Table 1. Parameters information of piezoelectric MFC plate [39].

Parameter	Values	Parameter	Values
Working mode	d_{31}	d_{31}	-171×10^{-12} C/N
Thickness	300 μm	Effective working length	56 mm
Electrode	Standard lead-free solder S-Sn99Cu1	Effective working width	28 mm
Capacitance	187 nF	Total length	67 mm
Upper limit of operating frequency	<1 MHz	Total width	31 mm
Elastic compliance coefficient s_{11}^E	16.4×10^{-12} m²/N	Permittivity ε_{33}^T	−0.9

According to the theory of structural mechanics, when the flexible beam is subjected to external forces, its dynamic model is [40]

$$E_b I_b \frac{\partial^4 y(x,t)}{\partial x^4} + \rho_b A_b \frac{\partial^2 y(x,t)}{\partial t^2} = \frac{\partial^2 M(x,t)}{\partial x^2} \quad (1)$$

where E_b, I_b, ρ_b and A_b are the elastic modulus, neutral axis moment of inertia, density and cross-sectional area of the flexible beam, respectively, but in which $y(x,t)$ is the deflection of the beam. $M(x,t)$ is the moment that acts on the beam.

When the flexible beam vibrates, the amount of charge generated by the sensing piezoelectric sheet attached to the upper surface of the beam is $Q(x,t)$. At this time, the voltage between the two surfaces of the piezoelectric plate is [41]

$$U_s(x,t) = \frac{Q(x,t)}{C_p} = \frac{bt_b d_{31} E_p}{2C_p} \sum_{i=1}^{n} [\varphi_i'(x_2) - \varphi_i'(x_1)] q_i(t) \quad (2)$$

where d_{31} is the piezoelectric constant, E_p is the elastic modulus of the piezoelectric plate, C_p is the capacitance of the piezoelectric plate, $\varphi_i(x)$ is the i-th order principal mode of the system, and $q_i(t)$ is the i-th order modal principal coordinate.

Letting

$$C_i = K_s [\varphi_i'(x_2) - \varphi_i'(x_1)] \quad (3)$$

where $K_s = bt_b d_{31} E_p / (2C_p)$.

When driving voltage U is applied to the actuated piezoelectric plate, the piezoelectric plate will generate stress due to the inverse piezoelectric effect:

$$\sigma_1 = \frac{E_p d_{31}}{t_p} U \quad (4)$$

where σ_1 is the stress generated by the piezoelectric plate.

The bending moment generated by the piezoelectric plate around the neutral axis of the flexible beam can be written as [42]

$$M = \int_{\frac{t_b}{2}}^{\frac{t_b}{2}+t_p} \sigma_1 b y \, dy = K_a U [h(x - x_1) - h(x - x_2)] \quad (5)$$

where $h(x)$ is the Heaviside function, K_a is the electromechanical coupling coefficient, and $K_a = b d_{31} E_p (t_b + t_p)/2$.

Substituting Equation (5) into Equation (1), the vibration differential equation of the beam under the action of the actuated piezoelectric plate can be expressed as

$$E_b I_b \frac{\partial^4 y(x,t)}{\partial x^4} + \rho_b A_b \frac{\partial^2 y(x,t)}{\partial t^2} = K_a U [\delta'(x - x_2) - \delta'(x - x_1)] \quad (6)$$

where $\delta'(x - x_i)$ is the first order derivative of the Dirac function. The Dirac function can be expressed as

$$\delta(x - x_i) = \begin{cases} 0, & (x \neq x_i) \\ 1, & (x = x_i) \end{cases} \quad (7)$$

According to the principle of modal superposition and considering the damping characteristics of the system, the vibration differential equation of the beam in modal coordinates can be written as

$$\ddot{q}_i(t) + 2\xi_i \omega_i \dot{q}_i(t) + \omega_i^2 q_i(t) = B_i U \quad (8)$$

where $B_i = K_a [\varphi_i'(x_2) - \varphi_i'(x_1)]$, ξ_i is the i-th mode damping ratio of the system, and ω_i is the natural frequency of the i-th mode of the system.

Introduce a state vector $x(t)$

$$x(t) = \{q, \dot{q}\}^T = \{q_1(t) \; q_2(t) \; \cdots \; q_n(t), \; \dot{q}_1(t) \; \dot{q}_2(t) \; \cdots \; \dot{q}_n(t)\}^T \quad (9)$$

Then, the state space equation of the piezoelectric flexible beam system can be expressed as

$$\begin{cases} \dot{x}(t) = \begin{bmatrix} 0_{n\times n} & I_{n\times n} \\ -\Omega & -2\Lambda \end{bmatrix} x(t) + [0_{1\times n} \; B_1 \; \cdots \; B_n]^T u(t) \\ y(t) = [C_1 \; \cdots \; C_n \; 0_{1\times n}] x(t) \end{cases} \quad (10)$$

where $\Omega = diag(\omega_1^2 \; \cdots \; \omega_n^2)$, $\Lambda = diag(\xi_1\omega_1 \; \cdots \; \xi_n\omega_n)$.

3. Piezoelectric Active Vibration Control System

3.1. LQR Controller Design

The controller is the core of the control system, and the controller's performance will directly affect the performance and stability of the control system. In this paper, a linear quadratic optimal controller, which has clear performance indicators and simple control rules, is designed based on the differential evolution algorithm is designed. The differential evolution algorithm is used to optimize the weighted matrix Q.

Define the system quadratic performance index function as

$$J = \frac{1}{2}\int_0^\infty [x^T Q x + u^T R u]dt \quad (11)$$

where the semi-positive definite matrix Q is the weighted matrix of the state variables, and the positive definite matrix R is the weighted matrix of the input variables.

The purpose of optimal control is to find an optimal input u so that the system quadratic performance index J is minimized.

The input law that determines the optimal control is

$$u^* = -Kx \quad (12)$$

where K is the state feedback gain matrix.

$$K = -R^{-1}B^T P \quad (13)$$

where P is a positive definite symmetric matrix, which is the solution of the Riccati algebraic equation:

$$PA + \dot{P} + A^T P - PBR^{-1}B^T P + Q = 0 \quad (14)$$

Then, the optimal control law can be expressed as

$$u^* = -Kx = -R^{-1}B^T P x \quad (15)$$

3.2. Selection of Weighted Matrices Q and R

In the quadratic performance index function J, the $x^T Q x$ term in the integrand mainly reflects the constraints or requirements of the state quantity during the system response, and the $u^T R u$ term mainly reflects the constraints or requirements put on the control input during the system response. Therefore, the quadratic performance index function is essentially a balance between the amount of state change in the system and the energy consumed. The weighted matrices Q and R represent the penalties for the state quantity and control input, respectively. Decreasing the matrix Q is equivalent to increasing the matrix R.

In LQR control, the weighting matrices Q and R represent the penalties for state quantity and control input, respectively. Their penalties are relative, i.e., increasing the matrix Q is equivalent to decreasing the matrix R. In this paper, the weighted matrix Q is optimized by differential evolution algorithm. When the matrix R is enlarged or reduced by different multiples on the basis of the identity matrix I, the element values in the matrix Q will be equally enlarged or reduced by the same multiples. However, the optimal control feedback gain matrix K will not be changed, nor will the vibration control effect be changed. Therefore, if $R = I$ is set, the elements in Q matrix are 10^7 orders of magnitude, and the J_c of optimization result is 10^4 orders of magnitude, which indicates that the numerical series is too large. Therefore, if $R = 0.01\ I$ is set, the elements in the matrix Q are reduced to 10^5 orders of magnitude. J_c is reduced to 10^2 orders of magnitude in the optimization result. At the same time, when $R = 0.01\ I$, the control effect will not be affected.

To simplify the calculation, set the weighted matrix R to $0.01\ I$ and the weighted matrix Q to be diagonal. Allow the weighted matrix Q to follow the form

$$Q = diag\begin{bmatrix} q(1) & q(2) & \cdots & q(2n) \end{bmatrix} \tag{16}$$

According to the characteristics of the piezoelectric flexible beam system, the optimization objective function of the system is defined as:

$$J_c = \frac{1}{2}\int_0^{ts} [x^T Q x + u^T R u] dt \tag{17}$$

where ts is the set time, which is determined according to the actual operating condition of the system.

In addition to considering the vibration suppression effect of flexible beams, the limitations of the system hardware should also be considered. Therefore, the following two constraints are proposed as

$$\left| \frac{y(ts)}{y(0)} \right| \leq \sigma\% \tag{18}$$

$$|u| \leq U_{\max} \tag{19}$$

where Equation (18) mainly considers the vibration suppression effect of the flexible beam, i.e., when the system applies the feedback control, the amplitude of the flexible beam should be less than $\sigma\%$ of the initial amplitude at t_s moment. Here, $y(ts)$ corresponds to the upper envelope curve formed by the vibration peak. Equation (19) mainly considers the input voltage range limitation of the actuated piezoelectric plate.

3.3. The Differential Evolution Algorithm

The differential evolution algorithm has the advantages of simple principles, strong robustness, fast convergence speed, and high precision. Therefore, the differential evolution algorithm is used to optimize the weighted matrix Q.

The principle of the differential evolution algorithm is similar to that of the genetic algorithm, being mainly composed of three operations: mutation, crossover and selection.

(1) Mutation operation

Using the difference strategy, the vector difference of any two individuals in the parent population is scaled and summed with a third individual to produce a mutant individual, and the expression of this is as follows

$$V_{i,G+1} = X_{r1,G} + F(X_{r2,G} - X_{r3,G}) \tag{20}$$

where the subscript G is the current algebra of the population, $r1$, $r2$ and $r3$ are random and mutually different integers between $[1, NP]$, NP is the population size and satisfies $NP \geq 4$, V_i is the i-th mutant individual in the population, and F is the mutation operator which usually takes the constant in the range $[0, 2]$.

(2) Cross operation

The method of probabilistic selection is used to form new candidates from the original target individuals and mutated individuals.

$$U_{ji,G+1} = \begin{cases} V_{ji,G+1}, r \leq CR \text{ or } j = i \\ X_{ji,G}, r > CR \text{ and } j \neq i \end{cases} ; j = 1, 2, \cdots, D \qquad (21)$$

where r is the random number in $[0, 1]$, CR is the cross operator, the value is $CR \in [0, 1]$, $U_{ji,G+1}$ is the individual generated after the crossover, j is the dimension position of the intersection operation within the individual, and D is the dimension within the individual.

(3) Select Actions

The greedy selection strategy is adopted to allow the target individual and the candidate compete to enable the selection of the individual with the better fitness function.

$$X_{ji,G+1} = \begin{cases} U_{ji,G+1}, f(X_{ji,G}) < f(U_{ji,G+1}) \\ X_{ji,G}, f(X_{ji,G}) \geq f(U_{ji,G+1}) \end{cases} \qquad (22)$$

where $X_{ji,G+1}$ is the new generation of individuals produced by the final selection, and f is the fitness function.

3.4. Fuzzy Controller

In order to compare the performance of the LQR controller when optimized by differential evolution algorithms, a fuzzy controller with dual input and single output is designed for comparison, and a fuzzy controller with dual input and single output is designed for the vibration active control of piezoelectric flexible beams. The basic working principle behind this is to convert the measured input into a fuzzy quantity that can be described through fuzzy control rules, perform fuzzy inference on a fuzzy output value, and then transform the fuzzy output value into an accurate value that can be used for actual control through defuzzification operation.

A common two-dimensional fuzzy logic controller is selected in this paper according to the vibration characteristics of the flexible beam. The first input is set as error E, which represents the difference between the end of the flexible beam and the beam balance position during the vibration process and further represents the amplitude of the beam end. The second input is set as the error change rate EC, which represents the velocity value of the flexible beam end in the process of vibration. The output U is set as the feedback control signal of the system.

In the process of input ambiguity, the discussion domain of input E and EC is set as $[-3, 3]$, and that of output U is set as $[-5, 5]$. Input/output adopts a seven-level fuzzy set, including descriptors of NL (negative large), NM (negative medium), NS (negative small), O (zero error), PS (positive small), PM (positive medium), and PL (positive large). In order to facilitate membership calculation, triangular membership functions are selected for input E, EC and output U. Their membership function curves are shown in Figure S1.

The research object of this paper is the vibration control of the beam. When the error E is larger, the beam amplitude will be larger. In order to make the beam return to the equilibrium position as soon as possible, the control quantity U should be larger, and vice versa. The greater the error rate EC, the greater the beam velocity. Using the Mamdani fuzzy inference method, the $7 \times 7 = 49$ fuzzy rules described by the matrix table are designed, as shown in Table S1.

Finally, the fuzzy output value is transformed into an accurate value which can be used for actual control by using the center of gravity method.

4. Frequency Sweeping Experiment

Frequency sweeping is one of the commonly used methods for harmonic response analysis in structural dynamics. Compared with performing modal tests such as the hammering method, the frequency response characteristics of the structure can be obtained efficiently and accurately through the sweeping vibration test and the natural frequency of the structure can be found. The procedure behind a sweep frequency experiment is shown in Figure S2.

The frequency sweep experiment of piezoelectric flexible beam system is carried out in this paper. The basic principle behind doing so is that the sweeping signal generated by the signal generator is amplified by the piezoelectric driver and then acts on the driving piezoelectric plate. The vibration of each mode of the flexible beam system is excited so that the sensing piezoelectric plate can generate induction signals. After processing the induced signal, the vibration response of the flexible beam system under the swept signal can be obtained. Because the vibration response of the beam is at a maximum at the natural frequency, the natural frequency of the flexible beam can be obtained through performing the sweep frequency experiment.

Since the vibration energy of the flexible beam is mainly distributed in the lower-order mode, the frequency sweep experiment is only used for the first two order modes of the flexible beam. The correct parameters for its use are as follows: set the frequency range of the sweep to 0–50 Hz, the sweep mode to linear, the signal amplitude to 5 V, the sweep time to 15 s, and the sampling frequency to 500 Hz. After the experiment, the output signal of the sensing piezoelectric plate is shown in Figure S3, and the frequency response curve is shown in Figure 2.

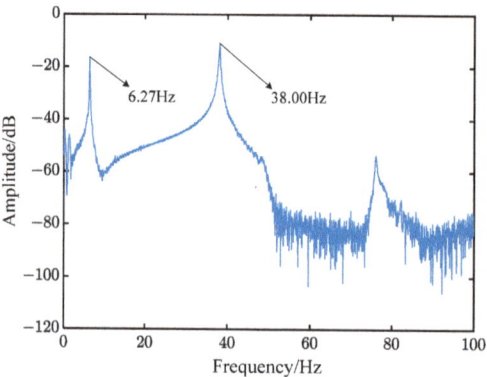

Figure 2. Results of frequency sweeping experiment.

The natural frequencies of the first two orders of the piezoelectric flexible beam system were obtained by performing an experiment and using the finite element method (FEM), as shown in Table 2. When simulated with FEM, the first two natural frequencies of the beam were 6.61 Hz and 39.04 Hz (see Figure S4). Additionally, the natural frequencies of the first two orders of the beam, obtained by using the frequency sweeping experiment, were 6.27 Hz and 38.00 Hz. It can be seen from Table 2 that the errors of the first two orders of natural frequency were 5.14% and 2.66%, respectively. The simulation results are consistent with the experimental results, which verify the correctness of the model.

Table 2. The natural frequency of piezoelectric flexible beams.

	FEM Results	Experiment Results	Errors (%)
First order natural frequency/Hz	6.61	6.27	5.14%
Second order natural frequency/Hz	39.04	38.00	2.66%

5. Simulation

5.1. Optimal Positions of Sensing/Actuating Piezoelectric Plates

In order to optimize the position of the sensing/actuating piezoelectric MFC plates in the flexible beam, scholars have proposed a variety of optimization criteria from different perspectives. Examples include the controllability/observability criterion, system energy criterion, system stability criterion, etc.

Compared with other criteria, the D optimization criterion proposed by Bayard is simple and has clear physical meaning. Because the damping of the flexible beam structure is small, the D optimization method can be simplified. The optimal position of the sensing plate can be determined by the structural modal information of the beam. The sensing plate can be arranged in the place where the modal strain of the structure is at its maximum, and the actuating voltage plate can be symmetrically arranged on the upper and lower surfaces of the beam by centering.

According to the boundary conditions, the modal shape function and natural frequency of the flexible beam can be derived as follows:

$$\varphi_i(x) = [\cosh \beta_i x - \cos \beta_i x + \gamma_i (\sinh \beta_i x - \sin \beta_i x)] \ (i = 1, 2, \cdots, n) \qquad (23)$$

$$w_i = (\beta_i L_b)^2 \sqrt{\frac{E_b I}{\rho A L_b^4}} \qquad (24)$$

where $\beta_1 = 1.875/L_b$, $\beta_2 = 4.694/L_b$, $\gamma_i = -(\cos\beta_i L_b + \cosh\beta_i L_b)/(\sin\beta_i L_b + \sinh\beta_i L_b)$.

The specific parameters of the flexible beam are shown in Table 3.

Table 3. Flexible beam parameters.

Item	Flexible Beams
Length/width/thickness/mm	330/32/0.8
Density ρ/(kg/m^3)	7850
Modulus of elasticity E/Gpa	198.6
Damping ratio ξ	0.004
1st round frequency/(rad/s)	38.9

According to Equation (24), the first two orders of modal shape curves of the flexible beam can be obtained by MATLAB simulation, as shown in Figure S5.

The first two-stage open-loop vibration curves of the flexible beam are shown in Figure S6. It can be seen from the figure that the first-order open-loop vibration amplitude of the flexible beam undergoes slow attenuation and has a long vibration duration. The second-order open-loop vibration amplitude of the flexible beam decays quickly, and it can be seen that most of the vibration energy of the flexible beam is concentrated in the first-order mode. Therefore, in this paper, only the first-order modal vibration of flexible beams is controlled by vibration suppression.

The maximum strain of first-order modal vibration of flexible beams is located at the root of the beam. Therefore, according to the D optimization criterion, the piezoelectric actuator should be arranged at the root of the flexible beam and, according to the principle of symmetry, the sensing/actuated piezoelectric plate should be aligned and pasted to the root of the beam.

5.2. Vibration Control Simulation

The differential evolution algorithm is combined with the Simulink model of the piezoelectric flexible beam system to optimize the weighted matrix Q. Because the objective function J_c is the minimum, the objective function J_c is used as the fitness function F. The process of optimizing Q using the differential evolution algorithm is shown in Figure 3.

The specific steps in the differential evolution algorithm are as follows:

(1) The differential differentiation algorithm generates the initial population;
(2) The population individuals are assigned to each element in the weighted matrix Q in turn, and the optimal control feedback gain matrix K is calculated;
(3) The Simulink is run to model the flexible beam system, obtain the state quantity, control output voltage value, and calculate the objective function value;

(4) Researchers determine whether the termination conditions are met; if so, the optimization process is over, and the output is the result; otherwise, the population continues the evolutionary operation to obtain the evolved population, and researchers can skip to step 2.

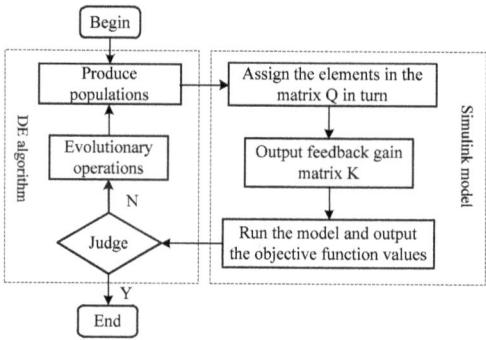

Figure 3. Algorithm optimization process.

In this paper, the first-order vibration model of the beam is selected as the research object. According to the established dynamic model, the state space equation of the piezoelectric flexible beam system can be obtained. In order to operate it, researchers should set the running time t_s = 6 s, attenuation ratio $\sigma\%$ = 1%, and the maximum voltage U_{max} = 360 V. The key parameter settings in the differential evolution algorithm are shown in Table 4.

Table 4. Key parameters of the differential evolution algorithm.

NP	F	CR	G
30	0.5	0.9	200

The evolution of fitness function value during differential evolution is shown in Figure S7. As the process of differential evolution advances, the individuals with low fitness in the population are gradually eliminated, while the number of individuals with high fitness gradually increases. When this evolves to approximately 50 generations, the fitness function converges to the optimal value until the end of the entire optimization process. At this point, the objective function value becomes J_c = 226.85, and the weighted matrix is Q = diag [6.19 × 10^5 3.83 × 10^5].

In order to verify the performance of the LQR controller based on the differential evolution algorithm, a dual-input single-output (DISO) fuzzy logic controller is also designed. The peak value of load voltage calculated by the two control algorithms should be equal.

In Simulink, an active vibration control simulation platform for the piezoelectric flexible beam is built, and vibration control experiments under different disturbances are carried out.

The value of the weighted matrix Q of the common LQR controller is artificially obtained through the empirical method or trial and error method after substantial verification, a process which takes a long time and which has poor numerical accuracy.

Figure 4 shows that the vibration control effect varies with weighted matrix Q, where the weighted matrix of LQR1 controller is artificially set as Q = diag [10^4 10^4], and the weighted matrix of LQR2 controller is Q = diag [10^5 10^5]. Therefore, the value of the weighted matrix Q needs to be adjusted artificially and continuously to induce the vibration control effect of the above constraints. The weighted matrix Q = diag [6.19 × 10^5 3.83 × 10^6] of the LQR controller in the figure is obtained by searching the optimal solution value within the constraints using the differential evolution algorithm; this process only takes a short period of time and has high numerical accuracy. Obviously, since Q is optimal, the vibration suppression effect under the LQR control shown in Figure 4 is the best. When the LQR control parameter deviates from the optimal solution value, the control effect cannot be guaranteed, and the amplitude of the LQR control cannot be guaranteed to be smaller than that of fuzzy control. Thus, it is possible for the amplitude of the LQR control to be larger than that of fuzzy control. Therefore, in order to achieve the best control LQR effect, the optimal weighting matrix Q should be obtained.

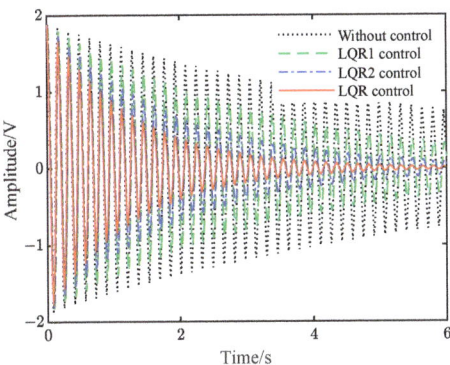

Figure 4. The vibration control effect varies with weighted matrix Q.

LQR control is essentially a balance between the amount of system state change and the energy consumed, i.e., better control performance is achieved with less energy consumption. Ordinary LQR control weighted matrices can only be obtained empirically or via pro forma methods. Thus, "optimal" control is actually artificial. The purpose of using the differential evolution algorithm is to replace artificial labor and find the most ideal weighted matrix values in a short period of time. In order to compare the LQR performances optimized by the differential evolution algorithm, fuzzy control is used to compare the performance of the LQR controller under the same loading voltage. This proves that the optimization of the LQR controller using a differential evolution algorithm is successful.

Figure 5 shows the vibration–response curve of the system under instantaneous disturbance. It can be seen from the figure that when there is no control, the vibration of the system decays slowly with time due to its damping. After applying LQR control and fuzzy control, the amplitude of the flexible beam is significantly suppressed over a short period of time, and the LQR control effect based on the differential evolution algorithm improves.

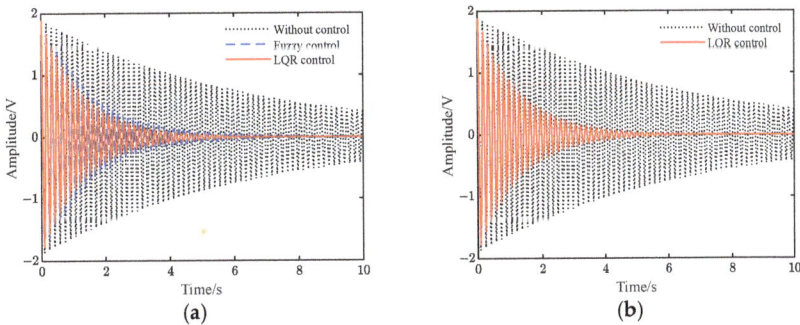

Figure 5. Vibration response under instantaneous disturbance. (**a**) Fuzzy and LQR control; (**b**) LQR control.

The results that correspond to a sinusoidal signal $z = 2\sin(2\pi ft)$ with a frequency of 6.27 Hz being applied as a disturbance signal are shown in Figure 6. It can be seen from the figure that when it is not controlled, the beam appears in the form of sinusoidal vibration. After the two controls are applied to the system, respectively, the vibration amplitude decreases first and then tends to become stable. Compared to the uncontrolled results, the amplitude is reduced by 68.7% and 63.8%. Additionally, the flexible beam under LQR control can reach a steady vibration suppression state faster. Compared to the fuzzy control, the LQR control is better.

In addition, the response of the beam system under the white noise signal is studied, as shown in Figure 7. In Figure 7, a white noise signal with a PSD height of 100 and a sampling time of 0.01 s is investigated as the vibration response to the system disturbance signal. Results show when it is not controlled, the system undergoes irregular vibration, with a peak value of approximately 2 V. After applying the two controls separately, the vibration peak of the system is reduced to approximately 1 V, and the average vibration suppression effect under LQR control is better than that of fuzzy control.

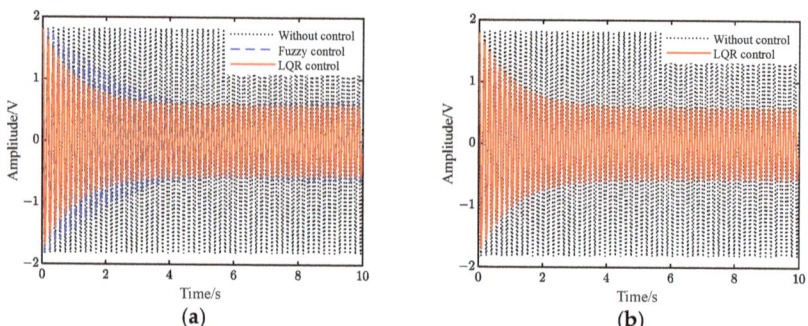

Figure 6. Vibration response under sinusoidal disturbance. (**a**) Fuzzy and LQR control; (**b**) LQR control.

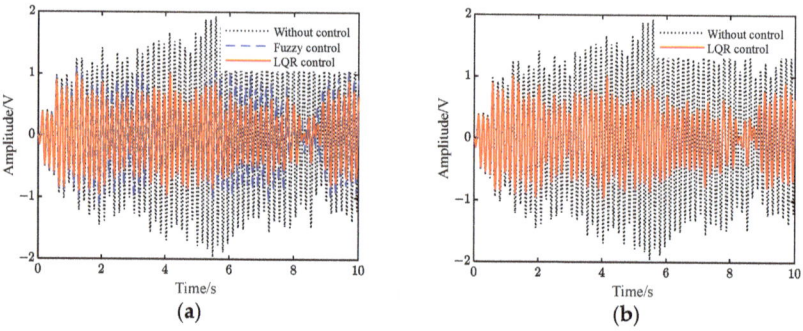

Figure 7. Vibration response under white noise disturbance. (**a**) Fuzzy and LQR control; (**b**) LQR control.

6. Experiments and Results

6.1. Vibration Control Experiment under Instantaneous Disturbance

A structural diagram of the piezoelectric flexible beam active vibration control system is shown in Figure S8. Additionally, the system hardware experimental platform is shown in Figure 8. The system consists of a signal acquisition unit, an active control unit, a feedback actuation unit, and an excitation unit.

The signal acquisition unit comprises a sensing piezoelectric sheet MFC-5628 and a charge amplifier YE5852, the latter of which converts the vibration signal of the collected flexible beam into an equal proportion of the voltage signal. The active control unit is composed of a data acquisition card NI USB-6002 and a computer installed with LabVIEW. Its function is to calculate the corresponding control signal through the active control algorithm. The feedback actuation unit is composed of piezoelectric driver E01 and actuating piezoelectric sheet MFC-5628. Its function is to make the piezoelectric plate generate deformation under the action of the control voltage. Then, the suppression of beam vibration is realized. The excitation unit is composed of three parts: signal generator FY8300, power amplifier CT5872 and exciter JZK-10. These components are responsible for interfering with the flexible beam and then making the beam generate vibration.

Figure 8. The experimental platform of the flexible beam system.

In this experiment, an external force is applied to the end of the flexible beam to deviate it from the equilibrium position. Additionally, after the external force is removed, the flexible beam vibrates. The two designed controllers are used to perform vibration control experiments under the instantaneous disturbance of flexible beams.

Figure 9 shows the vibration response of the flexible beam under the conditions of instantaneous disturbance. Transient disturbance refers to the sudden change in the force acting on the flexible beam, which makes the whole piezoelectric flexible beam system appear transient response. When the flexible beam is at rest, it should be located at the equilibrium position, and its state should be zero. In the experiment, an external force is used to offset the end of the beam by a fixed distance from the equilibrium position. During the experiment, when the moment the external force is removed, the beam will vibrate. Before and after the experiment, the beam will have a transient response due to the change in the external force.

Figure S9 shows the voltage when applied to the piezoelectric plate under different controls. It can be seen from Figure 9 and Figure S9 that when there is no control, the vibration of the flexible beam is slowly attenuated. Additionally, after the two controls are applied, the amplitude of the flexible beam is quickly suppressed to a significant extent. Furthermore, the amplitude of the beam under LQR control is smaller than that under fuzzy control.

Table 5 shows the amplitude comparison between the simulation and the experiment under instantaneous perturbation conditions. It can be seen from the table that after applying fuzzy control and LQR control, the beam amplitude drops to 0.10 V and 0.04 V in the simulation, which is a decrease of 88.6% and 95.5% compared with the uncontrolled 0.88 V. In the experiment, the beam amplitude decreases from an uncontrolled 0.90 V to 0.11 V and 0.05 V, a decrease of 87.8% and 94.4%, respectively. The simulation and experimental results are consistent.

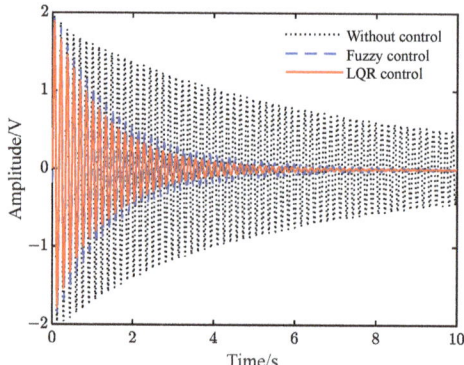

Figure 9. Vibration response under instantaneous disturbances.

Table 5. Comparison of simulation and experimental results under instantaneous disturbance.

	Without Control	Fuzzy Control	LQR Control
Simulate amplitude (V)	0.88	0.10	0.04
Experimental amplitude (V)	0.90	0.11	0.05

6.2. Vibration Control Experiment under Continuous Disturbance

The excitation unit is used to apply a sinusoidal signal with a frequency of 6.27 Hz and amplitude of 2 V to the root of the flexible beam to excite the first-order modal vibration of the flexible beam. The two designed controllers are used to carry out vibration control experiments on flexible beams under continuous disturbance.

Figure 10 shows the time-domain response under continuous disturbance. Figure S10 shows the voltage applied to the piezoelectric plate. It can be seen from the figure that when there is no control, the flexible beam displays sinusoidal vibration, with a vibration amplitude of 1.79 V. In our experiment, after the two controls were applied, the vibration amplitude of the flexible beam decreased to 0.62 V and 0.73 V, which was 65.4% and 59.2% lower than the uncontrolled amplitude, respectively. Additionally, the flexible beam under LQR control reached a stable vibration suppression state faster.

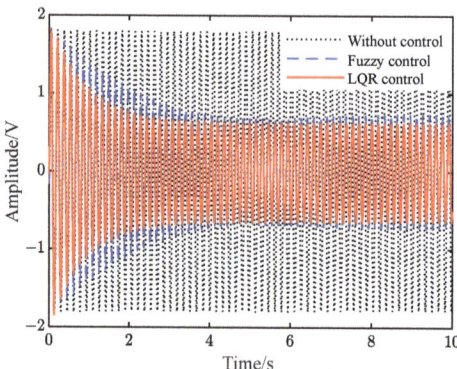

Figure 10. Vibration response under continuous disturbance.

Table 6 shows the amplitude comparison of the simulation and experiment results under sinusoidal disturbance. It can be seen that after we had applied the LQR control and fuzzy control, the beam amplitude decreased from uncontrolled 1.82 V to 0.57 V and

0.66 V, which marked decreases of 68.7% and 63.8%, respectively. In the experiment, the beam amplitude decreased from an uncontrolled 1.79 V to 0.62 V and 0.73 V, which marked decreases of 65.4% and 59.2%, respectively. Hence, the simulation and experimental results were basically consistent.

Table 6. Comparison of simulation and experimental results under sinusoidal disturbance.

	Without Control	Fuzzy Control	LQR Control
Simulate amplitude (V)	1.82	0.66	0.57
Experimental amplitude (V)	1.79	0.73	0.62

6.3. Vibration Control Experiment under White Noise Disturbance

The white noise signal with a PSD height of 100 and a sampling time of 0.01 s is applied to the root of the flexible beam. The vibration control experiment under the white noise disturbance is carried out on the flexible beam using the two controllers.

Figure 11 shows the time domain response under white noise disturbance conditions. Figure S11 presents the voltage applied to the piezoelectric plate under white noise disturbance conditions. In general, the amplitude of the flexible beam is suppressed after the control is applied. Additionally, the suppression effect of the LQR control is slightly better than that of the fuzzy control.

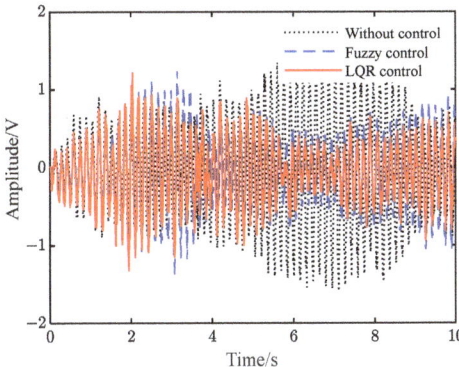

Figure 11. Vibration response under white noise disturbance.

Table 7 shows the amplitude of frequency response under white noise disturbance. It can be seen that after the two controls were applied in our experiment, the peak value of the amplitude at the first-order resonance of the flexible beam was reduced from 0.74 V to 0.38 V and 0.58 V, respectively. Additionally, the vibration of the flexible beam was suppressed. Compared with the fuzzy control, the amplitude of the beam under LQR control was smaller, and the control effect was better. This was consistent with the simulation results, which verifies the correctness of the simulation results.

Table 7. The amplitude of frequency response under white noise disturbance.

	Without Control	Fuzzy Control	LQR Control
Amplitude (V)	0.74	0.58	0.38

6.4. Innovation and Comparative Analysis

In this paper, the piezoelectric active vibration control of flexible beam is realized by using LQR controller based on difference algorithm. The innovation of this paper is mainly reflected in the following three aspects:

(1) The optimization of weighted matrix Q in LQR controller is traditionally obtained by using a genetic algorithm. However, genetic algorithms have the disadvantages of complex structure, amounts of setting parameters, and low search efficiency. However, the differential evolution algorithm has the advantages of simple principles, few setting parameters and high search efficiency. Compared with the genetic algorithm, the differential evolution algorithm can complete the optimization of the weighted matrix Q in a shorter time.

(2) As shown in Figure 10, when the weighted matrix Q reaches the optimal value, the average fitness of the LQR controller based on the difference algorithm becomes stable after 40 iterations. However, the LQR controller based on genetic algorithm achieved the stability of the average fitness after more than 60 iterations [43]. Furthermore, compared with the genetic algorithm, LQR controller based on differential evolution algorithm is more efficient.

(3) In the studies [20,21,30], maximum vibration attenuation amplitudes of 80%, 60% and 50% were achieved by piezoelectric active control in combination with other algorithms. In this paper, the differential evolution algorithm is used to achieve the maximum vibration attenuation amplitude of 94.4% of the flexible beam under instantaneous disturbance with LQR control. These results further show that the control effect obtained based on differential evolution algorithm is better.

7. Conclusions

In this paper, active control based on soft piezoelectric macro-fiber composite (MFC) consisting of polyimide (PI) sheet and lead zirconate titanate (PZT) is used to reduce the vibration. A dynamic coupling model of the piezoelectric flexible beam system is established. A linear quadratic optimal controller (LQR) is designed based on the optimal control theory. In order to verify the performance of the LQR controller optimized by the differential evolution algorithm, the fuzzy controller is used as a comparison, and simulation experiments are carried out in Simulink for different disturbance states. A systematic hardware experimental platform is established, and experimental research on active vibration control of flexible beams under different disturbances is carried out. The results show that:

(1) The design control algorithm has a good control effect on the vibration of flexible beams under different disturbance states. The vibration suppression effect of LQR control optimized based on the differential evolution algorithm is better than that of fuzzy control.

(2) The amplitudes of the piezoelectric flexible beams are reduced by 94.4% and 65.4% under instantaneous and continuous disturbances with LQR control.

(3) In the frequency sweeping experiment, the errors of the first two orders of natural frequencies are 5.14% and 2.77%, respectively.

Supplementary Materials: The following supporting information can be downloaded at: https://www.mdpi.com/article/10.3390/polym15081819/s1, Figure S1: Membership function graph of fuzzy controller; Table S1: Fuzzy control rule; Figure S2: The procedure of sweep frequency experiment; Figure S3: The output signal of the sensing piezoelectric plate; Figure S4: FEM analysis results of natural frequencies; Figure S5: Flexible beam modal curves; Figure S6: Open-loop vibration curves; Figure S7: Evolution of fitness function values; Figure S8: Schematic diagram of the experimental system; Figure S9: The voltage applied to the piezoelectric plate under instantaneous disturbance; Figure S10: The voltage applied to the piezoelectric plate under continuous disturbance; Figure S11: The voltage applied to the piezoelectric plate under white noise disturbance.

Author Contributions: Conceptualization, C.L.; Investigation, L.S.; Validation, L.S.; Writing—original draft, J.F.; Writing—review and editing, J.S. All authors have read and agreed to the published version of the manuscript.

Funding: This research was funded by Qing Lan Project of Jiangsu Province (No. 1024902201) and China Postdoctoral Science Foundation (No. 2018M640515).

Institutional Review Board Statement: Not applicable.

Data Availability Statement: Not applicable.

Conflicts of Interest: The authors declare no conflict of interest.

References

1. Liang, K.; Li, C.; Tong, Y.J.; Fang, J.; Zhong, W. Design of a Low-Frequency Harmonic Rotary Piezoelectric Actuator. *Actuators* **2021**, *10*, 4. [CrossRef]
2. Li, C.; Liang, K.; Zhong, W.; Fang, J.; Sun, L.; Zhu, Y. Electrochemical Coupled Analysis of a Micro Piezo-Driven Focusing Mechanism. *Micromachines* **2020**, *11*, 26. [CrossRef]
3. Li, C.; Zhong, W.; Fang, J.; Sun, L. Nonlinear Vibration of a Micro Piezoelectric Precision Drive System. *Micromachines* **2019**, *10*, 159. [CrossRef]
4. Huang, Z.; Peng, H.; Wang, X.; Chu, F. Modeling and Vibration Control of Sandwich Composite Plates. *Materials* **2022**, *16*, 896. [CrossRef] [PubMed]
5. Rahman, N.U.; Alam, M.N.; Ansari, J.A. An experimental study on dynamic analysis and active vibration control of smart laminated plates. *Mater. Today Proc.* **2021**, *46*, 9550–9554. [CrossRef]
6. Sun, L.F.; Li, W.J.; Wu, Y.Z.; Lan, Q. Active vibration control of a conical shell using piezoelectric ceramics. *J. Low Freq. Noise Vib. Act. Control* **2018**, *36*, 366–375. [CrossRef]
7. He, J.; Xing, T.; Wang, T.; Wu, X.; He, H.; Chen, G. Reduction of structural Vibrations with the Piezoelectric Stacks Ring. *Int. J. Appl. Electromagn. Mech.* **2020**, *64*, 729–736. [CrossRef]
8. Meng, D.; Xia, P.Q.; Song, L.S. MIMOMH feed-forward adaptive vibration control of helicopter fuselage by using piezoelectric stack actuators. *J. Vib. Control* **2018**, *24*, 5534–5545. [CrossRef]
9. Simões, R.C.; Steffen, V. Modal Active Vibration Control of a Rotor Using Piezoelectric Stack Actuators. *J. Vib. Control* **2007**, *13*, 45–64. [CrossRef]
10. Li, J.; Ma, Z.; Wang, Z.; Narita, Y. Random Vibration Control of Laminated Composite Plates with Piezoelectric Fiber Reinforced Composites. *Acta Mech. Solida Sin.* **2016**, *29*, 316–327. [CrossRef]
11. Panda, S. Performance of a short piezoelectric fiber-reinforced composite actuator in vibration control of functionally graded circular cylindrical shell. *J. Intell. Mater. Syst. Struct.* **2016**, *27*, 2774–2794. [CrossRef]
12. Dubey, M.K.; Panda, S. Shear actuation mechanism and shear-based actuation capability of an obliquely reinforced piezoelectric fibre composite in active control of annular plates. *J. Intell. Mater. Syst. Struct.* **2019**, *30*, 2447–2463. [CrossRef]
13. Matko, V.; Milanovic, M. Detection principles of temperature compensated oscillators with reactance influence on piezoelectric resonator. *Sensors* **2020**, *20*, 802. [CrossRef]
14. Matko, V.; Milanovic, M. Temperature-compensated capacitance-frequency converter with high resolution. *Sens. Actuators A Phys.* **2014**, *220*, 262–269. [CrossRef]
15. Hosseini, S.M.; Kalhori, H.; Al-Jumaily, A. Active vibration control in human forearm model using paired piezoelectric sensor and actuator. *J. Vib. Control* **2021**, *27*, 2231–2242. [CrossRef]
16. Huang, Z.; Mao, Y.; Dai, A.; Han, M.; Wang, X.; Chu, F. Active Vibration Control of Piezoelectric Sandwich Plates. *Materials* **2022**, *15*, 3907. [CrossRef]
17. Sohn, J.W.; Choi, S.B.; Lee, C.H. Active vibration control of smart hull structure using piezoelectric composite actuators. *Smart Mater. Struct.* **2009**, *18*, 074004. [CrossRef]
18. Watanabe, T.; Kazawa, J.; Uzawa, S.; Keim, B. Numerical and experimental study of active flutter suppression with piezoelectric device for transonic cascade. In Proceedings of the 2008 Proceedings of the ASME Turbo Expo: Power for Land, Sea, and Air, Berlin, Germany, 9–13 June 2008; pp. 849–859.
19. Duffy, K.P.; Choi, B.B.; Provenza, A.J.; Min, J.B.; Kray, N. Active Piezoelectric Vibration Control of Subscale Composite Fan Blades. *J. Eng. Gas Turbines Power* **2013**, *135*, 011601. [CrossRef]
20. Callipari, F.; Sabatini, M.; Angeletti, F.; Iannelli, P.; Gasbarri, P. Active vibration control of large space structures: Modelling and experimental testing of offset piezoelectric stack actuators. *Acta Astronaut.* **2022**, *198*, 733–745. [CrossRef]
21. Hashemi, A.; Jang, J.; Hosseini-Hashemi, S. Smart Active Vibration Control System of a Rotary Structure Using Piezoelectric Materials. *Sensors* **2022**, *22*, 5691. [CrossRef]
22. Lu, Q.; Wang, P.; Liu, C. An analytical and experimental study on adaptive active vibration control of sandwich beam. *Int. J. Mech. Sci.* **2022**, *232*, 107634. [CrossRef]
23. Wang, S.; Yang, Y.; Li, G.; Du, H.L.; Wei, Y.D. Microscopic vibration suppression for a high-speed macro-micro manipulator with parameter perturbation. *Mech. Syst. Signal Process.* **2022**, *179*, 109332. [CrossRef]
24. Padoin, E.; Santos, I.F.; Perondi, E.A.; Menuzzi, O.; Gonçalves, J.F. Topology optimization of piezoelectric macro-fiber composite patches on laminated plates for vibration suppression. *Struct. Multidiscip. Optim.* **2019**, *59*, 941–957. [CrossRef]
25. Dubey, M.K.; Panda, S. Shear-based vibration control of annular sandwich plates using different piezoelectric fiber composites: A comparative study. *J. Sandw. Struct. Mater.* **2021**, *23*, 405–435. [CrossRef]
26. Mohammed, U.Q.; Wasmi, H.R. Active Vibration Control of Cantilever Beam by Using Optimal LQR Controller. *J. Eng.* **2018**, *24*, 1–17. [CrossRef]
27. Lu, J.; Wang, P. Active vibration control of thin-plate structures with partial SCLD treatment. *Mech. Syst. Signal Process.* **2017**, *84*, 531–550. [CrossRef]
28. Ezzraimi, M.; Tiberkak, R.; Melbous, A.; Rechak, S. LQR and PID Algorithms for Vibration Control of Piezoelectric Composite Plates. *Mechanics* **2018**, *24*, 20645. [CrossRef]
29. Huang, Z.; Huang, F.; Wang, X.; Chu, F. Active Vibration Control of Composite Cantilever Beams. *Materials* **2023**, *16*, 95. [CrossRef]

30. Grzybek, D. Control System for Multi-Input and Simple-Output Piezoelectric Beam Actuator Based on Macro Fiber Composite. *Energies* **2022**, *15*, 2042. [CrossRef]
31. Her, S.C.; Chen, H.Y. Vibration Excitation and Suppression of a Composite Laminate Plate Using Piezoelectric Actuators. *Materials* **2022**, *15*, 2027. [CrossRef] [PubMed]
32. Lin, J.; Malakooti, M.H.; Sodano, H.A. Thermally stable poly (vinylidene fluoride) for high-performance printable piezoelectric devices. *ACS Appl. Mater. Interfaces* **2020**, *12*, 21871–21882. [CrossRef] [PubMed]
33. Gong, S.; Zhang, B.; Zhang, J.; Wang, Z.L.; Ren, K. Biocompatible poly (lactic acid)-based hybrid piezoelectric and electret nanogenerator for electronic skin applications. *Adv. Funct. Mater.* **2020**, *30*, 1908724. [CrossRef]
34. Yazdani, A.; Manesh, H.D.; Zebarjad, S.M. Piezoelectric properties and damping behavior of highly loaded PZT/polyurethane particulate composites. *Ceram. Int.* **2023**, *49*, 4055–4063. [CrossRef]
35. Sappati, K.K.; Bhadra, S. Piezoelectric polymer and paper substrates: A review. *Sensors* **2018**, *18*, 3605. [CrossRef]
36. Shao, J.; Liao, X.; Ji, M.; Liu, X. A Modeling Study of the Dielectric Property of Polymeric Nanocomposites Based on the Developed Rayleigh Model. *ACS Appl. Polym. Mater.* **2021**, *3*, 6338–6344. [CrossRef]
37. Wang, J.; Tong, Y.J.; Li, C.; Zhang, Z.; Shao, J. A Novel Vibration Piezoelectric Generator Based on Flexible Piezoelectric Film Composed of PZT and PI Layer. *Polymers* **2022**, *14*, 2871. [CrossRef]
38. Bryant, W.W.K.; Fox, R.G.; Hellbaum, R.L.; High, J.W.; Jalink, A., Jr.; Little, B.D.; Mirick, P.H. Method Of Fabricating A Piezoelectric Composite Apparatus. U.S. Patent 6,629,341, 7 October 2003.
39. MFC Macro Fiber Composite. Available online: http://www.coremorrow.com/en/proshow-7-423-1.html (accessed on 1 March 2023).
40. Bailey, T. Distributed piezoelectric polymer active vibration control of a cantilever beam. *Guid. Control* **1985**, *8*, 605–611. [CrossRef]
41. Narayanana, S.; Balamurugan, V. Finite element modeling of piezolaminated smart structures for active vibration control with distributed sensors and actuators. *J. Sound Vib.* **2003**, *26*, 529–562. [CrossRef]
42. Aldrihem, O.J. Optimal size and location of piezoelectric actuator/sensors practical consideration. *J. Guid. Control Dyn.* **2000**, *23*, 509–515. [CrossRef]
43. Kang, J.Y.; Bi, G.; Su, S.B. Experimental Identification and Active Vibration Control of Piezoelectric Flexible Manipulator. *J. Vib. Meas. Diagn.* **2021**, *41*, 90–95.

Disclaimer/Publisher's Note: The statements, opinions and data contained in all publications are solely those of the individual author(s) and contributor(s) and not of MDPI and/or the editor(s). MDPI and/or the editor(s) disclaim responsibility for any injury to people or property resulting from any ideas, methods, instructions or products referred to in the content.

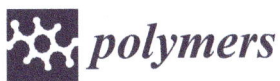

Article

Investigations on Five PMMA Closed Types of Piezo Actuators as a Cooling Fan

Rong-Tsu Wang [1,*] and Jung-Chang Wang [2,*]

[1] Department and Graduate Institute of Information Management, Yu Da University of Science and Technology, Miaoli County 36143, Taiwan
[2] Department of Marine Engineering (DME), National Taiwan Ocean University (NTOU), Keelung 202301, Taiwan
* Correspondence: rtwang@ydu.edu.tw (R.-T.W.); jcwang@ntou.edu.tw (J.-C.W.); Tel.: +886-2-24622192 (ext. 7109 or 7139) (J.-C.W.)

Abstract: There are five closed types of piezo actuators (closed type of PA, closed PA) as a cooling fan relative to those different PAJs of the previous work (open type of PAJ, open PAJ) for analysis in the present study. Closed PA was composed of circular piezoelectric ceramics (PCs) and acrylic (PMMA) plates and investigated on five different types at operating conditions. The results show that the noise of the closed PA is quieter than that of the open PAJ by about 10 dB. When the closed PA is deposed at a suitable distance of 10 to 20 mm from the heat source, averting sucking back the high-temperature fluids around that, the thermal convection coefficient is above 120% more than that of the conventional rotary fan. The cooling performances of these five closed PAs were evaluated by thermal analysis technique, and the convection thermal resistance of the best closed PA can be decreased by over 15%. In terms of energy consumption, a monolithic closed PA was less than 10% than that of a rotary fan. Among these five closed PAs, the best one has the essential qualities that the diameter of the piezoelectric sheet is 41 mm, the opening length is 4 mm, and the outer opening length is 10 mm. Moreover, the best operating conditions are a voltage frequency of 300 Hz and a release distance of 15 mm in the present study.

Keywords: PMMA; acrylic; actuator; PA; piezoelectric ceramic; thermal analysis

Citation: Wang, R.-T.; Wang, J.-C. Investigations on Five PMMA Closed Types of Piezo Actuators as a Cooling Fan. *Polymers* **2023**, *15*, 377. https://doi.org/10.3390/polym15020377

Academic Editor: Roman A. Surmenev

Received: 10 November 2022
Revised: 31 December 2022
Accepted: 5 January 2023
Published: 10 January 2023

Copyright: © 2023 by the authors. Licensee MDPI, Basel, Switzerland. This article is an open access article distributed under the terms and conditions of the Creative Commons Attribution (CC BY) license (https://creativecommons.org/licenses/by/4.0/).

1. Introduction

Chang et al. [1] utilized a PMMA plate with two circular piezoelectric ceramics (PCs) as an actuation jet named by the open PAJ for dissipating the heat of electronic devices in the previous study. The results showed that the heat convection effect was greater than that of a traditional rotary fan and the power consumption of a single open PAJ was less than 10% of that of a rotary fan. However, a flapping cantilever crossbeam constitution employing the piezoelectric matter for heat dissipation was first achieved in 1978 by Toda [2–5], in which integrated polyvinylidene fluoride resin 2 (PVF2) as the fan blade. There has been consequential attentiveness forwarded in the polymer of PVDF (PVF2 or polyvinylidene difluoride) based on the best piezoelectric behavior among these trade polymers. PVDF (homo- and co-polymers) is familiarly blended and polymerized between 5 and 160 °C and between 5 and 350 atm. Moreover, piezoelectric fans are thin elastic beams whose vibratory motion is actuated by means of piezoelectric material bonded to the beam. These fans have found use as a means to enhance convective heat transfer while requiring only small amounts of power, which is to quantify the influence of each operational parameter and its relative impact on thermal performance through the vibration frequency and amplitude as well as the geometry of the vibrating cantilever beam [6]. Yorinaga et al. [7] introduced a fan consisting of a piezoelectric bimorph tipped with an additional flexible blade in 1985. The fan case has two air outlets which are directed oppositely each other. The fan with such outlets can supply an airflow of 4 m^3/h at an applied voltage of 140 V_{p-p}, which is suitable for low-voltage operation, and it can supply airflow of 1 m^3/hr at an applied voltage of 17 V_{p-p}. Liu et al. [8] presented an experimental work concerning the thermal performance of piezoelectric fans. A total of six piezoelectric fans with various blade geometries are made and tested. It is found that the heat transfer augmentation of the piezo fan comes from the entrained airflow during each oscillation cycle and the jet-like air stream at the fan tip, yet these two modes are of the same order of magnitude. Based on the dimensionless analysis of the test results for all six fans, a correlation applicable for $x/L = 0$ is proposed. The mean deviation is 4.8%, which can well describe the influence of geometrical parameters.

The disclosure and development of micro-generators during the last decades has expressed a fine harvesting powering solution. It has also created a great background in engineering energy microsystems involving fabrication methods, system concepts, and optimal functionality. A brief representation of the major transduction mechanisms employed, namely the piezoelectric, electrostatic, electromagnetic, and triboelectric harvesting concepts. The mechanical structures used as motion translators include the employment of a proof mass, cantilever beams, the role of resonance, unmorphed structures, and linear/rotational motion translators. In this direction, the evolution of broadband electromechanical oscillators and the combination of environmental harvesting with power transfer operating schemes could unlock a widespread use of micro-generation in microsystems such as micro-sensors and micro-actuators [9]. Dau and Dinh [10] reported the numerical and experimental study of a valveless micro blower actuated by a lead zirconate titanate (PZT) diaphragm. The flow rate of the device of an air generator can be up to 0.7 l/m, and the developed back pressure is 300 Pa. Park et al. [11] conducted to analyze the effects of the freestream and simulated the unsteady flow on the 3-D piezoelectric fan. The counter-rotating vortices from the fan tip are more significantly affected by the freestream than the vortices from the fan side. Finally, two critical values of the Strouhal number were found in this study for determining the performance of a vibrating plate within the freestream. Revathi and Padmanabhan [12] proposed a piezoelectric polymer composite-based micropump that is biocompatible and inexpensive. Lead Zirconate Titanate (PZT) powder is dispersed in polyvinylidene difluoride (PVDF) to form a novel PZT/PVDF composite film material, which was used as the actuator. The piezoelectric polymer-based valveless micropump, thus, fabricated is subjected to performance testing. The micropump is found to deliver fluid in small quantities in a controlled manner, and sufficient back pressure and flow rate can be achieved at low applied voltage and frequency. Gil [13] investigated experimentally the heat transfer enhancement of an air-cooled heat sink using multiple synthetic jet actuators. A correlation of thermal resistance as a function of Reynolds number and dimensionless stroke length based on the experimental results was presented and discussed. The synthetic jet piezoelectric air pump is a potential miniature device for electronic cooling. The convective heat transfer coefficient of the synthetic jet piezoelectric pump is 28.8 W/(m$^2 \cdot °$C), which can prove that the device has a better heat dissipation capability [14].

Yoo et al. [15] fabricated various kinds of piezoelectric fans with lead zirconate titanate (PZT-5) raw material for cooling applications having AC 110 and 220 V, 60 Hz sources. The experiments showed the highest values of displacement and wind velocity, 35.5 mm and 3.1 m/s, respectively, at 220 V and 60 Hz. Zhang et al. [16] investigated that synthetic polymer jets driven by cantilever PZT bimorphs were fabricated and their cooling performance on a heat sink fin tip surface. Geometrical parameters of the synthetic jets, including cavity size, cavity depth, orifice size, orifice length, and diaphragm thickness, were optimized for increased jet velocity and high cooling performance using the Taguchi test method. Based on the test results, a synthetic jet with an optimized structure was fabricated. Measurements showed that the optimized jet could produce a peak air velocity of 50 m/s at 900 Hz from a round orifice 1.0 mm in diameter. The power consumption of the jet in this condition is 0.69 W, and the total mass is 6 g. Using the optimized synthetic jet, a heat transfer coefficient of 576 W/m^{2}K was achieved on the fin tip, indicating an increase of 630% over natural convection values. Wang et al. [17] utilize an acrylic (PMMA) plate with circular piezoelectric ceramics (PCs) as an actuator to design and investigate five different types of piezo actuation jets (PAJs) with operating conditions. The present study greatly improves the constructions of the open PAJ, which employs the circular piezoelectric ceramics (PCs) into the acrylic (PMMA) plates preventing the noise and enhancing the performances of the closed PA as a cooling fan. In addition, only one single circular piezoelectric ceramic (PC) has been applied in the closed PA. In other words, the difference between open PAJ and closed PA was the vibration effect resulting from the position and number of the PC. There are five closed PAs manufactured for various parameters, including the sheet spacing, the size, and the opening area, to set up the performance test methods and conduct cooling experiments on the heat dissipation of high-power LEDs to determine the best closed PA as combining several closed PAs in series. Connecting the closed PAs in series strengthened the overall amount of wind and permitted for the addition or subtraction of closed PAs according to the area of the heat source to effectively control the volume of the closed PA and achieve the best dissipation effect.

2. Research Methods

The closed type of piezo actuator (Closed PA) is composed of a circular piezoelectric sheet and acrylic (PMMA) in the present paper, which the materials of PMMA called ACRYREX® CM-207 with excessive gloss, distinct transparency, and great smoothness were furnished by Chi Mei Corporation (Taiwan, China). The closed-type structural design patterns discussed in this study included the opening size, spacing, and area. Figure 1 exhibits the noise of the closed PA is quieter than that of the open PAJ by about 10 dB. The best performance of the closed PA device was selected through the high-power LED (HI-LED) [18–23] experimental process, which can be combined in series to increase the overall air volume and wind speed, and the number of series can be increased or decreased according to the area of the heat source effectively control the volume of the closed PA device and achieve the best heat dissipation performance. The experimental flow chart is revealed in Figure 2, which involves structural design verified by measuring procedures and properties. These performance measurements contained noise, displacement, and wind speed. A closed PA device was investigated in HI-LED thermal performance module experiments through these parameters and through thermal resistance analysis.

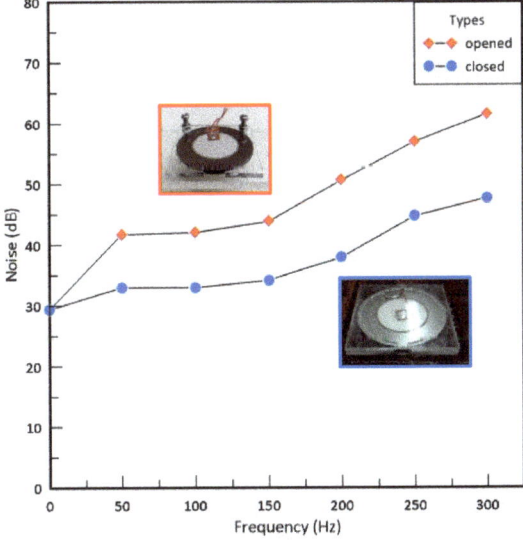

Figure 1. Nosie of PA and PAJ.

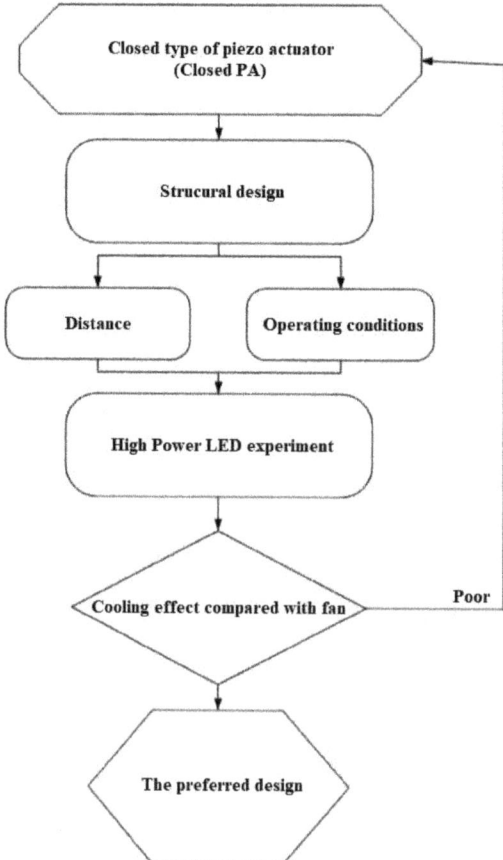

Figure 2. Experimental flow chart.

2.1. Structural Design Patterns

The Ref. [1] of open PAJ verified the diameter of the piezoelectric sheet of 41 mm, which straightly impacted the performance of the device of the closed type of piezo actuator (Closed PA) as a cooling fan. And the present experiment employed the five PMMA structural design patterns to explore the performance of the closed PA device. The base material of a piezoelectric sheet in this experiment was an iron–nickel alloy; the basic properties were Young's modulus of 141 GPa and Poisson's ratio of 0.29. Figure 3 displayed the XRD (X-Ray Diffraction) patterns of the circular piezoelectric sheet of Lead Zirconate Titanate (PZT), which revealed the chemical formula of Pb $(Zr_{0.44}Ti_{0.56})O_3$ as same as the Ref. [1]. The present study utilized the matter of the PMMA with a density of 1.16 g/cm^3, a melting point of 135 °C, a glass temperature of 102 °C, thermal conductivity of 0.23 W/(mk), and a tensile strength of 78 MPa, which were fabricated by the skill of the insert injection molding process of the vapor chamber (VCRHCS) [24–27] as same as the Ref. [1] and the strength can be improved outstandingly and lower the defect of the welding lines.

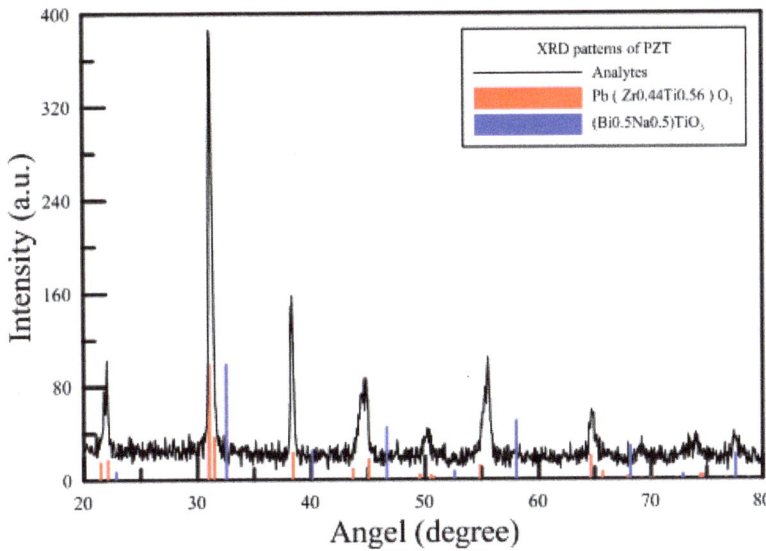

Figure 3. XRD patterns of the present PZT [1].

Table 1 displays the detailed specifications of a closed PA device, including six parameters. Ref. [1] examined the flared jet channel with an opening length of 4 mm and a spacing of 2 mm. Therefore, the opening part of the closed PA was designed with a gradually expanding flow channel. Fix the inside length (L_{inside}) of 4 mm and alter the outside length ($L_{outside}$) from 6 mm, 8 mm, 10 mm, 12 mm, to 15 mm, as shown in Figure 4a. That guided the air volume flow from the cavity to the outside, and the heat dissipation area was increased. All acrylic sheets had 1 mm thickness, which was mainly made in multiple layers for the convenience of subsequent series connection and low cost. The area of the closed PA device was only less than 4% of the current commercial rotary fan. The five tested real closed PA devices are shown in Figure 4b.

Table 1. Specifications of closed PA devices.

Case No.	Diameter (mm)	Cavity Volume (mm^3)	Spacing d_c (mm)	Inside Length L_{inside} (mm)	Outside Length $L_{outside}$ (mm)	Opening Area (mm^2)
Case (1)	41	2268.2	2	4	6	12.5
Case (2)	41	2268.2	2	4	8	15
Case (3)	41	2268.2	2	4	10	17.5
Case (4)	41	2268.2	2	4	12	20
Case (5)	41	2268.2	2	4	15	23.75

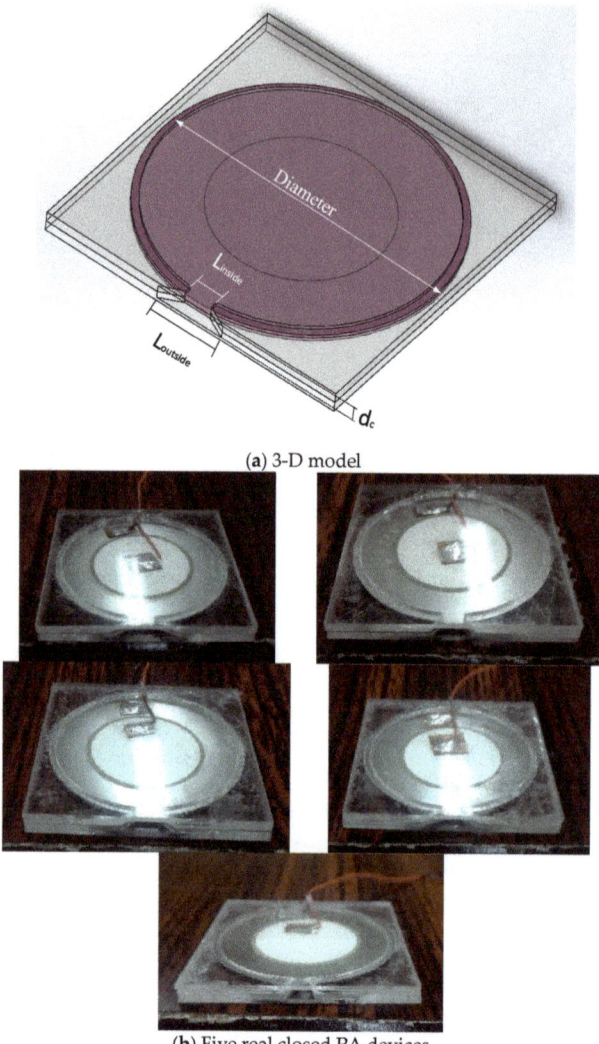

(a) 3-D model

(b) Five real closed PA devices

Figure 4. Schematic diagram of closed PA.

2.1.1. Noise

Figure 5 exhibited the schematic diagram of the noise experiment of the device of the closed type of piezo actuator (Closed PA) as a cooling fan, in which a sound-resistant box made by hand was adopted in this experimentation. A rotary fan of 12 cm evaluated in the consecutive tests of the sound-resistant box was compared with the closed PA device. The sound-resistant box sheet constructed by acrylic has a thickness of 5 mm and a size of $250 \times 250 \times 350$ mm^3, and the surface is covered with wave-shaped soundproof cotton at a thickness of 50 mm. The noise was measured by a DSL-333 decibel meter with a measuring scope between 30 and 130 dB according to the 0.1 dB resolution, where the error is within ± 1.5 dB, and the frequency response is between 30 Hz and 8 kHz. The rotary fan makes a noise value of 62.4 dB outside the sound-resistant box when yet put inside the box with a noise value of 50.7 dB. Equation (1) revealed the volume of sound correction of the background. The background volume of the experimental room was estimated to be

41.9 dB; however, the value was revised to 50.1 dB according to the standard noise control Equation (1) after correction. This demonstrates that the sound resistance box has well soundproofing ability.

$$L_0 = 10\log\left(10^{0.1L_1} - 10^{0.1L_2}\right) \quad (1)$$

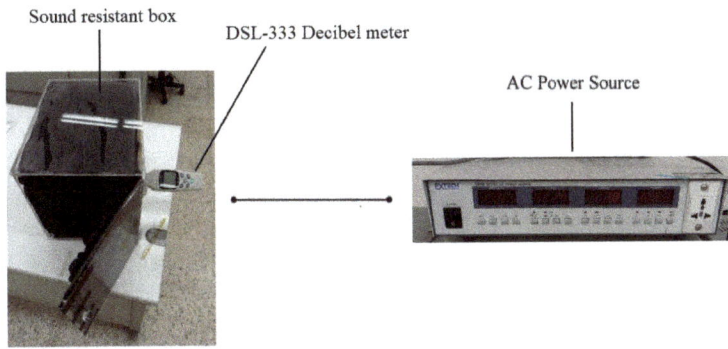

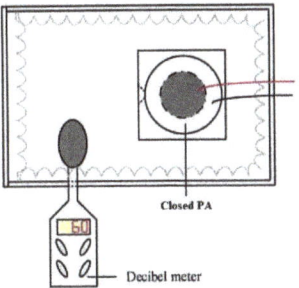

Figure 5. Noise experiment.

L_0: Measured value of the intended sound source;
L_1: Measured value of the total volume of sound;
L_2: Measured value of background volume of sound.

The noise of traditional fans is caused by the wind shearing sound produced by the blade end, the air turbulence, and the bearing rotating friction. The closed PA device uses metal sheets to replace the blades, which can eliminate the need for maintenance and no wind shearing effect. The source of its noise is the sound produced by the vibration of the metal sheet, and the sound intensity depends on the input voltage and frequency. The steps of this experiment are as follows. (1) Confirm that there is no external noise interference and movement of people to control the ambient noise below 45 dB. (2) Insert the decibel meter of DSL-333 into the sound-resistant box and stick soundproofing cotton at the gap to achieve complete sound insulation, and Install the draft shield. (3) Connect the closed PA device to the AC power supply, and the driving voltage and the initial frequency are 30 V and 50 Hz, respectively. (4) The frequency is every 50 Hz as an interval, and the measurement frequency range is between 50 and 450 Hz. (5) Record the measured decibel reading and draw a chart after the experiment.

2.1.2. Displacement

In this experiment, the piezoelectric ceramic (PC) sheet is the actuation source, which transfers electrical energy to the metal sheet and converts it into mechanical energy resulting in vibrating the metal sheet and regularly squeezing the air inside the PMMA cavity to form a jet. The displacement of the PC sheet mainly depends on the voltage and frequency by way of employing a three-leg fixture to clip the closed PA device and sensor head, as revealed in Figure 6. The frequency response and resolution for the EX-305 sensing head are, respectively, 18 kHz and 0.4 μm with operating temperatures between −10 °C and 60 °C. L = 1 mm means the measurement distance between the EX-305 and piezoelectric patches. The degree of interference of the sensor head impedance by the metal sheet and the high-frequency magnetic field caused as high-frequency current were converted into the voltage output. Consequently, the voltage signal was transported to a digital oscilloscope to create a waveform, and the displacement or strain of the metal sheet can be evaluated from the amplitude of the Y-axis of the waveform.

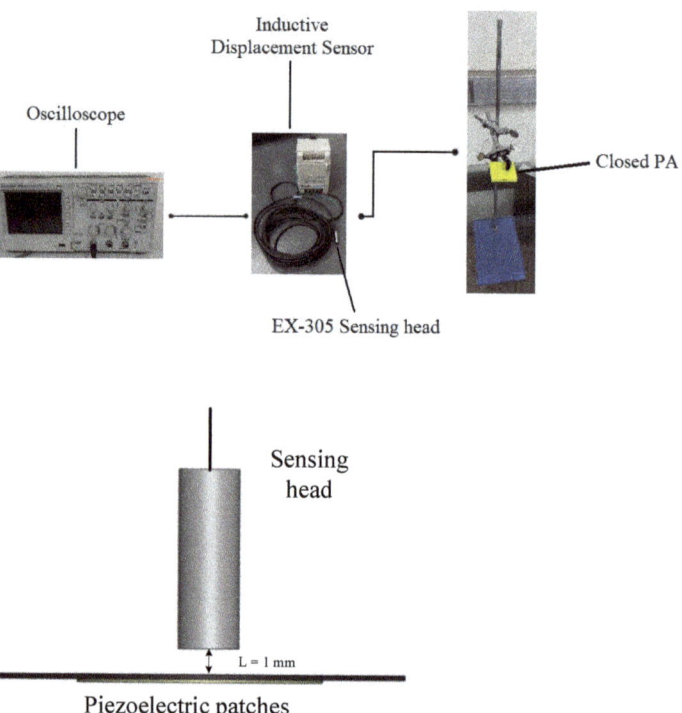

Figure 6. Schematic diagram of the jet path.

The steps of this experiment are as follows. (1) Open a circular hole with a diameter of 5.5 mm in the acrylic below the closed PA device. (2) Connect the closed PA device to the AC power supply with a driving voltage of 30 V and an initial frequency of 50 Hz. (3) Place the sensor head in the circular hole and adjust the measurement distance between them to be 1 mm. (4) Employ a DC power supply to drive the eddy current displacement sensor and connect the output end to a digital wave device. (5) The frequency is every 50 Hz as an interval, and the measurement frequency range is 50~300 Hz. (6) Capture the graph displayed by the digital oscilloscope and convert the voltage value measured by its voltage axis into the displacement. (7) Replace the piezoelectric sheets of different sizes and repeat the above steps.

2.1.3. Wind Speed

The objective of this experiment is to explore the position of the closed PA device from the heat source and find out the best placement between the closed PA and the heat source through this experiment to improve the heat dissipation effect. Generally, the air volume flow rate transmitted by the traditional rotary fan to the heat source is inversely proportional to the placement position, and the wind speed decreases with the increase in the distance. Accordingly, when the distance between them is too close, the flow resistance increases, and the heat dissipation effect may be worse than that of free convection. Thus, it is known that the placement distance is an important parameter that cannot be ignored. In the present experiment, a closed PA device was fixed by a tripod fixture, and the distance D_{HWA} from the hot wire anemometer named TES-1341 was changed at distances of 5, 10, 15, 20, and 25 mm, respectively, to measure the wind speed. The schematic diagram of the wind speed experiment is shown in Figure 7. The measured wind velocity ranges of TES-1341 are from 0 to 30 m/s with a 0.01 m/s resolution and an error within ±3%.

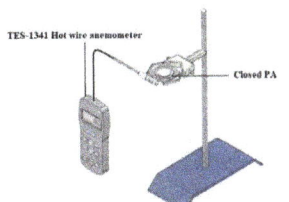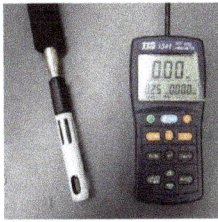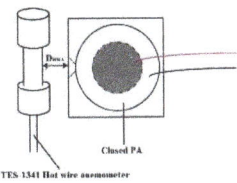

Figure 7. Hot wire anemometer.

The steps of this experiment are as follows. (1) Calibrate the hot wire anemometer based on the instructions in the manual. (2) Place the hot wire anemometer 5 mm in front of the closed PA device. (3) Connect the closed PA device to the AC power supply; the driving voltage is 30 V, and the initial frequency is 50 HZ. (4) The placement position is every 5 mm as an interval, and the farthest measurement is 25 mm. (5) Change the input voltage frequency under the same placement position, and the frequency is every 10 Hz as an interval, and the measurement frequency range is 50 to 450 Hz. (6) Record the average wind speed measured and draw a chart after the experiment.

2.2. Thermal Resistance Network Analysis

The thermal resistance value is mainly used to assess the heat dissipation qualification of the HI-LED package when designing HI-LED heat dissipation. The dimensions and power of this HI-LED are 36 × 34 × 2.6 mm³ and 10 W, respectively. The heat dissipation capability of the closed PA device as a cooling fan can also be used to judge by detecting the level of thermal resistance. That is, the larger the thermal resistance value, the poorer the heat dissipation effect, and, therefore, the overall chip temperature is higher. The smaller the thermal resistance value, the better the closed PA device capability. Equation (2) exhibits the definition of thermal resistance.

$$R_T = \frac{T_j - T_a}{W} \qquad (2)$$

R_T: Total thermal resistance, T_j: Junction temperature, T_a: Ambient temperature, W: Power Dissipation

The thermal performance test methods instructed the cooling experiments in confidence that the closed PA device as a cooling fan could be employed in the heat dissipation of HI-LEDs. Figure 8 displayed this experiment that was two parts. The first part discussed the thermal resistance between natural convection and forced convection for HI-LEDs without fins. The second part was the influence of heat dissipation fins on the same HI-LEDs. The thermal resistance of the present HI-LED can be resolved in two parts based on thermal performance modules, including the thermal resistance of the heat diffusion ($R_{L,1}$) and the thermal resistance of the natural convection ($R_{a,1}$), as revealed in Equation (3). The heat diffusion rate is spreader due to the thermal conductivity of the material because of the greater surface area of the LED substrate than the LED heat source such that the diffusion resistance is generated.

$$R_{L,1} = \frac{T_{L,1} - T_{M,1}}{Q_{in}} \tag{3}$$

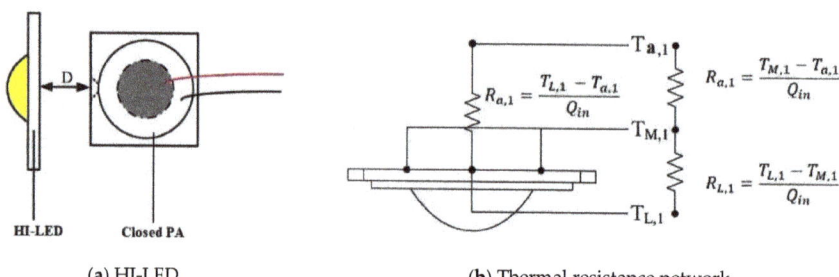

(a) HI-LED

(b) Thermal resistance network

Figure 8. LED Module network.

The thermal resistance $R_{a,1}$ describes that the temperature difference between the average temperature of the substrate interface ($T_{M,1}$) and the ambient temperature ($T_{a,1}$) is divided by the power (Q_{in}) as shown in Equation (4), in which the energy transmission generated through the density difference between the LED substrate temperature and the air is called free convection. A fan can also be mounted to raise the convection effect resulting in a forced convection phenomenon. The thermal resistance of heat convection exhibits the transfer process that heat capacity dispersed into the ambient via convection.

$$R_{a,1} = \frac{T_{M,1} - T_{a,1}}{Q_{in}} \tag{4}$$

To simplify the diagram of the network analysis, $R_{L,1} + R_{a,1} = R_{T,1}$ of the thermal resistance, $R_{T,1}$ is the total thermal resistance, as shown in Equation (5).

$$R_{T,1} = \frac{(T_{L,1} - T_{a,1})}{Q_{in}} \tag{5}$$

3. Results and Discussion

The present work observed the attributes of the device of the closed type of piezo actuator (Closed PA) as a cooling fan regarding the heat dissipation capacity with a high-power LED (HI-LED) module. The high temperature around the HI-LED heat source will be reflowed, resulting in raising the chamber temperature and reducing the cooling efficacy when the closed PA device is too close to the HI-LED. The closed PA device possesses a constricted region to enlarge the cooling area and promote the jet stream wind speed, which can crucially lower the thermal convection resistance. These operating situations, structural designs, and positions impact thermal performances and cooling functions. Therefore, we dominated to reform the operating performance and draw the best design of a closed PA device as a cooling fan.

3.1. Experiment of Performance Measurement

A piezoelectric sheet is a kind of sound-emitting element that transforms electric power to tone, in which its pitch and intensity depend on the voltage and frequency. The voltage was regularized at the maximum rated voltage of 30 V, and the frequency measurement range was 50 to 450 Hz. Simultaneously, the relationship between the voltage, frequency, and displacement of the metal sheet was discussed.

3.1.1. Noise

The present experiment employed a self-made soundproof box, in which the background noise of the lab was 41.9 dB, while inside, the empty soundproof box was 32.8 dB. Figure 9 shows the noise generated by the five cases at 50~450 Hz, in which the growth trends of these five cases were similar. The volume increased sharply after 150 Hz and started to slow down at 350 Hz. The red line segment is a rotating fan. The decibel value measured by the rotating fan in the soundproof box was 51.2 dB, which is similar to the decibel value of the closed PA at 300 Hz. The experimental results showed that the noise generated by the five closed PAs at 300 Hz was similar to that of the rotary fan and was lower than that of the rotary fan when the frequency was 50 Hz to 200 Hz, and the closed PA had no wind shear. If the low-frequency operation was used, the noise could be effectively reduced. After the formula was revised, the noise decibel was shown in Table 2. The results showed that the errors were all within 1.4%, indicating that the self-made soundproof box had a good sound insulation effect and was not disturbed by external background noise. After discussion, the following factors that may affect the noise measurement data and improvement methods were proposed. When the rotary fan rotates, the noise generated by the wind shear effect and bearing friction results in a slight difference from the decibel number marked on the product. The closed PA is fixed on the fixed frame. Due to the acrylic force squeezed to the upper and lower sides, the sound of the welding point on the piezoelectric sheet and the acrylic force colliding with each other can be heard between the fixed jaw and the closed two metal sheets are added to the PA to make the force point even and reduce the noise. There will be some protrusions of the welding wire on the piezoelectric plate, and this protrusion will cause noise when the piezoelectric sheet vibrates and hits the acrylic, so use a thinner wire for welding, and the protrusion will be lower, which can make the probability of impact is reduced, and noise generation is reduced. Coating a layer of resin on the metal sheet can effectively reduce its noise value and can also paste sound-absorbing cotton around it to absorb the sound emitted by the metal sheet.

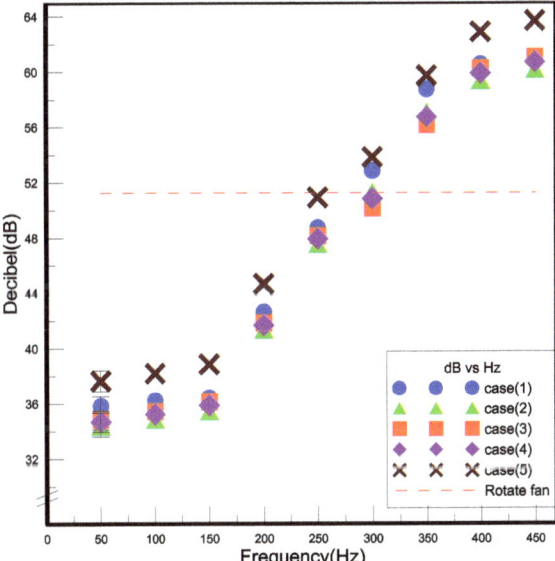

Figure 9. The results of noise measurement.

Table 2. The revised noise decibel.

Object	Before (dB)	After (dB)	Revised Percent (%)
Rotating Fan	51.2	50.6	1.17
case (1)	52.8	52.4	0.76
case (2)	51.3	50.8	0.97
case (3)	50.1	49.4	1.39
case (4)	50.8	50.2	1.18
case (5)	52.1	51.6	0.96

3.1.2. Displacement

The employed adhesives and the attached probes will be moderately not the same in view of hand error as constructing a device of closed PA in this paper. However, the piezoelectric sheets used are all the same, so this part of the experiment mainly finds out the frequency relationship between the piezoelectric sheet deformation and the input voltage. Moreover, it can be deduced from the disclosures that the same types of piezoelectric sheets have similar vibration behaviors. Figure 10 exhibited the relationship between the voltage frequency and the deformation distance of the piezoelectric ceramic metal sheet of the five closed PAs at 50 to 450 Hz. In the data, the displacement of the piezoelectric sheet is slightly different due to the amount of adhesive applied during the production of the closed PA and the human error when fixing the measuring head. The measure of deformation distance increased significantly when the input frequency was above 150 Hz. The amount of displacement and the vibration of the metal sheet is proportional to the input voltage frequency. When the frequency is 100, 250, and 400 Hz, there is a weakening phenomenon in which the natural frequency of the metal sheet does not match the supply frequency, so the metal sheet cannot resonate with the supply frequency.

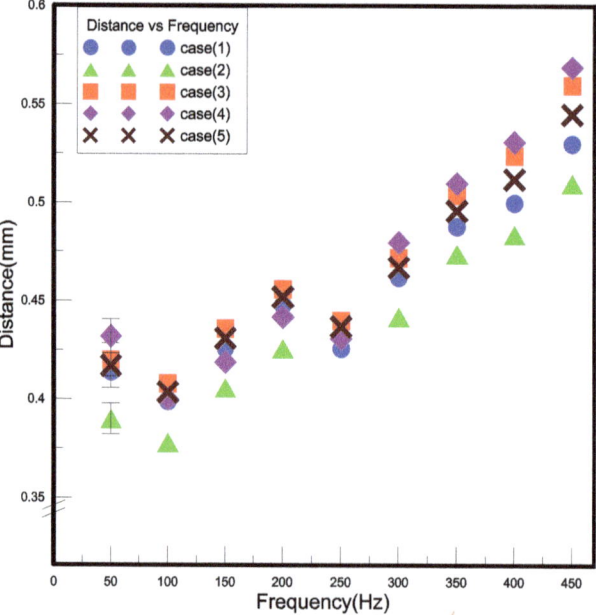

Figure 10. The results of displacement measurement.

3.1.3. Wind Speed

In general, the wind speed generated by the rotary fan and the closed PA is not high, and the distance between the placement position and the heat source is too close, which will make it difficult for the airflow to reach the heat source for heat exchange so that the effect of forced convection cannot be effectively exerted. This section is based on the strength of the wind speed as the basis for judging the heat dissipation effect, and the strength of the wind velocity disturbance can be regarded as an indicator of the turbulence strength. Figure 11a is the wind speed change measured by the rotary fan and the hot wire anemometer at different distances; Figure 11b–f is the closed piezoelectric jet fan with different external openings and the Wind speed measured by a hot wire anemometer at different distances. The measurement results show that the wind speed of the rotary fan and the piezoelectric fan is inversely proportional to the placement position. Among them, when the five cases are at 5 mm, the wind speed is the strongest. However, when the placement position is changed from 5 mm to 10 mm, the wind speed will increase rapidly due to the increase in the distance. However, when the placement position is between 10 mm and 30 mm, the distance of the wind speed decreases gradually, mainly because the eddy current is generated in this area, which drives the nearby air disturbance, so the measured wind speed will not vary greatly due to the increase in the distance; however, when the frequency is increased from 50 Hz to 250 Hz, the wind speed is proportional to the frequency, but when the frequency is increased from 250 Hz to 300 Hz, the wind speed is only slightly increased, indicating that the wind speed generated by the vibration frequency will reach the highest value. Figure 12a–e is represented by the Reynolds number; the Reynolds number of case (1)–case (5) is between 200 and 1800. Among them, in case (1), the static pressure is the highest, but the Reynolds number is the lowest; on the contrary, case (5) has the lowest static pressure and the highest Reynolds number, which shows that the main changing factor of the Reynolds number is the size of the outer opening. The higher the Reynolds number, the shorter the outer opening, and the lower the Reynolds number.

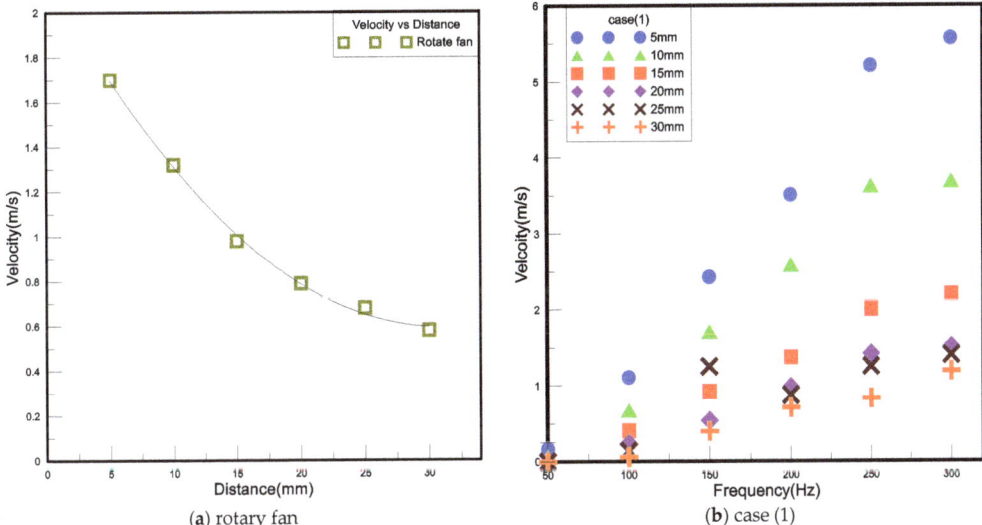

Figure 11. *Cont.*

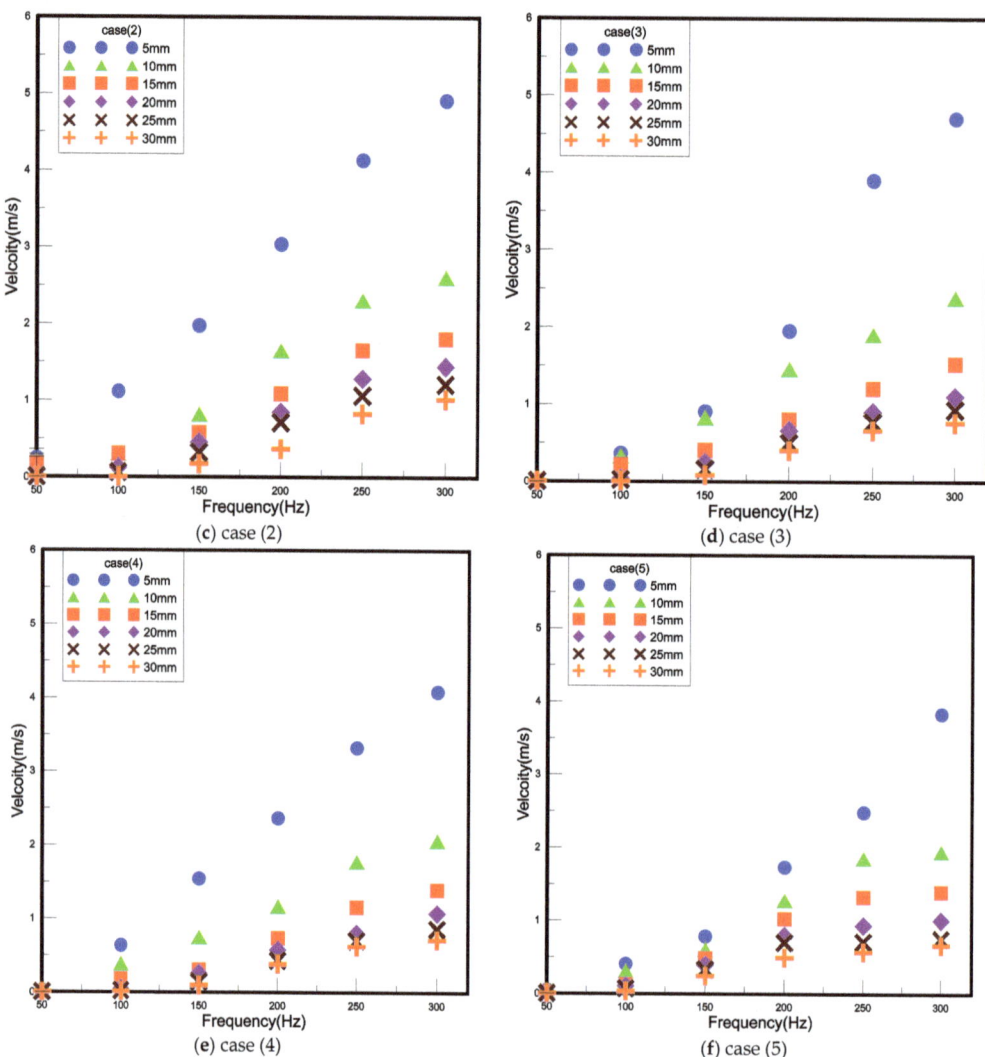

Figure 11. Each device when the wind speed and placement diagrams are under different frequencies.

3.2. LED Thermal Performance Experiment

Taking a single LED as an example, its working temperature is about −40 °C to 80 °C, but the temperature of its light-emitting crystal is about 120 °C. High temperature has a great influence on the service life of the LED, so the temperature must be controlled below 80 °C. It takes about 20 min for the temperature of the LED module to reach a steady state, so the experiment takes every 25 min as the recording time. In this experiment, the supplied wattage is designed to be 1 W, 2 W, and 3 W. When the supply reaches 4 W, the temperature under natural convection exceeds 80 °C, so the supplied wattage will not be further increased. Since the power supply can only adjust the voltage and current, the wattage needs to be calculated and can only be measured with a relatively close wattage. The energy supplied by the LED is not completely converted into light energy. At present, the efficiency of the white LED light is about 30%, and the other 70% is in the form of heat. This part is divided into the finless type and finned type, of which the finless type is further divided into horizontal type, upward blowing type, and downward blowing type, to discuss the heat dissipation situation of the closed PA in different directions and using the traditional rotation fan dissipates heat under the same wattage. The thermal resistance and thermal conductivity are calculated according to the experimental results. The performance of the closed PA and the traditional rotary fan was evaluated by the thermal resistance and thermal conductivity. The closed PAs used in this experiment were operated under the same conditions, the ambient temperature was 23.5 °C, the input voltage was 30 V, the voltage frequency was 300 Hz, and the placement distances were 5 mm, 10 mm, 15 mm, 20 mm, and 25 mm. The change of each thermal resistance judges the effect of each heat dissipation method.

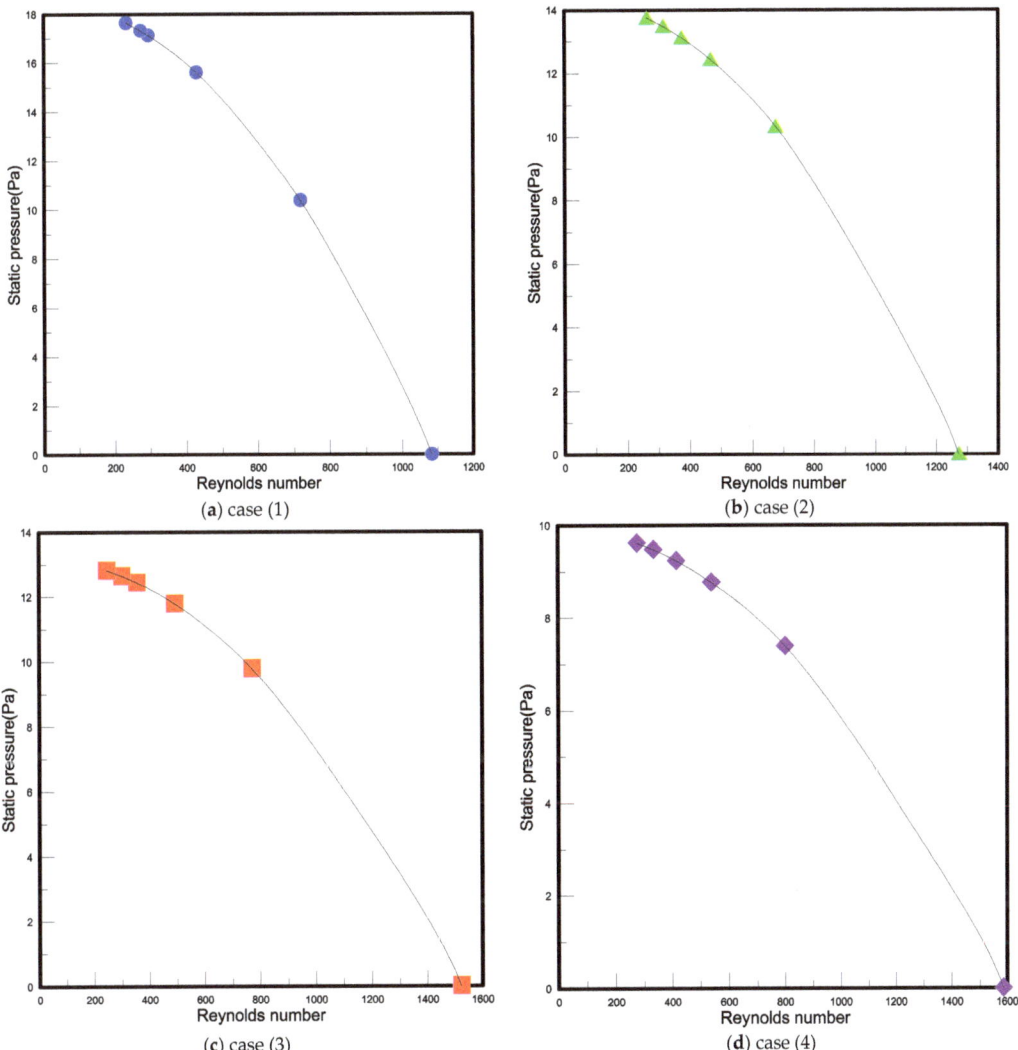

Figure 12. *Cont.*

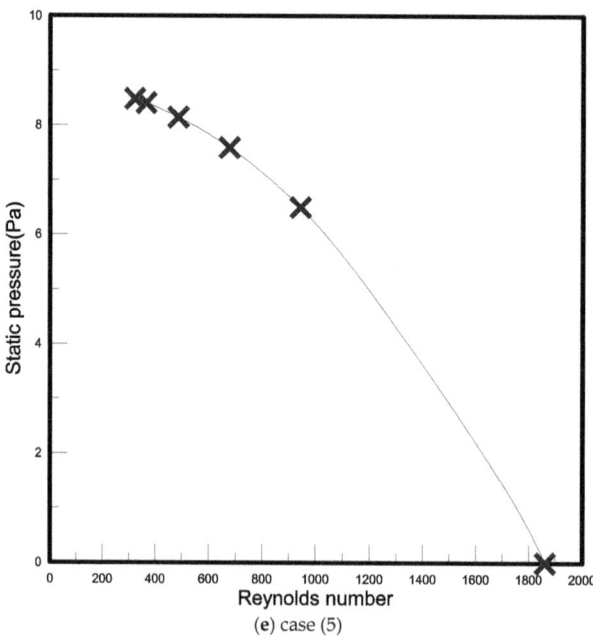

(**e**) case (5)

Figure 12. Each device of Reynolds number.

In the experiment, it was found that when the closed PA is dissipating heat in a steady state, the central temperature is lower than the temperature on both sides. When the thermal resistance analysis is performed by using the diffusion thermal resistance, the calculation will generate negative thermal resistance, so in this part, the temperature of the three points is calculated. Take the average value and the convection thermal resistance calculated from the ambient temperature for analysis. The CH6 point measured under the rotating fan and natural convection is the ambient temperature, and the closed PA is measured at the exit temperature (One near the air outlet and one on the air outlet). 5 mm away from the heat source, the experimental results when the supply wattage is 1 W are shown in Table 3. Figure 13 is a comparison diagram of the thermal resistance of the 1 W placement position of 5 mm. The heat dissipation effect of the traditional rotary fan is better than that of the closed PA. It is excellent. Compared with natural convection, the temperature drops by about 31% to 32%. Although the heat dissipation effect of case (1) is the worst among the five forms, compared with natural convection, the temperature drops by about 9 to 11%; The temperature of case (2) and case (4) drops by about 20–24%; case(3) has the best heat dissipation of the five forms, and the temperature drops by about 22–26%, while case (5) is between case (1) and case (2), the temperature drops about 12–16%. In the experiment, it was found that the heat dissipation area of the five cases will be affected by the size of the opening. The temperature of the center point with a small opening is lower than that of the left and right sides. As the opening becomes larger, the temperature difference will decrease, or even the temperature of the center point will exceed the temperature on both sides. The air volume of (1) is concentrated at the CH2 point in the center, so it cannot have good heat dissipation for the overall LED; case (5) is because the outer opening is too large so that the air volume cannot be concentrated, then case (3) is the best opening size. The closed PA has zero mass flux, and it will be sucked in and discharged at the same time. If it is too close to the heat source, during the vibration process, the fluid around the heat source or the fluid that has been discharged to the outer flow field will inevitably be sucked back into the cavity, which will not only affect the fluid flow during discharge but also bring the fluid with a higher temperature near the heat source back into the cavity. As time increases, the temperature of the fluid in the cavity will increase, which will not be conducive to heat dissipation. When the distance from the heat source is 10 mm, the temperature of the rotary fan after heat dissipation is not much different from the temperature of 5 mm. The closed PA has a slight improvement in the heat dissipation effect, mainly because the measurement distance increases, and the PA is in the vibration process. It can suck in the cold air on both sides of the outlet and push the air to the heat source for heat exchange.

Table 3. Temperatures of LED for free and forced convection under 1 W and 5 mm.

Flow Patterns	Temperatures of LED			
	CH1	CH2	CH3	CH6
Natural Convection	39.6 °C	40.5 °C	39.6 °C	23.5 °C
Rotate fan	27.5 °C	27.8 °C	27.5 °C	23.5 °C
case (1)	36.5 °C	35.9 °C	36.6 °C	24.4 °C
case (2)	31.3 °C	31 °C	31.2 °C	24.3 °C
case (3)	30.6 °C	29.8 °C	30.6 °C	24.3 °C
case (4)	31.0 °C	30.5 °C	30.9 °C	23.9 °C
case (5)	34 °C	34 °C	34 °C	23.8 °C

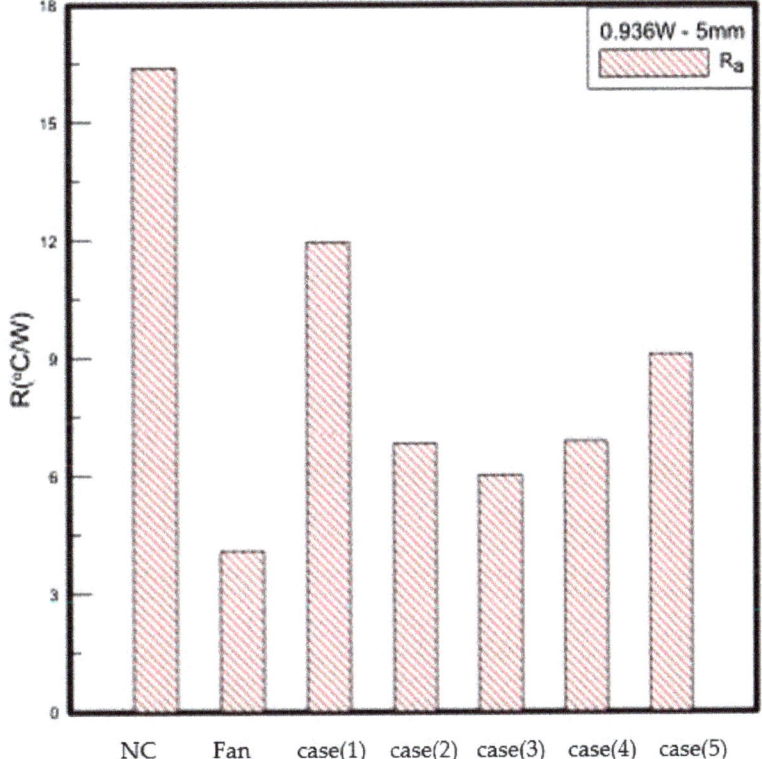

Figure 13. 1 W LED experimental data.

After the LED heat dissipation die is assembled with fins, the supply wattage is increased to 1–5 W. The experimental results under natural convection are shown in Figure 14. The thermal contact resistance (Ri) is the factor between the LED substrate and the heat dissipation fins. The gap makes it difficult for thermal convection, thus, forming thermal resistance. This thermal resistance is not directly related to the wattage supplied and the heat dissipation method. Generally, it can be applied by pressurizing or applying a thermally conductive medium to the gap. The material used in this experiment is a thermal paste, and to reduce to avoid experimental error, try to control the contact thermal resistance within a fixed value, and the contact thermal resistance measured in the experiment is between 0.08–0.15 °C/W. The material thermal resistance (R_M) is the thermal resistance caused by the material and the thickness of the material itself. The experimental results show that the material thermal resistance of the flat fin is about 1.4 °C/W. After adding heat dissipation fins, the LED heat dissipation module does have a cooling effect. The temperature of the LED heat dissipation module with flat heat dissipation fins at 5 W is 61.8 °C, which is higher than the temperature when no fins are installed to provide 3 W. Low, it mainly strengthens the convection effect and reduces the convection thermal resistance (Ra).

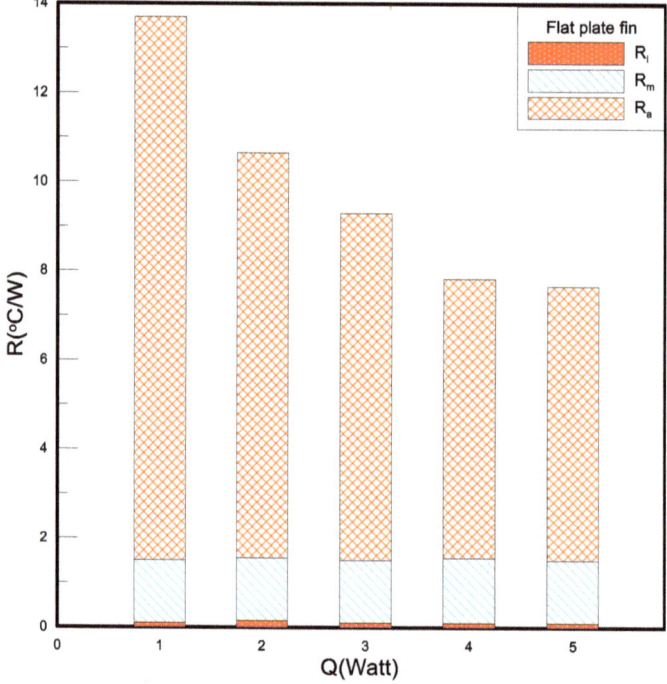

Figure 14. Thermal resistances of nature convection.

4. Conclusions

This paper is devoted to designing a new type of closed PA as a cooling fan and discussing the input conditions. According to the different parameters, such as piezoelectric sheet spacing and opening area, five groups of closed PAs are made. The research results show that the better type design and operating conditions can enable the fan to give full play to its performance, and the heat dissipation effect is better than that of the traditional rotary fan, which is a new way of heat dissipation. The summary is divided into two parts for discussion. The first part is the results of type design and performance analysis, and the second part is the experimental results of LED module heat dissipation. (1) It can be known from the experiment that the wind speed, displacement of the piezoelectric sheet, and noise of the closed piezoelectric jet fan are proportional to the frequency, but considering the noise factor, the frequency is set to 300 Hz for heat dissipation experiments. (2) The displacement of the piezoelectric sheet will change with the supply voltage frequency, which is roughly proportional. When the supply frequency is 100 Hz, 250 Hz, and 400 Hz, the displacement of the piezoelectric sheet will decrease, indicating that the supply frequency Resonance cannot be achieved with the natural frequency of the piezoelectric sheet. (3) The size of the opening will affect the flow rate. A large opening can provide stronger air volume, but the wind speed is low; a small opening can increase the jet airspeed, and the air volume is weak. Case (3) is a form of both air volume and wind speed. (4) The volume of the enclosed piezoelectric jet fan is 8100 mm^2, which is conducive to placing it in electronic products for heat dissipation. (1) When the closed piezoelectric jet fan is placed too close to the heat source, the high-temperature fluid around the heat source will be sucked back, causing the temperature of the fluid in the cavity to rise and reducing the heat dissipation effect. Therefore, the fan should be kept 10–20 degrees from the heat source. The distance of mm reduces the temperature of the fluid sucked into the cavity and achieves a better heat dissipation effect. If the distance is too far, the enclosed piezoelectric jet fan cannot send the air to the heat source, nor can it effectively dissipate heat. (2) According to the experiment, in the heat dissipation experiment without fins, it is found that case (3) has a good heat dissipation capacity. When the placement position is 15 mm, the heat dissipation effect capacity is 120% of that of the traditional rotary fan. (3) After adding heat dissipation fins, the LED can effectively reduce the temperature, which can increase the wattage by 3–4 watts. Coupled with the closed piezoelectric jet fan for heat dissipation, it can be increased by about 2–3 watts. The piezoelectric jet fan will still suck the high-temperature fluid between the fins due to the short distance, which is the same as the situation without fins. (4) When the heat dissipation device is equipped with a single cooling fin or is replaced with a combined piezoelectric jet fan, its optimal placement will be affected. The best placement of the closed piezoelectric jet fan is 15–20 mm, and in the case of adding fins at the same time, the placement position is 30 mm or later. (5) Find out the empirical formula of case (2)–case (5) by dimensional analysis, and its error value is within 28%, which shows the reliability of this empirical formula.

From the comparison results of power consumption, it can be seen that the power consumption of the closed PA in the same reduction of 1 °C is only 10–20% of that of the rotary fan, and the cost is low while achieving low cost and low energy consumption. The principle of the closed PA studied in this paper is similar to that of the cantilever beam piezoelectric fan, but through the difference in structure and material, a new heat dissipation mechanism has been invented, and its heat dissipation efficiency is better than that of the traditional rotary fan in some cases. In terms of power consumption, the closed piezoelectric jet fan is indeed lower than the traditional rotary fan, which can effectively achieve energy saving. In terms of volume, the closed PA is thin and low in cost, and a new invention patent has been applied for.

Author Contributions: Conceptualization, J.-C.W.; data curation, R.-T.W. and J.-C.W.; investigation, J.-C.W. and R.-T.W.; methodology, J.-C.W.; supervision, R.-T.W. All authors have read and agreed to the published version of the manuscript.

Funding: This research received no external funding.

Institutional Review Board Statement: Not applicable.

Informed Consent Statement: Not applicable.

Data Availability Statement: All data are offered by the authors for reasonable request, and the novel device of the piezo actuation jet is available from the authors.

Conflicts of Interest: The authors declare no conflict of interest.

References

1. Chang, Y.T.; Wang, R.T.; Wang, J.C. PMMA Application in Piezo Actuation Jet for Dissipating Heat of Electronic Devices. *Polymers* **2021**, *13*, 2596. [CrossRef] [PubMed]
2. Toda, M. Theory of air flow generation by a resonant type PVF2 bimorph cantilever vibrator. *Ferroelectrics* **1978**, *22*, 911–918. [CrossRef]
3. Toda, M. High field dielectric loss of PVF2 and the electromechanical conversion efficiency of a PVF2 fan. *Ferroelectrics* **1978**, *22*, 919–923. [CrossRef]
4. Toda, M. Elastic properties of piezoelectric PVF2. *J. Appl. Phys.* **1980**, *51*, 4673–4677. [CrossRef]
5. Toda, M.; Osaka, S. Vibrational fan using the piezoelectric polymer PVF 2. *Proc. IEEE* **1979**, *67*, 1171–1173. [CrossRef]
6. Kimber, M.; Garimella, S.V. Measurement and prediction of the cooling characteristics of a generalized vibrating piezoelectric fan. *Int. J. Heat Mass Transf.* **2009**, *52*, 4470–4478. [CrossRef]
7. Yorinaga, M.; Makino, D.; Kawaguchi, K.; Naito, M. A piezoelectric fan using PZT ceramics. *Jpn. J. Appl. Phys.* **1985**, *24*, 203. [CrossRef]
8. Liu, S.F.; Huang, R.T.; Sheu, W.J.; Wang, C.C. Heat transfer by a piezoelectric fan on a flat surface subject to the influence of horizontal/vertical arrangement. *Int. J. Heat Mass Transf.* **2009**, *52*, 2565–2570. [CrossRef]
9. Kiziroglou, M.E.; Yeatman, E.M. Micromechanics for energy generation. *J. Micromech. Microeng.* **2021**, *31*, 114003. [CrossRef]
10. Dau, V.T.; Dinh, T.X. Numerical study and experimental validation of a valveless piezoelectric air blower for fluidic applications. *Sens. Actuators B Chem.* **2015**, *221*, 1077–1083. [CrossRef]
11. Park, S.H.; Oh, M.H.; Kim, Y.H.; Choi, M. Effects of freestream on piezoelectric fan performance. *J. Fluids Struct.* **2019**, *87*, 302–318. [CrossRef]
12. Revathi, S.; Padmanabhan, R. Fabrication and Testing of PDMS Micropump Actuated by Piezoelectric Polymer Composite. *ECS Trans.* **2022**, *107*, 1811.
13. Gil, P. Experimental investigation on heat transfer enhancement of air-cooled heat sink using multiple synthetic jets. *Int. J. Therm. Sci.* **2021**, *166*, 106949. [CrossRef]
14. Li, X.; Liu, X.; Dong, L.; Sun, X.; Tang, H.; Liu, G. A High-Performance Synthetic Jet Piezoelectric Air Pump with Petal-Shaped Channel. *Sensors* **2022**, *22*, 3227. [CrossRef] [PubMed]
15. Yoo, J.H.; Hong, J.I.; Park, C.Y. Characteristics of piezoelectric fans using PZT ceramics. In Proceedings of the 5th International Conference on Properties and Applications of Dielectric Materials, Seoul, Republic of Korea, 25–30 May 1997; Volume 2, pp. 1075–1081.
16. Zhang, M.; Simon, T.W.; Huang, L.; Selvi VB, A.; North, M.T.; Cui, T. A polymeric piezoelectric synthetic jet for electronic cooling. In Proceedings of the ASME 2011 International Mechanical Engineering Congress and Exposition, Denver, CO, USA, 11–17 November 2011; Volume 54969, pp. 235–239.
17. Wang, R.T.; Wang, J.C.; Chen, S.L. Design and Testing of a Bearing less Piezo Jet Micro Heat Sink. *J. Mar. Sci. Technol.* **2022**, *30*, 180–191. [CrossRef]
18. Wang, R.T.; Wang, J.C. Optimization of heat flow analysis for exceeding hundred watts in HI-LEDs projectors. *Int. Commun. Heat Mass Transf.* **2015**, *67*, 153–162. [CrossRef]
19. Wang, J.C.; Wang, R.T.; Chang, T.L.; Hwang, D.S. Development of 30 Watt high-power LEDs vapor chamber-based plate. *Int. J. Heat Mass Transf.* **2010**, *53*, 3990–4001. [CrossRef]
20. Wang, J.C. Thermoelectric transformation and illuminative performance analysis of a novel LED-MGVC device. *Int. Commun. Heat Mass Transf.* **2013**, *48*, 80–85. [CrossRef]
21. Tucker, R.; Khatamifar, M.; Lin, W.; McDonald, K. Experimental investigation of orientation and geometry effect on additive manufactured aluminium LED heat sinks under natural convection. *Therm. Sci. Eng. Prog.* **2021**, *23*, 100918. [CrossRef]
22. Wang, J.C. Thermal investigations on LED vapor chamber-based plates. *Int. Commun. Heat Mass Transf.* **2011**, *38*, 1206–1212. [CrossRef]
23. Wang, R.T.; Wang, J.C. Analysis of thermal conductivity in HI-LEDs lighting materials. *J. Mech. Sci. Technol.* **2017**, *31*, 2911–2921. [CrossRef]
24. Wang, J.-C.; Li, A.-T.; Tsai, Y.-P.; Hsu, R.-Q. Analysis for Diving Regulator Applying Local Heating Mechanism of Vapor Chamber in Insert Molding Process. *Int. Commun. Heat Mass Transf.* **2011**, *38*, 179–183. [CrossRef]
25. Wang, R.-T.; Wang, J.-C.; Chen, S.-L. Investigations on Temperatures of the Flat Insert Mold Cavity using VCRHCS with CFD Simulation. *Polymers* **2022**, *14*, 3181. [CrossRef] [PubMed]
26. Wang, J.-C.; Tsai, Y.-P. Analysis for Diving Regulator of Manufacturing Process. *Adv. Mater. Res.* **2011**, *213*, 68–72. [CrossRef]
27. Tsai, Y.-P.; Wang, J.-C.; Hsu, R.-Q. The Effect of Vapor Chamber in an Injection Molding Process on Part Tensile Strength. *Exp. Tech.* **2011**, *35*, 60–64. [CrossRef]

Disclaimer/Publisher's Note: The statements, opinions and data contained in all publications are solely those of the individual author(s) and contributor(s) and not of MDPI and/or the editor(s). MDPI and/or the editor(s) disclaim responsibility for any injury to people or property resulting from any ideas, methods, instructions or products referred to in the content.

Article

Swelling and Collapse of Cylindrical Polyelectrolyte Microgels

Ivan V. Portnov [1,2], Alexandra A. Larina [1], Rustam A. Gumerov [1] and Igor I. Potemkin [1,3,*]

1. Physics Department, Lomonosov Moscow State University, 119991 Moscow, Russia
2. A. N. Nesmeyanov Institute of Organoelement Compounds, Russian Academy of Sciences, 119991 Moscow, Russia
3. National Research South Ural State University, 454080 Chelyabinsk, Russia
* Correspondence: igor@polly.phys.msu.ru

Abstract: In this study, we propose computer simulations of charged cylindrical microgels. The effects of cross-linking density, aspect ratio, and fraction of charged groups on the microgel swelling and collapse with a variation in the solvent quality were studied. The results were compared with those obtained for equivalent neutral cylindrical microgels. The study demonstrated that microgels' degree of swelling strongly depends on the fraction of charged groups. Polyelectrolyte microgels under adequate solvent conditions are characterized by a larger length and thickness than their neutral analogues: the higher the fraction of charged groups, the longer their length and greater their thickness. Microgels' collapse upon solvent quality decline is characterized by a decrease in length and non-monotonous behavior of its thickness. First, the thickness decreases due to the attraction of monomer units (beads) upon collapse. The further thickness increase is related to the surface tension, which tends to reduce the anisotropy of collapsed objects (the minimum surface energy is known to be achieved for the spherical objects). This reduction is opposed by the network elasticity. The microgels with a low cross-linking density and/or a low enough aspect ratio reveal a cylinder-to-sphere collapse. Otherwise, the cylindrical shape is preserved in the course of the collapse. Aspect ratio as a function of the solvent quality (interaction parameter) demonstrates the maximum, which is solely due to the electrostatics. Finally, we plotted radial concentration profiles for network segments, their charged groups, and counterions.

Keywords: cylindrical microgels; collapse; electrostatics; computer simulations

Citation: Portnov, I.V.; Larina, A.A.; Gumerov, R.A.; Potemkin, I.I. Swelling and Collapse of Cylindrical Polyelectrolyte Microgels. *Polymers* **2022**, *14*, 5031. https://doi.org/10.3390/polym14225031

Academic Editor: Jung-Chang Wang

Received: 21 October 2022
Accepted: 14 November 2022
Published: 20 November 2022

Publisher's Note: MDPI stays neutral with regard to jurisdictional claims in published maps and institutional affiliations.

Copyright: © 2022 by the authors. Licensee MDPI, Basel, Switzerland. This article is an open access article distributed under the terms and conditions of the Creative Commons Attribution (CC BY) license (https://creativecommons.org/licenses/by/4.0/).

1. Introduction

Polymer microgels are soft, porous, colloidally stable macromolecular objects that reveal the properties of polymers, (nano-)microparticles, and surfactants [1]. They have a network-like internal structure and their size ranges between tens of nanometers and tens of microns (the upper limit is determined by stability towards the precipitation of the single macromolecules). Such as for hydrogels, the most spectacular property of the microgels is their ability to drastically swell and collapse under external stimuli, e.g., temperature [2], pH [3], magnetic field [4], etc., and significantly change their size, porosity, and characteristics of their interaction with each other (from repulsion in the swollen state to attraction in the collapsed one). However, in contrast to the macroscopic gels, the microgels' stimuli response is considerably faster and makes them very promising for many applications. In particular, they can be used as carriers for guest molecules, which can be released on demand [5,6]; as porous and functional alternatives to solid particles for emulsion stabilization [7,8]; in catalysis [9,10]; as scavengers [11]; in membrane technologies [12]; in tissue engineering [13]; and many others.

The common methods of microgel synthesis include precipitation [14], (mini)emulsion and [15] template [16] polymerization, as well as microfluidics-based synthesis. [17] In the case of precipitation and (mini)emulsion polymerization the obtained microgels are spherical in shape due to the presence of surface tension, the minimum of which is only achieved in the spherical geometry. Indeed, the growing chains in precipitation polymerization are surrounded by a poor quality solvent while microgels in (mini)emulsion polymerization reproduce the shape of the droplets that also have surface tension and a spherical shape [18]. On the contrary, microfluidics provides control of the microgel shape, varying it from spherical to cylindrical, and the latter's aspect ratio is controlled by droplet volume with respect to the diameter of the channel [19]. More sophisticated shapes, including an anisotropic core-shell or hollow [20–22], ribbon-like microgels, can be obtained via template polymerization [23]. However, a pretty large size in some cases could be considered a disadvantage of this technique. The design of nanoscale anisotropic microgels is still challenging.

Given that the properties of the spherical microgels are studied well enough [1], much less is known about anisotropic microgels [21,22]. In particular, it was recently found that anisotropic microgels adsorbed at oil–water and air–water interfaces upon compression demonstrate self-assembly into liquid-crystalline and more sophisticated structures, caused by excluded volume and capillary forces [22]. The incorporation of magnetic nanoparticles into anisotropic microgels can lead to monovalent cationic charged groups of the microgel being randomly distributed over the network. The same amount of counterions is added to the system to provide a macroscopic electric neutrality of the system. The fraction of charged groups f was varied between 0 and 50%. Note that such high charge content can be obtained in reality, for instance, via their previous orientation in the magnetic field, forming a strongly anisotropic medium, which is a perspective for directed cell growth [13,24]. Liquid-crystalline ordering in solutions of the cylindrical microgels is predicted for their high enough concentrations depending on their aspect ratio and cross-linking density [25]. The swelling and collapse of neutral cylindrical microgels demonstrates peculiar behavior. Depending on microgels' cross-linking density, aspect ratio, and molecular weight, one can observe either self-similar (cylinder-to-cylinder) or non-self-similar collapse, leading to cylinder-to-sphere transformation [25]. Therefore, in addition to microgel size and porosity, their aspect ratio in the solution and self-assembled structures can be finely tuned by external stimuli.

In this work, we continued the study of the swelling and collapse of the cylindrical microgels with computer simulations. The main aim was to reveal the effect of electrostatics on the swelling and collapse of the single microgels. Both effects of cross-linking density and fraction of charged groups were investigated. We demonstrate how microgels' length and thickness (as well as aspect ratio) change upon solvent quality decline (temperature increase for thermoresponsive gels). A comparison with equivalent neutral microgels is provided here. The radial concentration profiles of monomer units and charged groups were analyzed. The effect of the fraction of charged groups on the distribution of counterions are demonstrated here.

2. Model and Simulation Methods

We have used a conventional Brownian dynamics technique of simulations within a coarse-grained model and with an implicit solvent. All structural units of the microgel (charged and neutral monomer units) and counterions are modeled as beads of equal radius σ and mass m, interacting with each other through the truncated Lennard-Jones potential [26]:

$$U_{LJ} = 4\varepsilon\left[(\sigma/r_{ij})^{12} - (\sigma/r_{ij})^{6}\right], \quad r_{ij} \leq r_{cut}, \quad (1)$$

where r_{ij} is the distance between two interacting beads and r_{cut} is a cut-off radius beyond which the potential is equal to zero. The cut-off radii for interaction of the monomer units with each other and counterions with the monomer units and each other were selected as $r_{cut} = 2.5\ \sigma$ and $r_{cut} = 2^{1/6}\ \sigma$, respectively. The solvent quality is determined by the interaction parameter ε of the potential acting between microgel beads. The increase in this parameter corresponds to the solvent quality decline. In our simulations, ε varies between 0.01 k_BT (favorable solvent quality) and 1.4 k_BT (poor solvent quality) [27]. A similar parameter describing interactions of the counterions with each other and the monomer units was fixed to the value 1 k_BT in all simulations. Electrostatic interactions between any pair of the charged species were described by Coulomb potential. The Bjerrum length was chosen as $l_B = 1\ \sigma$ and corresponds to the aqueous solutions [28,29].

The cylindrical microgels were designed as follows. Fully stretched subchains of an ideal microgel (all subchains have equal length) were connected through tetrafunctional cross-links in such a way that repeats a unit cell of the diamond crystal lattice. Then, a parallelepiped microgel template consisting of $10 \times 10 \times 100$ modified unit cells, mentioned above, was constructed. The subchain length M was considered to be 10 and 20 corresponding to approximately 5% and 2.5% of cross-links, respectively. To provide a cylindrical shape to the microgel particles, a cylinder was inscribed into the template in such a way that the cylinder axis passes through a set of the cross-links belonging to a straight. Then, all beads outside the cylinder were "cut off" (Figure 1). The cylinder is characterized by the initial aspect ratio A, i.e., the ratio of the length L_0 to its diameter $2R_0$, and $A = L_0/2R_0$ under preparation conditions. Annealing of the microgel and a change in the solvent quality can lead to the change in the aspect ratio. Two aspect ratio values, $A = 4$ and 10, were considered. The number of beads (the molecular weight) in the different microgels N was fixed at 50 and 100 thousand.

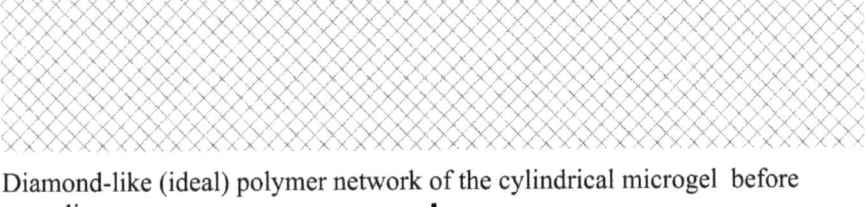

Diamond-like (ideal) polymer network of the cylindrical microgel before annealing.

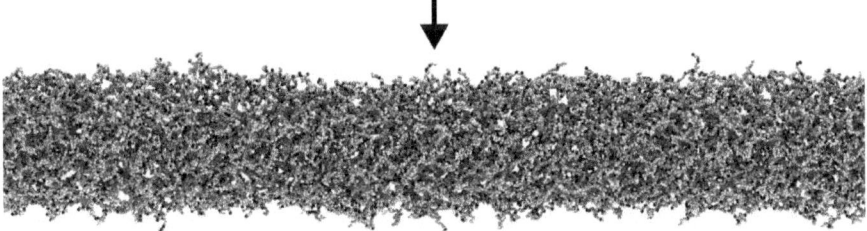

Equilibrated microgel under good solvent conditions. Random distribution of cationic charged groups.

 $\varepsilon = [0.01 \ldots 1.4]k_BT$

Figure 1. Internal structure (thin layer near the axis) of the cylindrical microgel before and after annealing. The charged groups (red beads) are randomly distributed over the network. Interactions between network beads are quantified by the interaction parameter ε.

Monovalent cationic charged groups of the microgel are randomly distributed over the network (Figure 1). The same amount of counterions is added to the system to provide a macroscopic electric neutrality of the system. The fraction of charged groups f was varied between 0 and 50%. Note that such high charge content can be obtained in a reality, for instance, via the post modification of network monomers [30]. adjacent bonded beads of a chain are connected by a finitely extensively, nonlinear, elastic (FENE) spring potential: [31]

$$U_{FENE} = -\frac{k}{2}R_0^2 \ln\left[1 - \left(\frac{r}{R_0}\right)^2\right], \, r < R_0 \qquad (2)$$

with parameters $k = 30 \, k_B T/\sigma^2$ and $R_0 = 1.5 \, \sigma$. The simulations were performed using the open-source software LAMMPS [32] in NVT ensemble with imposed periodic boundary conditions; the linear sizes of the simulation box were $L_y = L_z = 500 \, \sigma$ and $L_z = 1500 \, \sigma$. The electrostatic interactions between charged beads were calculated using the PPPM algorithm [33] with an accuracy of 10^{-5}. The equations of motion were integrated with a time step of $\Delta t = 0.005 \, \tau$, where τ is the standard time unit for a Lennard-Jones fluid. Initially, single microgels with fully stretched chains and counterions uniformly distributed throughout the box were annealed (equilibrated) for 8×10^6 steps under adequate solvent conditions, $\varepsilon = 0.01 \, kT$. The statistics were gathered within additional 2×10^6 steps. Then, the interaction parameter was increased until the value $\varepsilon = 1.4 \, kT$ was reached. For each case of solvent quality, the amount of simulation time (equilibration + statistics) was the same.

3. Results and Discussion

First, let us demonstrate how the presence of charges affects the swelling behavior of the cylindrical microgels. In our research, interaction parameter ε describes the attraction between microgel segments and quantifies solvent quality: the larger the parameter, the poorer the solvent quality. Figure 2a,b shows contour length L and thickness $2R$ of single charged and neutral microgels as functions of ε. The contour length is determined as the sum of distances between neighbor cross-links, which the cylinder axis passes through under the preparation conditions [25]. It is nearly equal to the apparent length of the cylindrical microgels. In turn, the thickness (the diameter of a cylinder) is determined geometrically from a slab of a network cut from its central part (determined from a center of mass) with a length equal to 20% of L. Figure 2a clearly demonstrates that the presence of charged groups leads to a significant elongation of the microgel as compared to the neutral one. This effect is due to the osmotic pressure of those counterions, which are localized within the microgel, and electrostatic repulsion of similarly charged groups of the network. In turn, the latter is due to the local violation of the electric neutrality of the microgel caused by the partial release of the counterions [27,34,35] (see below). Both neutral and charged microgels shorten upon the worsening of the solvent quality. Such a behavior is caused by the attraction of neutral beads, leading to the microgel collapse. However, a charged microgel always has a longer length than a neutral one under the same value of the parameter ε (the same solvent quality). Equal lengths of the neutral and charged microgels can be achieved upon the stronger attraction of the beads in the charged microgel to overcome the exerted osmotic pressure of counterions and electrostatic repulsion. At the same time, the more interesting behavior is found for microgel thickness (Figure 2b). Under favorable solvent conditions, $\varepsilon = 0.01 \, kT$, the polyelectrolyte network is thicker than the neutral one and, keeping in mind the same ratio for the length, we can say that the charged microgel is more swollen than the equivalent neutral one. Such an effect is well-known for spherical microgels [36,37] and macroscopic gels [38]. Initially, the thickness decreases with an increase in ε for both neutral and charged networks (Figure 2b). This decrease is caused by microgel collapse, leading to the increase in polymer concentration within the networks (see below). However, a further increase in ε is responsible for microgel thickening (in both cases (Figure 2b)), which is also visible in the snapshots in Figure 3. This behavior is a feature of a finite-sized anisotropic object and is caused by the presence of surface tension. Indeed, if microgels did not have elasticity, an increase in surface tension (solvent quality decline) would lead to the transformation of the cylinder to the sphere because the latter has a smaller surface area than the former (under fixed volume). Therefore, in the presence of microgel elasticity, the cylinder shortening and thickening at high ε values is driven by surface tension. In contrast to the $L(\varepsilon)$ dependences (Figure 2a), $2R(\varepsilon)$ curves cross each other (Figure 2b). This means that, at a certain solvent quality, the thickness of the charged microgel can be less than that of the neutral one. This effect is a consequence of the anisotropy of the object and long-range electrostatic repulsion between the network's charged groups. An increase in polymer concentration within the charged microgel upon collapse proceeds via microgel shortening and thinning. However, thinning is electrostatically more favorable than shortening. Indeed, the electrostatic energy of a cylindrical object is proportional to Q^2/L, apart from a logarithmic factor [39]. Here, Q and L are a charge and a length of a cylinder, respectively. Therefore, the electrostatic energy decreases upon elongation (such as in the case of linear polyelectrolyte chains in dilute solution) [40]. That is why microgel length can less progressively decrease than thickness upon collapse. This behavior is detectable if we plot the aspect ratio function (Figure 2c). The $A(\varepsilon)$ dependence is non-monotonous for polyelectrolyte microgels. The initial increase in A is caused by electrostatics: the surface tension is too low to overcome the favorable electrostatic energy for an elongated object. At higher ε values, the aspect ratio decreases, meaning that surface tension dominates electrostatics (Figure 2c). Meanwhile, for the neutral microgel, A monotonously decreases with ε (Figure 2c).

Despite the difference in collapse behavior, the considered neutral and charged microgels are common in a sense of so-called "rod-to-rod" collapse: the final (under poor quality solvent conditions) values of the aspect ratio $A > 4$ (Figure 2c). On the contrary, a near "rod-to-sphere" transition is observed for both microgels if we take a shorter and smaller cylinder. Figure 4 demonstrates the transition for the microgels with the initial (under preparation condition) aspect ratio $A = 4$, molecular weight $N = 50,000$, and the same subchain length $M = 10$. For convenience, the A parameter is determined from the eigenvalues of a gyration tensor [41]. While for longer cylinders, this approach produces similar values as from the ratio of contour length and diameter, for shorter and smaller cylinders, it allows for more apparent tracking size changes.

Under adequate solvent conditions, $\varepsilon = 0.01\ kT$, both microgels are swollen, revealing a cylindrical shape, and the charged sample has a longer length and greater thickness (Figure 4c). A collapse of the neutral microgel is characterized by a more progressive shortening with respect to the thinning: the aspect ratio monotonously decreases with ε (Figure 4a). Such as for longer counterparts, the charged microgel first decreases thickness upon a more progressive collapse, which corresponds to the aspect ratio increase (Figure 4a). This effect also occurs due to the electrostatics, which stabilizes the elongated shape. However, the further solvent quality decline (an increase in ε) leads to the formation of a nearly spherical microgel. Furthermore, conformations of neutral and charged microgels are nearly equal at highest considered values of ε (Figure 4c). The aspect ratio curves for different microgels converge at these values as well. An asphericity parameter [41] shown in Figure 4b demonstrates the evolution of the microgels' shape: cylindrical and spherical shapes are quantified by the values > 0.6 and < 0.1, respectively.

The non-monotonous dependence of the aspect ratio on the solvent quality is explained by the presence of the uncompensated charge of the microgel. To prove this hypothesis, we plot a dependence of the fraction of counterions β, which are localized within the microgel, on the fraction of charged groups f in the network (Figure 5). It can be seen that the fraction of localized counterions depends on the fraction of the charged groups: the higher the f value, the larger the fraction of localized counterions. However, it does not mean that more charged microgels have lower net charge. On the contrary, the microgel net charge fraction increases with f due to the convex shape of the β-f dependence, $\beta \sim f^\alpha$, $\alpha < 1$. Indeed, if the microgel total net charge is introduced as Q, its ratio to the total number of beads (relative net charge) in the network N would take the form $Q/N = ef(1-\beta)$, where e is the elementary charge. Thus, for the 10%, 20%, and 40% microgels, the relative net charge grows as $Q/Ne \approx 1.8\%, 2.5\%, 3.6\%$, respectively (Figure 5).

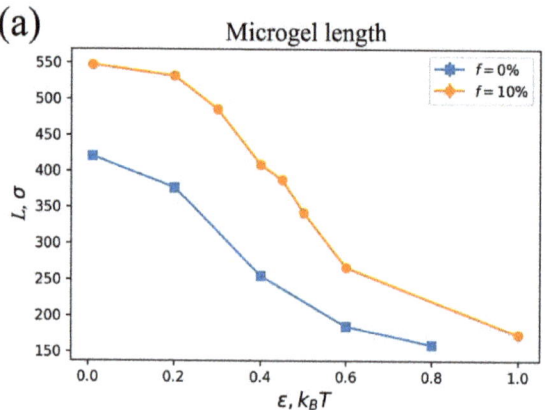

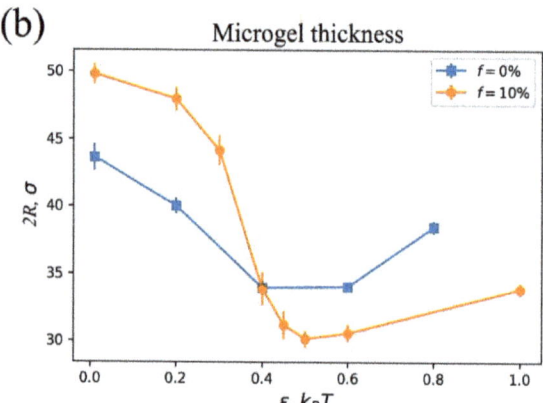

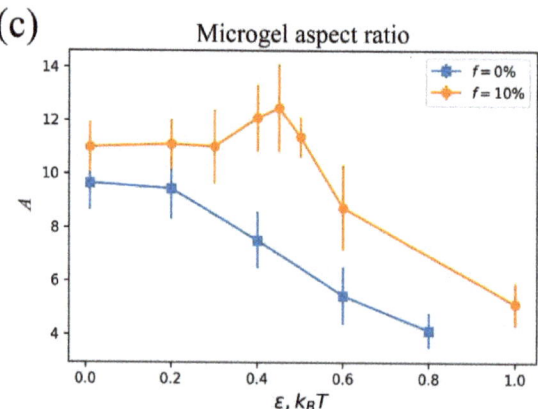

Figure 2. Contour length L (**a**), thickness $2R$ (**b**), and aspect ratio $A = L/2R$ (**c**) of equivalent neutral ($f = 0\%$) and charged ($f = 10\%$) microgels as functions of the interaction parameter ε. The molecular weight and subchain length of the microgels are $N = 100{,}000$ and $M = 10$, respectively. The initial aspect ratio of the microgels is $A = 10$.

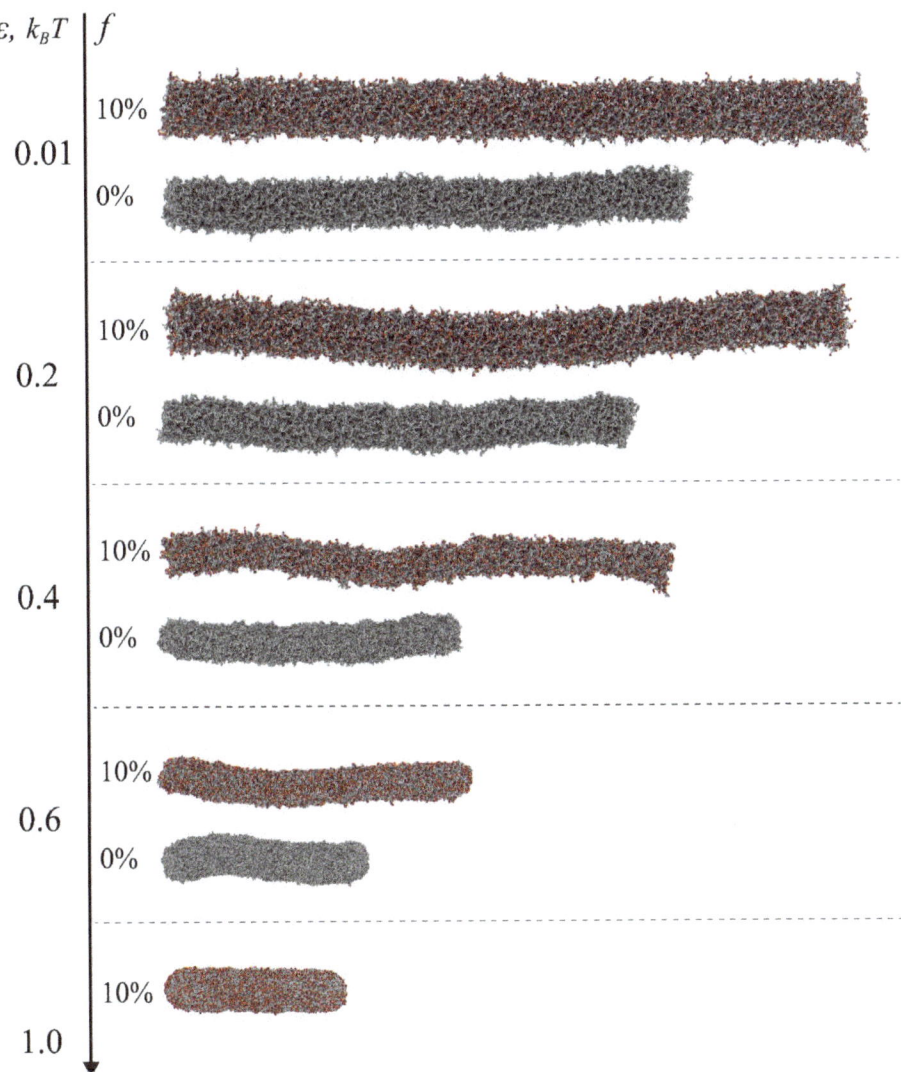

Figure 3. Side-view snapshots of equivalent neutral (f = 0%) and charged (f = 10%) microgels upon solvent quality decline. The charged groups were slightly enlarged for clarity. The molecular weight and subchain length of the microgel are N = 100,000 and M = 10, respectively. The initial aspect ratio of the microgels is A = 10.

The effect of the fraction of charged groups in the network on microgel collapse is demonstrated in Figure 6. The microgel degree of swelling under favorable solvent conditions strongly depends on the fraction of charged groups. Both the microgel contour length and thickness increase with f (Figure 6a,b, respectively). This is mainly due to the increase in the number of counterions, which creates exerted osmotic pressure. Additionally, for this reason, stronger attraction between monomer units is required to induce microgel collapse: the transition to the collapsed state shifts towards higher ε values (Figure 6). We can see that the collapse becomes more abrupt upon the increase in the fraction of charged groups. This effect is an indication of the polyelectrolyte nature: the collapse of macroscopic gels (a limiting case of an infinitely sized network) proceeds as a first order phase transition [42].

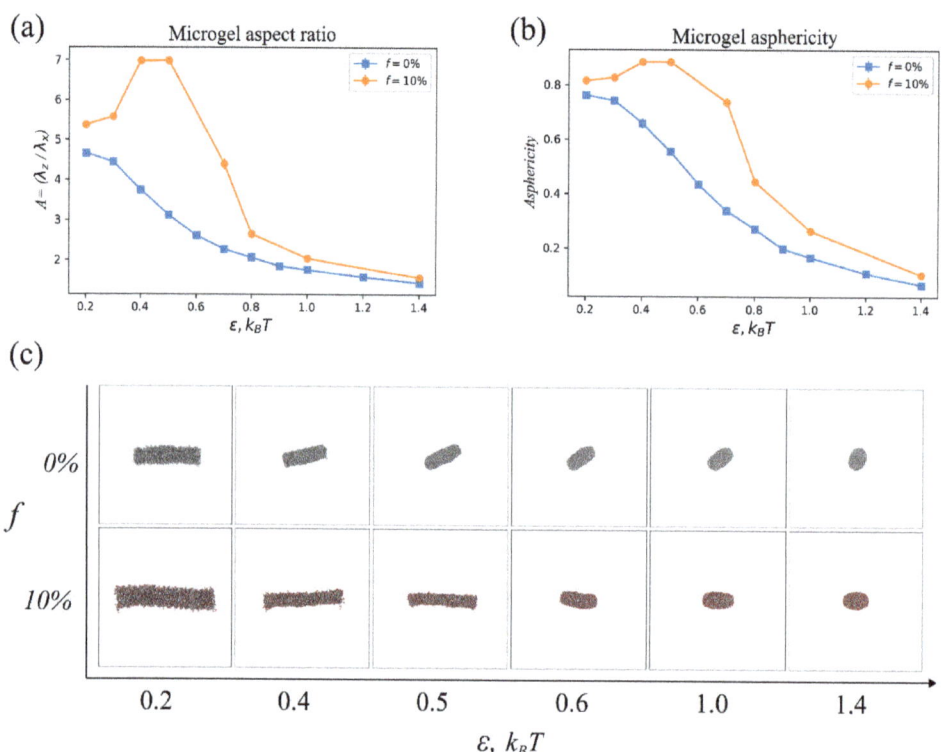

Figure 4. (**a**) Aspect ratio calculated from eigenvalues of gyration tensor λ_z/λ_x [41]. (**b**) Asphericity parameter and (**c**) snapshots of equivalent neutral ($f = 0\%$) and charged ($f = 10\%$) microgels vs. interaction parameter ε. The molecular weight and subchain length of the microgel are $N = 50{,}000$ and $M = 10$, respectively. The initial aspect ratio of the microgels is $A = 4$.

The radial concentration profiles for microgels differing in the fraction of charged groups under adequate solvent conditions are presented in Figure 7. They were obtained as follows. The microgel was divided into a few cylindrical segments of the length $l = 0.2L$ (similar to the determination of microgel thickness). The concentration profiles were calculated for each inner segment (excluding end segments) with further averaging over the segments and over their rotation with respect to the axis. The polymer volume fraction (dimensionless concentration of polymer beads) is presented in Figure 7a. The highest swelling of the microgel is characterized by the lowest polymer concentration and was achieved for $f = 50\%$. The decrease in f leads to the increase in the polymer concentration within the microgel (Figure 7a). We can also observe that the polymer volume fraction has a local maximum at the microgel periphery. This maximum is more distinct for the lowest fraction of charged groups and diminishes upon microgel charging (Figure 7a). Such an effect is known for the spherical microgels and is related to the partial redistribution of the charged groups towards the periphery to reduce the electrostatic energy [27]. The distribution of the charged groups of the network and counterions is presented in Figure 7b by solid and dashed lines, respectively. For all considered f values, we observed the local increase in the fraction of the network charged groups at the periphery. Here, we can also see a maximum violation of the electric neutrality: the counterion concentration profiles significantly deviate from those of the network charged groups at the microgel periphery.

This is caused by the entropy of counterions. Therefore, at the periphery, an electric double layer is formed.

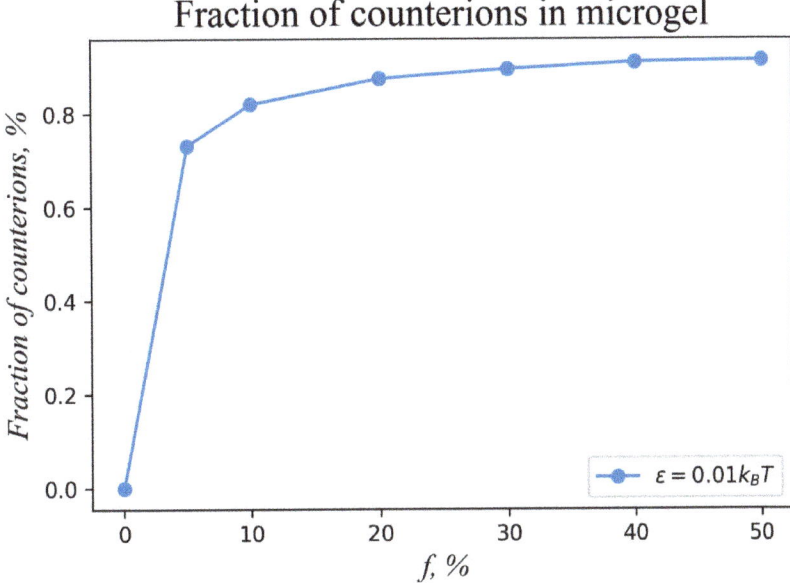

Figure 5. Fraction of counterions localized within the charged microgel as a function of the fraction of charged groups of the network f. The molecular weight and subchain length of the microgel are $N = 100{,}000$ and $M = 10$, respectively. The initial aspect ratio is $A = 10$. The microgel is modeled under favorable solvent conditions, $\varepsilon = 0.01kT$.

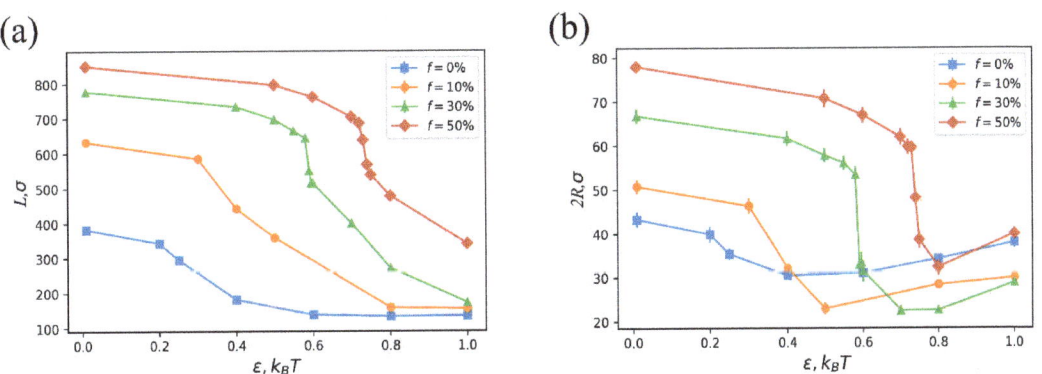

Figure 6. Microgel contour length L (**a**) and thickness $2R$ (**b**) vs. interaction parameter ε between beads. Different curves correspond to the different fraction of charged groups in the microgel. The molecular weight and subchain length of the microgel are $N = 50{,}000$ and $M = 20$, respectively. The initial aspect ratio is $A = 10$.

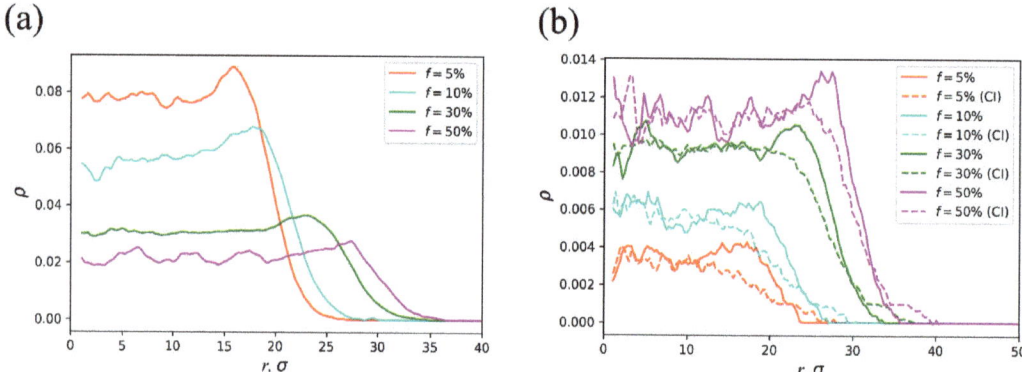

Figure 7. (**a**) Polymer volume fraction, (**b**) volume fraction of the network charged groups (solid) and counterions (dashed) as functions of the radial coordinate r at different f values and favorable solvent conditions, $\varepsilon = 0.01\ kT$. The molecular weight and subchain length of the microgel are $N = 100,000$ and $M = 10$, respectively. The initial aspect ratio is $A = 10$.

The evolution of the polymer concentration profile and the positions of charged groups and counterions upon solvent quality decline are presented in Figure 8. The polymer concentration gradually increases in the microgel interior with ε. Such as under favorable solvent conditions, a maximum concentration value is detectable at the periphery in a poor quality solvent, although the difference with respect to the inner part is not essential (Figure 8a). This effect is due to the violation of the electric neutrality in the peripheral layer, which therefore leads to a lower concentration of counterions and, further, to less exerted osmotic pressure (Figure 8b). Indeed, counterion concentration in the microgel interior is higher and gradually decreases towards the periphery (Figure 8b).

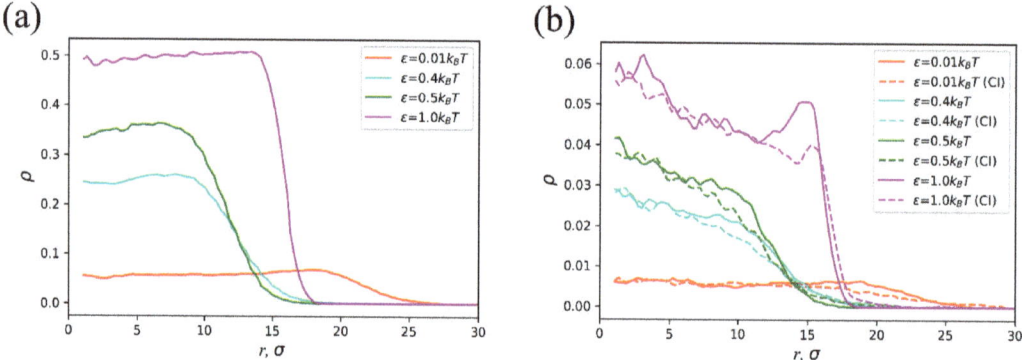

Figure 8. (**a**) Polymer volume fraction, (**b**) volume fraction of the network charged groups (solid) and counterions (dashed) as functions of the radial coordinate r at different ε values. The molecular weight and subchain length of the microgel are $N = 100,000$ and $M = 10$, respectively. The initial aspect ratio and the fraction of charged groups are $A = 10$ and $f = 10\%$.

4. Conclusions

In this study, we performed computer simulations of the swelling and collapse of cylindrical microgels bearing charged groups. Both aspect ratio and charged group fraction effects on microgel collapse were studied. A comparison with neutral counterparts is provided here. We have shown that under adequate solvent conditions, charged microgels have a longer length and greater thickness than their neutral counterparts. This effect primarily occurs due to the exerted osmotic pressure of counterions localized within the microgel to compensate for the electric charge of the network. Such as for macroscopic polyelectrolyte gels, the swelling degree of the cylindrical microgel depends on the fraction of charged groups: the higher the fraction, the greater the swelling (length and thickness of the microgel). We detected peculiarities in the microgel collapse upon solvent quality decline. An increase in the attraction between beads (interaction parameter ε) leads to microgel shortening. However, the microgel thickness demonstrates non-monotonous behavior. The initial decrease caused by the attraction between beads is followed by the microgel thickening, which is a consequence of the surface tension. Though, if it has a lower value for the less anisotropic object, the microgel would collapse into the sphere in the absence of elasticity. In addition, in the case of low cross-linking density or a low enough aspect ratio, we observed a "cylinder-to-sphere" collapse. Otherwise, the cylindrical shape is preserved upon collapse. The microgel aspect ratio (the length-to-diameter ratio) has a maximum in the course of collapse. The initial growth (more progressive thinning with respect to the shortening) is caused by electrostatics. The microgel releases some counterions due to entropic reasons and has uncompensated charge. The elongated charged objects have less electrostatic energy than their spherical counterparts (similar to polyelectrolyte chains in a dilute solution). Therefore, the microgel collapse accompanied by thinning is more favorable. The further decrease in the aspect ratio is driven by the surface tension. We have demonstrated that the characteristic of the collapse depends on the fraction of charged groups: the higher the fraction, the more abrupt the transition. The computer simulations allowed us to plot polymer concentration profiles and demonstrate the distribution of charged groups and counterions. Because of the relative novelty of the technique for synthesizing microgels of complex architectures [16,23,24], and so far, because of the current absence of methods of their preparation on a nanometer scale, there is little information in the literature on the experiments with anisotropic polyelectrolyte networks. Therefore, our research has both fundamental and practical significance. We believe that it can inspire the creation of functional materials with specified properties based on cylindrical microgels.

Author Contributions: Conceptualization, I.I.P.; methodology, I.V.P. and I.I.P.; software, I.V.P.; validation, I.I.P. and R.A.G.; formal analysis, I.V.P. and A.A.L.; investigation, I.V.P., A.A.L. and R.A.G.; data curation, I.I.P.; writing—original draft preparation, I.I.P.; writing—review and editing, A.A.L. and R.A.G.; visualization, I.V.P.; supervision, I.I.P.; project administration, I.I.P.; funding acquisition, I.I.P. All authors have read and agreed to the published version of the manuscript.

Funding: This research was funded by the Russian Science Foundation, project # 21-73-30013.

Institutional Review Board Statement: Not applicable.

Informed Consent Statement: Not applicable.

Data Availability Statement: The data presented in this study are available on request from the corresponding author.

Acknowledgments: The research was carried out using the equipment of the shared research facilities of HPC computing resources at Lomonosov Moscow State University [43]. The data processing was performed using the facilities of the Interlaboratory Computer Center at INEOS RAS supported by the Ministry of Science and Higher Education of Russian Federation.

Conflicts of Interest: The authors declare no conflict of interest.

References

1. Plamper, F.A.; Richtering, W. Functional Microgels and Microgel Systems. *Acc. Chem. Res.* **2017**, *50*, 131–140. [CrossRef] [PubMed]
2. Senff, H.; Richtering, W. Temperature Sensitive Microgel Suspensions: Colloidal Phase Behavior and Rheology of Soft Spheres. *J. Chem. Phys.* **1999**, *111*, 1705–1711. [CrossRef]
3. Schroeder, R.; Rudov, A.A.; Lyon, L.A.; Richtering, W.; Pich, A.; Potemkin, I.I. Electrostatic Interactions and Osmotic Pressure of Counterions Control the PH-Dependent Swelling and Collapse of Polyampholyte Microgels with Random Distribution of Ionizable Groups. *Macromolecules* **2015**, *48*, 5914–5927. [CrossRef]

4. Backes, S.; Witt, M.U.; Roeben, E.; Kuhrts, L.; Aleed, S.; Schmidt, A.M.; von Klitzing, R. Loading of PNIPAM Based Microgels with CoFe2O4 Nanoparticles and Their Magnetic Response in Bulk and at Surfaces. *J. Phys. Chem. B* **2015**, *119*, 12129–12137. [CrossRef] [PubMed]
5. Gelissen, A.P.H.; Scotti, A.; Turnhoff, S.K.; Janssen, C.; Radulescu, A.; Pich, A.; Rudov, A.A.; Potemkin, I.I.; Richtering, W. An Anionic Shell Shields a Cationic Core Allowing for Uptake and Release of Polyelectrolytes within Core–Shell Responsive Microgels. *Soft Matter* **2018**, *14*, 4287–4299. [CrossRef] [PubMed]
6. Dirksen, M.; Dargel, C.; Meier, L.; Brändel, T.; Hellweg, T. Smart Microgels as Drug Delivery Vehicles for the Natural Drug Aescin: Uptake, Release and Interactions. *Colloid Polym. Sci.* **2020**, *298*, 505–518. [CrossRef]
7. Wiese, S.; Tsvetkova, Y.; Daleiden, N.J.E.; Spieß, A.C.; Richtering, W. Microgel Stabilized Emulsions: Breaking on Demand. *Colloids Surf. A Physicochem. Eng. Asp.* **2016**, *495*, 193–199. [CrossRef]
8. Zhang, T.; Ngai, T. One-Step Formation of Double Emulsions Stabilized by PNIPAM-Based Microgels: The Role of Co-Monomer. *Langmuir* **2021**, *37*, 1045–1053. [CrossRef]
9. Yang, L.Q.; Hao, M.M.; Wang, H.Y.; Zhang, Y. Amphiphilic Polymer-Ag Composite Microgels with Tunable Catalytic Activity and Selectivity. *Colloid Polym. Sci.* **2015**, *293*, 2405–2417. [CrossRef]
10. Sabadasch, V.; Dirksen, M.; Fandrich, P.; Cremer, J.; Biere, N.; Anselmetti, D.; Hellweg, T. Pd Nanoparticle-Loaded Smart Microgel-Based Membranes as Reusable Catalysts. *ACS Appl. Mater. Interfaces* **2022**, *14*, 49181–49188. [CrossRef]
11. Berger, S.; Singh, R.; Sudha, J.D.; Adler, H.J.; Pich, A. Microgel/Clay Nanohybrids as Responsive Scavenger Systems. *Polymer* **2010**, *51*, 3829–3835. [CrossRef]
12. Saha, P.; Santi, M.; Emondts, M.; Roth, H.; Rahimi, K.; Großkurth, J.; Ganguly, R.; Wessling, M.; Singha, N.K.; Pich, A. Stimuli-Responsive Zwitterionic Core–Shell Microgels for Antifouling Surface Coatings. *ACS Appl. Mater. Interfaces* **2020**, *12*, 58223–58238. [CrossRef] [PubMed]
13. Rose, J.C.; Gehlen, D.B.; Haraszti, T.; Köhler, J.; Licht, C.J.; De Laporte, L. Biofunctionalized Aligned Microgels Provide 3D Cell Guidance to Mimic Complex Tissue Matrices. *Biomaterials* **2018**, *163*, 128–141. [CrossRef] [PubMed]
14. Balaceanu, A.; Verkh, Y.; Kehren, D.; Tillmann, W.; Pich, A. Thermoresponsive Core-Shell Microgels. Synthesis and Characterisation. *Z. für Phys. Chem.* **2014**, *228*, 253–267. [CrossRef]
15. Crespy, D.; Zuber, S.; Turshatov, A.; Landfester, K.; Popa, A.M. A Straightforward Synthesis of Fluorescent and Temperature-Responsive Nanogels. *J. Polym. Sci. Part A Polym. Chem.* **2012**, *50*, 1043–1048. [CrossRef]
16. Krüger, A.J.D.; Köhler, J.; Cichosz, S.; Rose, J.C.; Gehlen, D.B.; Haraszti, T.; Möller, M.; De Laporte, L. A Catalyst-Free, Temperature Controlled Gelation System for in-Mold Fabrication of Microgels. *Chem. Commun.* **2018**, *54*, 6943–6946. [CrossRef]
17. Seiffert, S.; Weitz, D.A. Microfluidic Fabrication of Smart Microgels from Macromolecular Precursors. *Polymer* **2010**, *51*, 5883–5889. [CrossRef]
18. Rudyak, V.Y.; Kozhunova, E.Y.; Chertovich, A.V. Towards the Realistic Computer Model of Precipitation Polymerization Microgels. *Sci. Rep.* **2019**, *9*, 13052. [CrossRef]
19. Krüger, A.J.D.; Bakirman, O.; Guerzoni, L.P.B.; Jans, A.; Gehlen, D.B.; Rommel, D.; Haraszti, T.; Kuehne, A.J.C.; De Laporte, L. Compartmentalized Jet Polymerization as a High-Resolution Process to Continuously Produce Anisometric Microgel Rods with Adjustable Size and Stiffness. *Adv. Mater.* **2019**, *31*, 1903668. [CrossRef]
20. Crassous, J.J.; Mihut, A.M.; Månsson, L.K.; Schurtenberger, P. Anisotropic Responsive Microgels with Tuneable Shape and Interactions. *Nanoscale* **2015**, *7*, 15971–15982. [CrossRef]
21. Nickel, A.C.; Scotti, A.; Houston, J.E.; Ito, T.; Crassous, J.; Pedersen, J.S.; Richtering, W. Anisotropic Hollow Microgels That Can Adapt Their Size, Shape, and Softness. *Nano Lett.* **2019**, *19*, 8161–8170. [CrossRef] [PubMed]
22. Nickel, A.C.; Kratzenberg, T.; Bochenek, S.; Schmidt, M.M.; Rudov, A.A.; Falkenstein, A.; Potemkin, I.I.; Crassous, J.J.; Richtering, W. Anisotropic Microgels Show Their Soft Side. *Langmuir* **2022**, *38*, 5063–5080. [CrossRef] [PubMed]
23. Wolff, H.J.M.; Linkhorst, J.; Göttlich, T.; Savinsky, J.; Krüger, A.J.D.; De Laporte, L.; Wessling, M. Soft Temperature-Responsive Microgels of Complex Shape in Stop-Flow Lithography. *Lab Chip* **2020**, *20*, 285–295. [CrossRef]
24. Rose, J.C.; Cámara-Torres, M.; Rahimi, K.; Köhler, J.; Möller, M.; De Laporte, L. Nerve Cells Decide to Orient inside an Injectable Hydrogel with Minimal Structural Guidance. *Nano Lett.* **2017**, *17*, 3782–3791. [CrossRef] [PubMed]
25. Zholudev, S.I.; Gumerov, R.A.; Larina, A.A.; Potemkin, I.I. Swelling, Collapse and Ordering of Rod-like Microgels in Solution: Computer Simulation Studies. *J. Colloid Interface Sci.* **2023**, *629*, 270–278. [CrossRef]
26. Toxvaerd, S.; Dyre, J.C. Role of the First Coordination Shell in Determining the Equilibrium Structure and Dynamics of Simple Liquids. *J. Chem. Phys.* **2011**, *135*, 134501. [CrossRef] [PubMed]
27. Rumyantsev, A.M.; Rudov, A.A.; Potemkin, I.I. Communication: Intraparticle Segregation of Structurally Homogeneous Polyelectrolyte Microgels Caused by Long-Range Coulomb Repulsion. *J. Chem. Phys.* **2015**, *142*, 171105. [CrossRef]
28. Stevens, M.J.; Kremer, K. The Nature of Flexible Linear Polyelectrolytes in Salt Free Solution: A Molecular Dynamics Study. *J. Chem. Phys.* **1995**, *103*, 1669–1690. [CrossRef]
29. Jeon, J.; Dobrynin, A.V. Molecular Dynamics Simulations of Polyelectrolyte-Polyampholyte Complexes. Effect of Solvent Quality and Salt Concentration. *J. Phys. Chem. B* **2006**, *110*, 24652–24665. [CrossRef]
30. Tiwari, R.; Hönders, D.; Schipmann, S.; Schulte, B.; Das, P.; Pester, C.W.; Klemradt, U.; Walther, A. A Versatile Synthesis Platform to Prepare Uniform, Highly Functional Microgels via Click-Type Functionalization of Latex Particles. *Macromolecules* **2014**, *47*, 2257–2267. [CrossRef]

31. Kremer, K.; Grest, G.S. Dynamics of Entangled Linear Polymer Melts: A Molecular-dynamics Simulation. *J. Chem. Phys.* **1990**, *92*, 5057–5086. [CrossRef]
32. LAMMPS Molecular Dynamics Simulator. Available online: http://Lammps.Sandia.Gov/ (accessed on 20 October 2022).
33. Hockney, R.; Eastwood, J. *Computer Simulation Using Particles*; CRC Press: Boca Raton, FL, USA, 2021. [CrossRef]
34. Kobayashi, H.; Halver, R.; Sutmann, G.; Winkler, R.G. Polymer Conformations in Ionic Microgels in the Presence of Salt: Theoretical and Mesoscale Simulation Results. *Polymers* **2017**, *9*, 15. [CrossRef] [PubMed]
35. Del Monte, G.; Ninarello, A.; Camerin, F.; Rovigatti, L.; Gnan, N.; Zaccarelli, E. Numerical Insights on Ionic Microgels: Structure and Swelling Behaviour. *Soft Matter* **2019**, *15*, 8113–8128. [CrossRef]
36. Kratz, K.; Hellweg, T.; Eimer, W. Influence of Charge Density on the Swelling of Colloidal Poly(N-Isopropylacrylamide-Co-Acrylic Acid) Microgels. *Colloids Surfaces A Physicochem. Eng. Asp.* **2000**, *170*, 137–149. [CrossRef]
37. Su, W.; Yang, M.; Zhao, K.; Ngai, T. Influence of Charged Groups on the Structure of Microgel and Volume Phase Transition by Dielectric Analysis. *Macromolecules* **2016**, *49*, 7997–8008. [CrossRef]
38. Quesada-Pérez, M.; Maroto-Centeno, J.A.; Forcada, J.; Hidalgo-Alvarez, R. Gel Swelling Theories: The Classical Formalism and Recent Approaches. *Soft Matter* **2011**, *7*, 10536. [CrossRef]
39. Potemkin, I.I.; Khokhlov, A.R. Nematic Ordering in Dilute Solutions of Rodlike Polyelectrolytes. *J. Chem. Phys.* **2004**, *120*, 10848–10851. [CrossRef] [PubMed]
40. Kundagrami, A.; Kumar, R.; Muthukumar, M. Simulations and Theories of Single Polyelectrolyte Chains. In *Modeling and Simulation in Polymers*; Wiley: Hoboken, NJ, USA, 2010; pp. 247–341. [CrossRef]
41. Noguchi, H.; Yoshikawa, K. Morphological Variation in a Collapsed Single Homopolymer Chain. *J. Chem. Phys.* **1998**, *109*, 5070–5077. [CrossRef]
42. Khokhlov, A.R.; Yu, A.; Grosberg, V.S.P. *Statistical Physics of Macromolecules*; AIP series in polymers and complex materials; AIP Press: Melville, NY, USA, 1994.
43. Voevodin, V.V.; Antonov, A.S.; Nikitenko, D.A.; Shvets, P.A.; Sobolev, S.I.; Sidorov, I.Y.; Stefanov, K.S.; Voevodin, V.V.; Zhumatiy, S.A. Supercomputer Lomonosov-2: Large Scale, Deep Monitoring and Fine Analytics for the User Community. *Supercomput. Front. Innov.* **2019**, *6*, 4–11.

Article

A Dual-Cavity Fiber Fabry–Pérot Interferometer for Simultaneous Measurement of Thermo-Optic and Thermal Expansion Coefficients of a Polymer

Cheng-Ling Lee [1,*], Chao-Tsung Ma [2,*], Kuei-Chun Yeh [1] and Yu-Ming Chen [1]

1 Department of Electro-Optical Engineering, National United University, Miaoli 360, Taiwan
2 Department of Electrical Engineering, National United University, Miaoli 360, Taiwan
* Correspondence: cherry@nuu.edu.tw (C.-L.L.); ctma@nuu.edu.tw (C.-T.M.); Tel.: +886-37-382568 (C.-L.L.); +886-37-382482 (C.-T.M.); Fax: +886-37-382555 (C.-L.L.); +886-37-382488 (C.-T.M.)

Abstract: This paper presents a novel method based on a dual-cavity fiber Fabry–Pérot interferometer (DCFFPI) for simultaneously measuring the thermo-optic coefficient (TOC) and thermal expansion coefficient (TEC) of a polymer. The polymer is, by nature, highly responsive to temperature (T) in that its size (length, L) and refractive index (RI, n) are highly dependent on the thermal effect. When the optical length of the polymer cavity changes with T, it is difficult to distinguish whether there is a change in L or n, or both. The variation rates of L and n with a change in T were the TOC and TEC, respectively. Therefore, there was a cross-sensitivity between TOC and TEC in the polymer-based interferometer. The proposed DCFFPI, which cascades a polymer and an air cavity, can solve the above problem. The expansion of the polymer cavity is equal to the compression of the air cavity with the increase in T. By analyzing the individual optical spectra of the polymer and air cavities, the parameters of TOC and TEC can be determined at the same time. The simultaneous measurement of TOC and TEC with small measured deviations of 6×10^{-6} (°C^{-1}) and 3.67×10^{-5} (°C^{-1}) for the polymer NOA61 and 7×10^{-6} (°C^{-1}) and 1.46×10^{-4} (°C^{-1}) for the NOA65 can be achieved. Experimental results regarding the measured accuracy for the class of adhesive-based polymer are presented to demonstrate the feasibility and verify the usefulness of the proposed DCFFPI.

Keywords: dual-cavity fiber Fabry–Pérot interferometer (DCFFPI); thermo-optic coefficient (TOC); thermal expansion coefficient (TEC); polymer

Citation: Lee, C.-L.; Ma, C.-T.; Yeh, K.-C.; Chen, Y.-M. A Dual-Cavity Fiber Fabry–Pérot Interferometer for Simultaneous Measurement of Thermo-Optic and Thermal Expansion Coefficients of a Polymer. *Polymers* **2022**, *14*, 4966. https://doi.org/10.3390/polym14224966

Academic Editor: Jung-Chang Wang

Received: 27 October 2022
Accepted: 15 November 2022
Published: 16 November 2022

Publisher's Note: MDPI stays neutral with regard to jurisdictional claims in published maps and institutional affiliations.

Copyright: © 2022 by the authors. Licensee MDPI, Basel, Switzerland. This article is an open access article distributed under the terms and conditions of the Creative Commons Attribution (CC BY) license (https://creativecommons.org/licenses/by/4.0/).

1. Introduction

An acrylic-based polymer can be developed into many kinds of optical adhesives that can be widely utilized in various technologies and industrial applications. There are various highly transparent polymer materials that can be employed as functional optical materials for applications in the optics field. However, there are few studies on the investigation and simultaneous measurement of the optical or mechanical parameters of the polymers. The optical parameter, the thermo-optic coefficient (TOC), and the mechanical parameter, the thermal expansion coefficient (TEC), are both very important values because they reveal how the thermal effect would affect the properties of optical materials. As the optical materials in optoelectronic devices are greatly influenced by the thermal effect of the surrounding environment, the parameters of TOC and TEC have to be precisely determined simultaneously in practical applications. However, studies on sensing these two parameters simultaneously are very rare. In 2017, a low-cost monochromatic single-arm double interferometer was developed for the simultaneous measurement of the linear thermal expansion and thermo-optic coefficients of transparent samples [1]. Similar work published in the literature on the simultaneous analysis of the optical and mechanical properties of polymers has shown the importance of the issues investigated in this paper [2–4]. Hossain et al. reported a simple method for the simultaneous measurement of the thermo-optic and stress-optic coefficients of polymer thin films using a prism coupler technique with two different kinds of polymers. The finite element method was used to predict the stresses of the polymer film to further obtain its real thermo-optic coefficient [2]. Furthermore, Shimamura, A., et al. presented the simultaneous analysis of the optical and mechanical properties of cross-linked azobenzene-containing liquid–crystalline polymer films. The photoinduced stress of the polymer films upon UV irradiation was investigated to gain a fundamental understanding of the complicated physical processes of the photomechanical response [4]. The above-mentioned methodology and implementation mechanisms are all effective; however, the measurement systems used were general bulk, which could limit their further application. In recent decades, fiber optical sensors have been developed rapidly because of their excellent sensing performances and capability to function under variable conditions. In the measurement of the TOC or TEC of materials using fiber-based sensors, Esposito et al. proposed a fiber Bragg grating (FBG) sensor to measure the TEC of polymers at cryogenic temperatures [5]. Other work has proposed a fiber-optic temperature sensor based on the difference in TEC between fused silica and metallic materials [6]. The above studies are the results of TEC measurement by fiber sensors. Regarding the precise measurement of TOC, the information regarding the temperature (T) and refractive index (RI) variations enables the TOC of materials to be accurately measured by the following fiber-optic sensors: in-line hollow-core fiber Fabry-Pérot interferometers (HCFFPIs) [7,8], FBG [9], the core-offset fiber Mach–Zehnder interferometer (FMZI) [10] and two-mode fiber interferometer (TMFI) [11]. However, the above structures are still unable to measure the parameters of TEC and TOC concurrently. One solution is to take multiplexing advantage of the fiber-optic sensors. In view of the distributed and multiplexed capabilities of the fiber-based sensors, simultaneous measurement would be the most attractive advantage, as multiple fiber sensing elements can be spliced and integrated in a single-fiber device [12–19], e.g., a fiber Fizeau interferometer (FZI) connected to a FBG for the measurement of the relative humidity (RH) and T [12], two long-period fiber gratings (LPGs) for strain (S) and T measurement [13], a LPG/photonic crystal fiber interferometer (PCFI) for sensing the T and RI [14], two polymer FZIs for measuring the RH and T [15], a hybrid structured fiber Fabry–Perot interferometer (FFPI) for sensing the S and T simultaneously [16], a fiber Michelson interferometer for wide-range curvature measurement with low T cross-sensitivity [17], a single optical fiber tip modified with a coating of poly(allylamine hydrochloride) (PAH)/silica nanoparticles (SiO_2 NPs) for RH measurement and then coated with thermochromic liquid crystal (TLC) for sensing T to achieve the simultaneous measurement [18], and a pendant polymer droplet-based fiber Fabry–Pérot interferometer (FFPI) for sensing the pressure and T [19]. The above sensing results based on multiple fiber sensing elements demonstrated the superiority of sensing characteristics in eliminating the possible cross-sensitivities problems; however, studies focused on the simultaneous sensing of the TOC and TEC are actually very limited.

In this study, we propose a novel configuration based on a dual cavity fiber Fabry–Pérot interferometer (DCFFPI) that can simultaneously measure the thermo-optic coefficient (TOC) and thermal expansion coefficient (TEC) of a polymer. The polymer is particularly sensitive to T, not only regarding its size (length, L), but also its refractive index. The variation in the size (L) and refractive index (RI) of polymers with the change in T denotes the parameters of TEC and TOC. As mentioned previously, there is a cross-sensitivity between TEC and TOC in polymer interferometers because the interference wavelength shifts as n or L changes with T. The DCFFPI that cascaded a polymer and an air cavity could obtain a superimposed signal of the multiple interferences. With the proposed sensor structure, the polymer expands and compresses the air cavity when T increases. By analyzing the individual optical spectra of the polymer and the air cavity, the parameters of TEC and TOC could be determined at the same time. Experimental results indicating the accuracy in the simultaneous measurement of the TEC and TOC with small measured errors are shown to verify the feasibility and the usefulness of the proposed DCFFPI.

2. Sensor Fabrication and Principle

The configuration of the dual-cavity fiber Fabry–Pérot interferometer (DCFFPI) is based on a polymer and an air cavity formed as a dual cavity, as presented in Figure 1. The fabrication process is described as follows. A hollow-core fiber (HCF_1) with an air-core diameter (D) of 50 µm was fusion-spliced to another HCF_2 with D = 10 µm. The endface of HCF_1 was cleaved for a certain length (preferably <150 µm) with a slight slant and followed by being spliced with a single-mode fiber (SMF). The fusion splicing could create a miniature tilted gap (2–3 µm) for the effective access of different monomers (polymers). A capillary action was used to fill the polymer into the core of HCF_1 without fully filling the core. Filling through capillary action only took seconds to accomplish the desired structure, as there was an open cavity at the HCF_2 endface. This provided a high fabrication efficiency. Here, R_1, R_2, and R_3 denote the Fresnel reflections from the interfaces of the structure. Light that propagates into the long HCF_2 section with a hollow core of 10 µm could not cause Fabry–Pérot interference, but would only excite some leaky modes, increasing the loss associated with the reflection of R_3 and reducing the visibility of the interference from the air cavity [4].

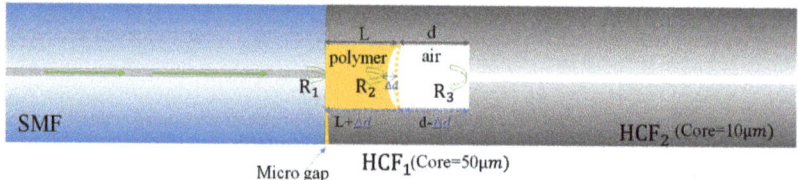

Figure 1. Configuration of the proposed DCFFPI with a dual cavity (polymer and air). Here, R_1, R_2, and R_3 denote the Fresnel reflections from the interfaces of the structure.

Figure 2 shows the micrographs of the proposed DCFFPIs with various lengths of the dual cavity. Figure 2a,b shows the HCF_1 of ~86 µm filled with the polymer NOA65 and HCF_1 of ~110 µm filled with the polymer NOA61, respectively. As mentioned previously, the monomer (polymer) was added to the HCF_1 and formed a controlled microcavity; after the filling step, the HCF_1 was exposed to UV light. In the UV-curing process, the monomer gradually transformed into a robust polymer cavity. The remaining unfilled section was the air cavity. It is worth noting the tiny air core of the HCF_2 as an opening for airflow.

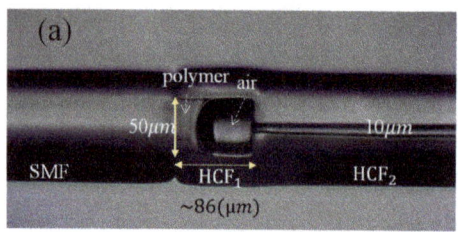 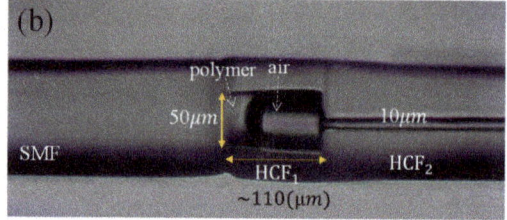

Figure 2. Micrographs of the proposed DCFFPIs with various structures with lengths of (**a**) HCF_1 = 86 µm and (**b**) HCF_1 = 110 µm. Here, HCF_1: (D = 50 µm) and HCF_2: (D = 10 µm).

It can improve the polymer filling and not inhibit the thermal expansion of the polymer, but it may slightly deform during the arc-discharging of the fusion splicing of fibers. Even so, the air core deformation condition did not affect the measurement results as long as HCF_2 was not closed. The experimental setup for the simultaneous measurement of TOC and TEC of a polymer is displayed in Figure 3. The proposed DCPFFI was positioned on a TE cooler for varying T with a fixed relative humidity in the surrounding environment. When a broadband light source (BLS, BLS-GIP Technology, New Taipei, Taiwan) with an optical circulator was propagated to the device, spectral interference reflections from the polymer and air cavities were superimposed and measured by an optical spectrum analyzer (OSA, Advantest Q8381 A, Las Vegas, NV, USA). The analysis of the optical responses of the combined interferences was accomplished using the fast Fourier transform (FFT) method, which is used to separate multiple interferences in spatial frequency into two individual spatial frequencies for a polymer and air cavities [12]. Once individual interference spectra for the corresponding cavity of the polymer and air were determined, we could study the thermal effect on the characteristics of a polymer in the DCFFPI. As can be seen in Figure 1, when the DCFFPI sensor was heated by the TE cooler, the polymer cavity expanded and generated a variation in the optical length (i.e., optical path) to shift the interference wavelengths. Meanwhile, the air cavity was compressed to reduce its optical path, with the wavelength shifting to a shorter wavelength region (blue-shifted). Here, the TEC of a silica fiber ($\alpha = +5.5 \times 10^{-7}\ °C^{-1}$) was an extremely small value that could be ignored. Therefore, the length of HCF_1 can be regarded as almost fixed.

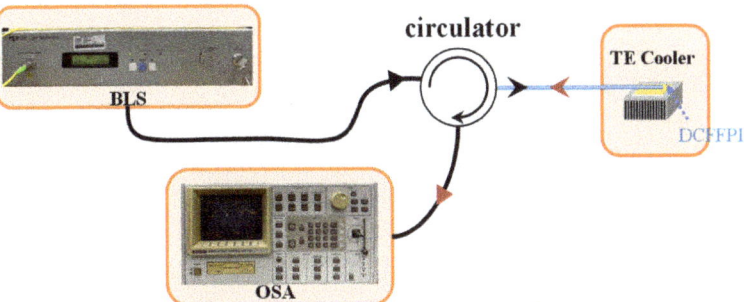

Figure 3. Experimental setup for simultaneously measuring the TEC and TOC of a polymer.

3. Experimental Results and Discussions

In the experiment, the T of the proposed DCFFPI device was controlled by a TE cooler with steps of 1 °C. Several polymers of the Norland Optical Adhesive (NOA) series by Norland Products Inc., e.g., NOA61, NOA65, and NOA146 [14] were utilized for the simultaneous measurement of their TOC and TEC. Figure 4 shows the results for the measurement of the NOA61. The original spectra obtained by the OSA under T = 20~30 °C are shown in Figure 4a. Figure 4a plots the superimposed spectra from the air and polymer cavities. The FFT method was utilized to separate multiple interferences in the spatial frequency (Figure 4b) into two individual spatial frequencies for the corresponding air and polymer cavities, respectively (as shown in Figure 4c,d). In this case, the polymer length (L = 26 μm) and air length (d = 85 μm) were measured, and Figure 5 shows the experimental results of the wavelength shifts due to the variation in T.

Figure 5a,b displays the optical spectra interferences of the proposed DCFFPI for the air and polymer cavities, respectively. The interference fringes were estimated to blue-shift, shifting to the shorter wavelength region, and red-shift, shifting to the longer wavelength region, for the air and polymer cavities, respectively, due to the increase in T. It can be seen that the polymer expanded to compress the air cavity from the heating process. Heating performances with high linear sensitivities of −0.4 nm/°C and +1.0057 nm/°C were shown for the air and NOA 61-polymer cavity, respectively. Based on the above results in Figure 5a,b, the TOC and TEC can be simultaneously determined by the description below.

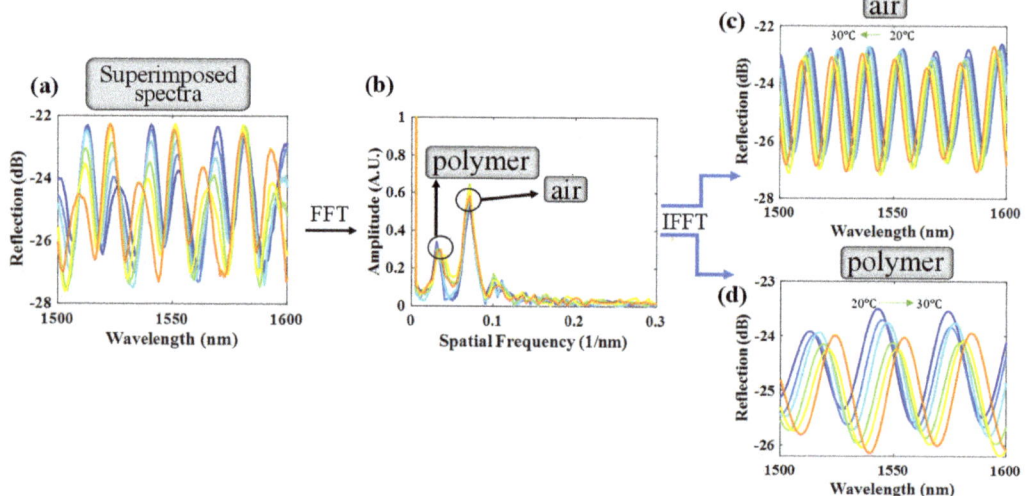

Figure 4. (a) Optical response of the superimposed interferences with variation in T; (b) superimposed spectra processed by FFT; separated reflection spectra of the (c) air cavity and (d) polymer cavity.

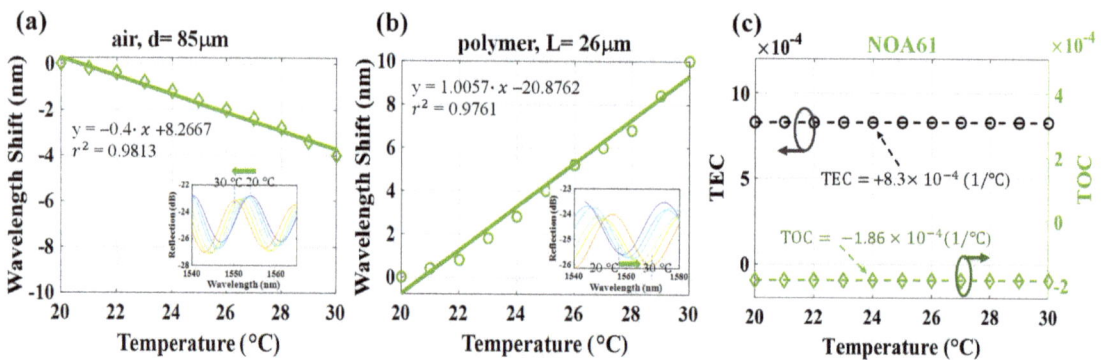

Figure 5. Wavelength shifts from 20 °C to 30 °C for the cavities of (a) air, $\Delta\lambda_a$, and (b) NOA61-polymer, $\Delta\lambda_p$; (c) measured TEC and TOC from the results of (a,b).

When the DCFFPI was heated by the TE cooler, the wavelength shifts per °C of the air ($\Delta\lambda_a$) and polymer ($\Delta\lambda_p$) cavities could be respectively estimated by using the following simultaneous Equations (1) and (2).

$$\frac{\Delta\lambda_a}{\lambda_a} = \frac{\Delta d}{d} \quad (air\ cavity) \tag{1}$$

$$\frac{\Delta\lambda_p}{\lambda_p} = \frac{\Delta n}{n} + \frac{\Delta L}{L} + \frac{\Delta n}{n} \cdot \frac{\Delta L}{L} = TOC + TEC\ (polymer\ cavity) \tag{2}$$

where λ_a and λ_p are the monitored interference wavelength dip at a specific T for the air and polymer cavities (generally set at 1550 nm), respectively. The symbols d, L and Δd, ΔL denote the original values and variations in the cavity length for the air and polymer, respectively, and n is the RI of the polymer. $\frac{\Delta n}{n} \cdot \frac{\Delta L}{L}$ is extremely small that is neglected. Here, in the proposed structure, $\Delta L = -\Delta d$, and the HCF_1 can be considered essentially fixed. Based on the experimental results, the wavelength shifts from 20 °C to 30 °C for the air ($\Delta\lambda_a$) and (b) polymer ($\Delta\lambda_p$) cavities are shown in Figure 5a,b, respectively. Then, the

TEC and TOC of the test polymer, NOA61 could be calculated by Equations (1) and (2). The results are displayed in Figure 5c and indicate that the values of the TOC and TEC of NOA61 were -1.86×10^{-4} (°C^{-1}) and $+8.3 \times 10^{-4}$ (°C^{-1}), respectively.

Figure 6 shows the results of the NOA65 polymer measurement. The results in Figure 6a,b also show the high linear sensitivity of -0.817 nm/°C and $+1.563$ nm/°C for the air and NOA 65-polymer cavity with cavity lengths of 59.8 µm and 26.5 µm, respectively. The TOC and TEC of NOA 65 could be simultaneously determined to be -1.9×10^{-4} (°C^{-1}) and $+11.9 \times 10^{-4}$ (°C^{-1}), respectively. Again, based on the results shown in Figure 7, thermal sensitivities of -0.64 nm/°C and $+1.1147$ nm/°C for the air and NOA146H polymer with cavity lengths of 65.1 µm and 24.86 µm were obtained, respectively. The simultaneous measurement of TOC and TEC yielded -3.59×10^{-4} (°C^{-1}) and $+10.78 \times 10^{-4}$ (°C^{-1}), respectively.

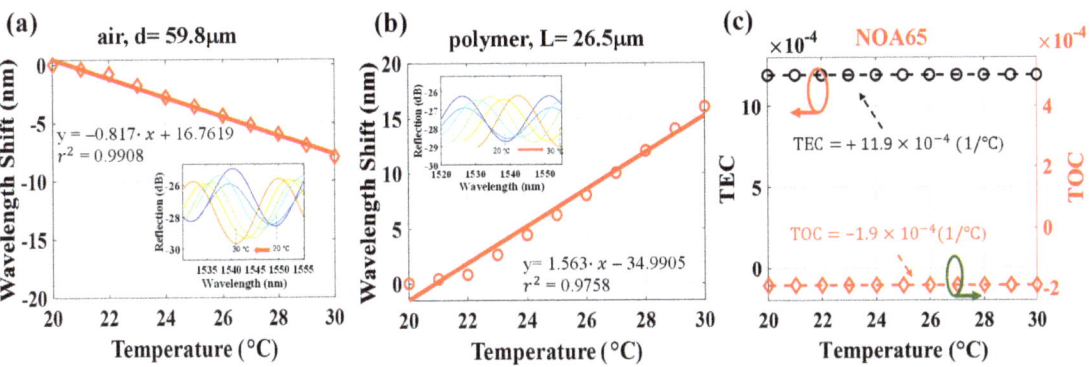

Figure 6. Wavelength shifts from 20 °C to 30 °C for the cavities of (**a**) air, $\Delta\lambda_a$, and (**b**) NOA65-polymer, $\Delta\lambda_p$; (**c**) measured TEC and TOC from the results of (**a**,**b**).

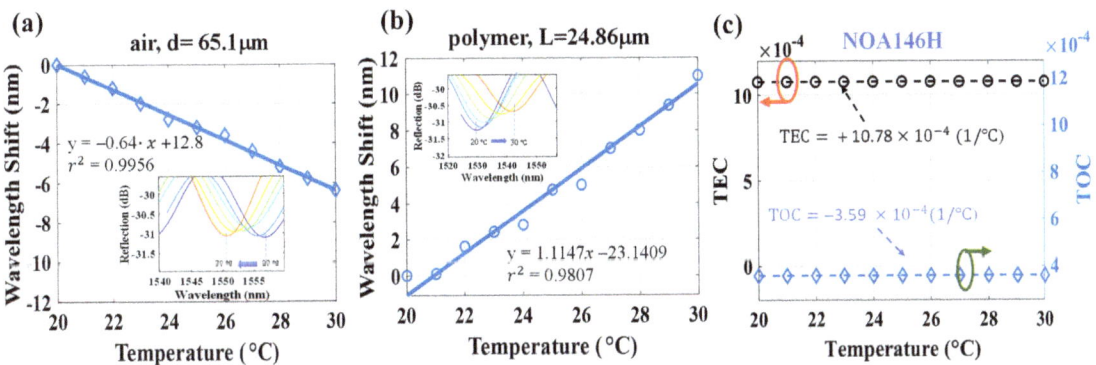

Figure 7. Wavelength shifts from 20 °C to 30 °C for the cavities of (**a**) air, $\Delta\lambda_a$, and (**b**) NOA146H-polymer, $\Delta\lambda_p$; (**c**) the measured TEC and TOC from the results of (**a**,**b**).

The above-determined TOCs and TECs obtained by the proposed approach are listed in Table 1 and compared with those in the reference data. We can see that the results of the obtained TOCs showed consistency. However, the values of the obtained TECs in this study were several times higher than those of the reference data. It can be easily understood that, in this study, with the polymer filled inside the cylindrical hollow fiber and adhered to the inner wall of the HCF, the radial expansion was almost zero to enhance the axial expansion of the polymers. Therefore, volume expansion degenerated into linear expansion to show that the linear expansion was further exacerbated. The volume expansion coefficient (γ) was demonstrated to be almost three times the linear expansion coefficient (α), i.e., $\gamma = 3\alpha$ [20]. The deviations in the TOC and TEC measurement via the proposed method could be estimated by the following relations: $TOC_d = |TOC_m - TOC_r|$ and $TEC_d = |TEC_m - TEC_r|$, respectively. TOC_d, TOC_m, and TOC_r, and TEC_d, TEC_m, and TEC_r are, respectively, the deviated, measured, and referred to values for the TOC and TEC. Here, we estimated that the TOC_d and TEC_d (linear) were 6×10^{-6} (°C^{-1}) and 3.67×10^{-5} (°C^{-1}) for the polymer NOA61, and 7×10^{-6} (°C^{-1}) and 1.46×10^{-4} (°C^{-1}) for NOA65, respectively. Consistent results show that, by using the proposed sensing scheme with a simple mathematical method, the proposed DCFFI could simultaneously and effectively sense the two parameters of TOC and TEC. We believe that the small deviations could be attributed to the measured error of the polymer/air cavity limited by the resolution of the microscope. It should be noted that, at the time of the writing of this paper, the NOA146H is a newly developed polymer; thus, there is no TOC and TEC-related information on the official website or other published references.

Table 1. Comparisons of the simultaneously measured TEC and TOC for three types of polymers between the experimental results and the reference data.

Polymers	NOA 61 (At 1550 nm)	NOA65 (At 1550 nm)	NOA146H (At 1550 nm)
TOC: (°C^{-1}) TEC: (°C^{-1}) (measured in the study)	-1.86×10^{-4} $+8.3 \times 10^{-4}$ (volume)	-1.9×10^{-4} $+11.9 \times 10^{-4}$ (volume)	-3.59×10^{-4} $+10.78 \times 10^{-4}$ (volume)
TOC: (°C^{-1}) TEC: (°C^{-1}) (reference data)	-1.8×10^{-4} $+2.4 \times 10^{-4}$ (linear) [21]	-1.83×10^{-4} $+2.5 \times 10^{-4}$ (linear) [19]	
TOC_d: (°C^{-1}) $TEC_{d, linear}$: (°C^{-1})	6×10^{-6} 3.67×10^{-5}	7×10^{-6} 1.46×10^{-4}	

4. Conclusions

This study demonstrated a new sensing configuration based on a dual-cavity fiber Fabry–Pérot interferometer (DCFFPI) for simultaneously measuring the thermo-optic coefficient (TOC) and thermal expansion coefficient (TEC) of a polymer. The polymer is naturally highly responsive to temperature (T); thus, the variation in the length (L) or refractive index (n), or both, can generate an optical path difference to shift the interference spectra. The proposed DCFFPI with a cascaded polymer and an air cavity can resolve the cross-sensitivity between the TOC and TEC. By analyzing the individual optical spectra of the polymer and air cavities, the parameters of TOC and TEC could be determined simultaneously. A set of experimental results showed the simultaneous measurement of the TOC and TEC with small measured deviations of 6×10^{-6} (°C^{-1}) and 3.67×10^{-5} (°C^{-1}) for the polymer NOA61 and 7×10^{-6} (°C^{-1}) and 1.46×10^{-4} (°C^{-1}) for NOA65. Both the feasibility and effectiveness of the proposed DCFFPI sensor have been fully demonstrated for the class of adhesive-based polymers. We believe that the dual-cavity fiber Fabry–Pérot interferometer presented in this study can be further applied in the simultaneous measurement of the thermal optical and thermal–mechanical parameters of various materials with liquid, adhesive, or gel types. It is particularly worth mentioning that, with the proposed DCFFPI sensing mechanism, only a picoliter volume of the materials to be measured is required. This is a great merit in the investigation of precious and rare materials.

Author Contributions: This work was carried out in collaboration between all authors. The first author and corresponding author C.-L.L., proposed the sensing concept, verified the mathematical method, analyzed the data, wrote and revised the manuscript. Another corresponding author C.-T.M., also proposed the sensing concept, verified the mathematical method, and revised the manuscript. Author, K.-C.Y., an undergraduate student in the Department of EOE, finished the experiments, managed figures and data. Author, Y.-M.C., the graduated student in the Department of EOE, performed the first stage experiments, managed figures, and data. All authors have read and agreed to the published version of the manuscript.

Funding: This work was supported by the Ministry of Science and Technology of Taiwan, MOST 110-2221-E-239-025-MY2.

Institutional Review Board Statement: Not applicable.

Informed Consent Statement: Not applicable.

Data Availability Statement: The data presented in this study are available on request from the corresponding author.

Conflicts of Interest: The authors declare no conflict of interest.

References

1. Jose, F.M.D.; Andrade, A.A.; Pilla, V.; Zilio, S.C. Simultaneous measurement of thermo-optic and thermal expansion coefficients with a single arm double interferometer. *Opt. Express* **2017**, *25*, 313–319. [CrossRef]
2. Hossain, M.; Chan, H.; Uddin, M. Simultaneous measurement of thermo-optic and stress-optic coefficients of polymer thin films using prism coupler technique. *Appl. Optics* **2010**, *49*, 403–408. [CrossRef] [PubMed]
3. Van Aken, J.A.; Janeschitz-Kriegl, H. Simultaneous measurement of transient stress and flow birefringence in one-sided compression (biaxial extension) of a polymer melt. *Rheol. Acta* **1981**, *20*, 419–432. [CrossRef]
4. Shimamura, A.; Priimagi, A.; Mamiya, J.I.; Ikeda, T.; Yu, Y.; Barrett, C.J.; Shishido, A. Simultaneous Analysis of Optical and Mechanical Properties of Cross-Linked Azobenzene-Containing Liquid-Crystalline Polymer Films. *ACS Appl. Mater. Interfaces* **2011**, *3*, 4190–4196. [CrossRef] [PubMed]
5. Esposito, M.; Buontempo, S.; Petriccione, A.; Zarrelli, M.; Breglio, G.; Saccomanno, A.; Szillasi, Z.; Makovec, A.; Cusano, A.; Chiuchiolo, A.; et al. Fiber Bragg Grating sensors to measure the coefficient of thermal expansion of polymers at cryogenic temperatures. *Sens. Actuators A Phys.* **2013**, *189*, 195–203. [CrossRef]
6. Li, X.; Lin, S.; Liang, J.; Zhang, Y.; Oigawa, H.; Ueda, T. Fiber-optic temperature sensor based on difference of thermal expansion coefficient between fused silica and metallic materials. *IEEE Photonics J.* **2012**, *4*, 155–162. [CrossRef]
7. Lee, C.L.; Ho, H.Y.; Gu, J.H.; Yeh, T.Y.; Tseng, C.H. Dual hollow core fiber-based Fabry–Perot interferometer for measuring the thermo-optic coefficients of liquids. *Opt. Lett.* **2015**, *40*, 459–462. [CrossRef] [PubMed]
8. Lee, C.L.; Lu, Y.; Cheng, C.H.; Ma, C.T. Microhole-pair hollow core fiber Fabry-Perot interferometer micromachining by a femtosecond laser. *Sens. Actuators A Phys.* **2020**, *302*, 111798. [CrossRef]
9. Kamikawachi, R.C.; Abe, I.; Paterno, A.S.; Kalinowski, H.J.; Muller, M.; Pinto, J.L.; Fabris, J.L. Determination of Thermo-Optic Coefficient in Liquids with Fiber Bragg Grating Refractometer. *Opt. Commun.* **2008**, *281*, 621–625. [CrossRef]
10. Yao, Q.; Meng, H.; Wang, W.; Xue, H.; Xiong, R.; Huang, B.; Tan, C.; Huang, X. Simultaneous Measurement of Refractive Index and Temperature Based on a Core-Offset Mach–Zehnder Interferometer Combined with a Fiber Bragg Grating. *Sens. Actuators A Phys.* **2014**, *209*, 73–77. [CrossRef]
11. Kim, Y.H.; Park, S.J.; Jeon, S.W.; Ju, S.; Park, C.S.; Han, W.T.; Lee, B.H. Thermo-Optic Coefficient Measurement of Liquids Based on Simultaneous Temperature and Refractive Index Sensing Capability of a Two-Mode Fiber Interferometric Probe. *Opt. Express* **2012**, *20*, 23744–23754. [CrossRef] [PubMed]
12. Lee, C.L.; You, Y.W.; Dai, J.H.; Hsu, J.M.; Horng, J.S. Hygroscopic polymer microcavity fiber Fizeau interferometer incorporating a fiber Bragg grating for simultaneously sensing humidity and temperature. *Sens. Actuators B Chem.* **2016**, *222*, 339–346. [CrossRef]
13. Wang, L.; Zhang, W.; Wang, B.; Chen, L.; Bai, Z.; Gao, S.; Li, J.; Liu, Y.; Zhang, L.; Zhou, Q.; et al. Simultaneous Strain and Temperature Measurement by Cascading Few-Mode Fiber and Single-Mode Fiber Long-Period Fiber Gratings. *Appl. Opt.* **2014**, *53*, 7045. [CrossRef] [PubMed]
14. Hu, D.J.J.; Lim, J.L.; Jiang, M.; Wang, Y.; Luan, F.; Shum, P.P.; Wei, H.; Tong, W. Long Period Grating Cascaded to Photonic Crystal Fiber Modal Interferometer for Simultaneous Measurement of Temperature and Refractive Index. *Opt. Lett.* **2012**, *37*, 2283. [CrossRef]
15. Ma, C.T.; Chang, Y.W.; Yang, Y.J.; Lee, C.L. A Dual-Polymer Fiber Fizeau Interferometer for Simultaneous Measurement of Relative Humidity and Temperature. *Sensors* **2017**, *17*, 2659. [CrossRef] [PubMed]
16. Zhou, A.; Qin, B.; Zhu, Z.; Zhang, Y.; Liu, Z.; Yang, J.; Yuan, L. Hybrid structured fiber-optic Fabry–Perot interferometer for simultaneous measurement of strain and temperature. *Opt. Lett.* **2014**, *39*, 5267. [CrossRef] [PubMed]

17. Shao, M.; Cao, Z.; Gao, H.; Hao, M.; Qiao, X. Large measurement-range and low temperature cross-sensitivity optical fiber curvature sensor based on Michelson interferometer. *Opt. Fiber Technol.* **2022**, *72*, 102990. [CrossRef]
18. He, C.; Korposh, S.; Correia, R.; Liu, L.; Hayes-Gill, B.R.; Morgan, S.P. Optical fibre sensor for simultaneous temperature and relative humidity measurement: Towards absolute humidity evaluation. *Sens. Actuators B Chem.* **2021**, *344*, 130154. [CrossRef]
19. Sun, B.; Wang, Y.; Qu, J.; Liao, C.; Yin, G.; He, J.; Zhou, J.; Tang, J.; Liu, S.; Li, Z.; et al. Simultaneous measurement of pressure and temperature by employing Fabry-Perot interferometer based on pendant polymer drople. *Opt. Express* **2015**, *23*, 1906–1911. [CrossRef] [PubMed]
20. Available online: https://www.sciencedirect.com/topics/engineering/linear-coefficient-of-expansion (accessed on 22 October 2022).
21. Available online: https://www.norlandprod.com/adhesiveindex2.html (accessed on 10 October 2022).

Article

A Flexible Multifunctional PAN Piezoelectric Fiber with Hydrophobicity, Energy Storage, and Fluorescence

Qisong Shi [1,*], Rui Xue [1], Yan Huang [1], Shifeng He [1], Yibo Wu [1] and Yongri Liang [2,*]

1. College of New Materials and Chemical Engineering, Beijing Institute of Petrochemical Technology, Beijing 102617, China
2. State Key Lab of Metastable Materials Science and Technology, School of Materials Science and Engineering, Yanshan University, Qinhuangdao 066004, China
* Correspondence: shiqisong@bipt.edu.cn (Q.S.); liangyr@ysu.edu.cn (Y.L.)

Citation: Shi, Q.; Xue, R.; Huang, Y.; He, S.; Wu, Y.; Liang, Y. A Flexible Multifunctional PAN Piezoelectric Fiber with Hydrophobicity, Energy Storage, and Fluorescence. *Polymers* 2022, *14*, 4573. https://doi.org/10.3390/polym14214573

Academic Editor: Jung-Chang Wang

Received: 10 October 2022
Accepted: 24 October 2022
Published: 28 October 2022

Publisher's Note: MDPI stays neutral with regard to jurisdictional claims in published maps and institutional affiliations.

Copyright: © 2022 by the authors. Licensee MDPI, Basel, Switzerland. This article is an open access article distributed under the terms and conditions of the Creative Commons Attribution (CC BY) license (https://creativecommons.org/licenses/by/4.0/).

Abstract: Lightweight, flexible, and hydrophobic multifunctional piezoelectric sensors have increasingly important research value in contemporary society. They can generate electrical signals under the action of pressure and can be applied in various complex scenarios. In this study, we prepared a polyacrylonitrile (PAN) composite fiber doped with imidazolium type ionic liquids (ILs) and europium nitrate hexahydrate (Eu $(NO_3)_3 \cdot 6H_2O$) by a facile method. The results show that the PAN composite fibers had excellent mechanical properties (the elongation at break was 114% and the elastic modulus was 2.98 MPa), hydrophobic self-cleaning ability (water contact angle reached 127.99°), and can also emit light under UV light irradiation red fluorescence. In addition, thanks to the induction of the piezoelectric phase of PAN by the dual fillers, the composite fibers exhibited efficient energy storage capacity and excellent sensitivity. The energy density of PAN@Eu-6ILs reached a maximum of 44.02 mJ/cm^3 and had an energy storage efficiency of 80%. More importantly, under low pressure detection, the sensitivity of the composite fiber was 0.69 kPa^{-1}. The research results show that this PAN composite fiber has the potential to act as wearable piezoelectric devices, energy storage devices, and other electronic devices.

Keywords: piezoelectricity; polyacrylonitrile; ionic liquid; energy storage; sensors

1. Introduction

Wearable electronic products have developed rapidly in recent decades [1], and they are mostly used in the medical and health field to monitor human body signals or as self-powered generators as new energy devices in the future [2,3]. Compared with traditional electrical materials, which are generally rigid and unsuitable for flexible devices [4], nanofiber-based electronic products can achieve light weight and stretchable flexibility, which ensures structural integrity under a certain tensile deformation [5]. Flexible and wearable pressure sensors can respond to external pressure stimuli and convert them into electrical signals [6], and the capacitive piezoelectric sensor, as the one of the most dazzling of the new generations of piezoelectric sensors, has high-quality features such as simple and efficient device structure, high sensitivity, low loss, and fast response speed [7,8]. It can continuously accumulate the internal charge of the material under pressure to achieve self-powering functionality [9]. The distance between electrodes and the high dielectric constant of the dielectric layer has a crucial impact on capacitive piezoelectric sensors [10]. At the same time, the products prepared by electrospinning technology are often disturbed by the influence of moisture in the air and sweat produced by the human body [11]. Therefore, the development of a capacitive flexible piezoelectric sensor with high dielectric constant, high sensitivity, and hydrophobicity has become a research hotspot. To meet the requirements of energy and environmental sustainability [12], there is an urgent need to develop multipurpose sensors with energy storage or self-powering. There are recent research reports. For example, Khalifa [13] et al., reported a poly(vinylidene fluoride) PVDF/mica nanosheet composite (PMNCs) prepared by solution casting method with a high sensitivity of 3.2 N^{-1} and capable of generating a maximum output voltage of 32 V under a pressure of 5 N, which was used in wearable energy storage and piezoelectric sensors. Wang [14] et al. reported a hybrid nanogenerator with self-powered and simultaneous energy storage through a power management circuit. This hybrid nanogenerator had a maximum output power of 1.7 mW when the load resistance was 10 MΩ, and the energy storage efficiency was as high as 112%.

Polyacrylonitrile (PAN) is a kind of piezoelectric polymer. PAN nanofibers prepared by electrospinning technology have the advantages of appropriate piezoelectric properties, good mechanical properties, and high piezoelectric sensitivity, which makes it widely used in the field of flexible piezoelectric sensors [15]. PAN is an amorphous vinyl-type polymer containing a cyano group (-CN) in each repeating unit [16]. PAN has two typical conformations, planar zigzag conformation and 3^1 helical conformation [17], and the planar zigzag conformation has a total transformation (TTTT) structure [18]. Compared with the most popular piezoelectric polymer polyvinylidene fluoride (PVDF), it has a stronger dipole moment of 3.5 Debye than the β phase (2.1 Debye) of PVDF [19]. PAN also has smaller dielectric loss and good thermal stability [20]. Therefore, PAN is expected to be a better piezoelectric material than PVDF. Ionic liquid (ILs) is a conductive material composed entirely of ions [21] that has high ionic conductivity, electrochemical stability, and other properties [22]. It was reported that good dielectric properties and improved mechanical properties were obtained by incorporating ionic liquids into polymer systems [23]. In addition, mono-, binary, or ternary sensor systems combined with ionic liquids have also been widely developed and utilized. Jiang [24] et al. reported a strain sensor of TPU/ILs. The sensor had a maximum tensile limit of 400%, a high sensitivity of 1.28 and a fast response of 67 ms under high strain conditions. Zhang [25] et al. prepared a PVDF dielectric composite material doped with ILs and graphene. When the concentrations of graphene and ILs both reached 2 wt%, the composite material achieved the best dielectric properties of about 7, which was nearly six times that of PVDF, and the dielectric loss was lower than 0.2. Yuan [26] et al. designed a ternary composite of RGO@ILs/PBO. The mechanical properties and dielectric properties of the composite material were greatly enhanced. Its dielectric constant at 10^3 Hz and 200 °C was 35.51, and the dielectric loss was 0.09. Meanwhile, its application in energy storage had an energy density as high as 2.38 J/cm^3 and a breakdown strength of 122.96 kV/mm. In addition, fluorescent nanofibers were prepared by doped fluorescent fillers with piezoelectric polymers, which could be applied to self-powered optoelectronic devices. Fu [27] et al. prepared a polyvinylidene fluoride hexafluoropropylene (PVDF-HFP) composite fibers with addition of Eu $(TTA)_3(TPPO)_2$ and $BaTiO_3$ by electrospinning technology. The results showed that it can display stable red fluorescence and had a piezoelectric sensitivity of 0.49 kPa^{-1} and an energy storage density of 30.45 mJ/cm^3.

PAN composite fibers doped with ILs and (Eu $(NO_3)_3 \cdot 6H_2O$) were prepared, and a flexible multifunctional PAN piezoelectric fiber with hydrophobicity, fluorescence, and energy storage was obtained through the synergistic effect of the dual fillers. It can be used in fields such as flexible piezoelectric sensors and energy storage devices. The results show that the content of the planar zigzag phase in the dual-filler PAN@Eu-ILs composite fiber was as high as 94.74%, which demonstrated its high piezoelectric phase content. At the same time, the small interplanar spacing was also proof of the high piezoelectric phase content. The double-filled PAN composite fibers also had a good energy storage density, reaching an energy storage density of 44.02 mJ/cm^3 under the action of an electric field of 420 kV/cm^3, which was 1.64 times of the energy storage density of pure PAN (26.84 mJ/cm^3). More importantly, the dual-filler PAN composite fibers also had a larger tensile deformation range, softer properties, and can exhibit red fluorescence. In addition, the water contact angle of the single-component PAN-ILs composite fiber reached 127.99°, and its excellent hydrophobic properties and self-cleaning ability can avoid the influence of environmental humidity on the sensor. Finally, the PAN composite fiber has a high sensitivity up to 0.69 kPa^{-1} in the pressure range of <1 kPa, and it has a high sensitivity for low pressure detection.

2. Materials and Methods

2.1. Materials

The chemicals used in this study include polyacrylonitrile (PAN), MW = 85,000, Shanghai MACKLIN Company, Shanghai, China; 1-allyl-3-butylimidazolium tetrafluoroborate [AMIm][BF$_4$] (ILs), Lanzhou Institute of Chemical Physics, Chinese Academy of Sciences, Lanzhou, China; europium nitrate hexahydrate (Eu (NO$_3$)$_3$·6H$_2$O), Beijing Warwick Chemical Co., Ltd., Beijing, China; N, N dimethylformamide (DMF), Beijing Chemical Plant, Beijing, China. All samples were used directly.

2.2. Preparation of PAN@Eu-ILs Multifunctional Composite Fibers

A total of 1.4 g of PAN was added to 10 mL of DMF solution. The mass fraction of PAN in the solution was 14% (w/v). It was heated and stirred at 60 °C until the PAN was completely dissolved and the solution was pale yellow. Then, 0.14 g of Eu (NO$_3$)$_3$·6H$_2$O was added to the solution, and the solution was heated and stirred until the Eu (NO$_3$)$_3$·6H$_2$O was completely dissolved. Finally, ILs were added to the solution, where IL levels were 0%, 3%, 6%, and 9% of the mass fraction of PAN, which was named PAN@Eu-XILs (X was the mass fraction of ILs in PAN). The solution was heated and stirred for electrospinning. A 5 mL syringe was used to extract an appropriate amount of spinning solution. The process parameters were as follows. A 20 G metal needle was used on the syringe, and the injection speed was 0.2 mm/min. The spinning distance was 15 cm, the applied spinning voltage was a positive voltage of 16 kV and a negative voltage of 2 kV, and the rotational speed of the drum was 120 r/min. The ambient humidity was 45–50%. A schematic diagram of the experimental preparation is shown in Figure 1.

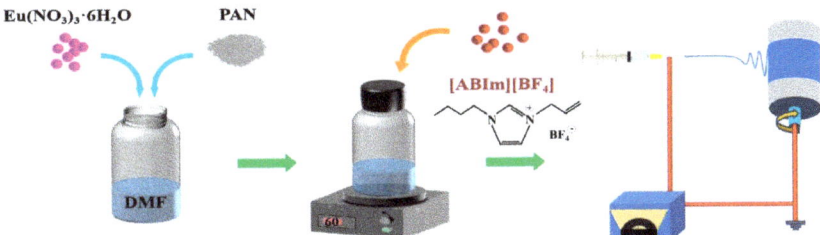

Figure 1. Experimental flow chart.

2.3. Characterization

The SSX-550 model SEM (Shimadzu Co., Kyoto, Japan) was used to study the surface morphology and orientation of nanofibers. The diameters of the nanofibers in the SEM images were statistically calculated by Image J software. EDS images were tested by a Sigma 300 model Oxford Energy Spectrometer (Carl Zeiss Co., Jena, Germany). All samples were treated with gold spray. Nicolet-380 Fourier Transform infrared spectrometer (Thermoelectric Co., West Chester, PA, USA) was used to determine the chemical structure of nanofibers, and the vibration bands areas were calculated by OMINC software analysis. The FS5 spectrofluorometer (Edinburgh Instruments, SCT, UK) was used to analyze the fluorescence properties of composite fibers. The crystal structure of composite fibers was studied by XRD-7000 X-ray diffractometer (Shimadzu Co., Kyoto, Japan). The SL200KS optical contact Angle and surface tensionmeter (Solon Information Technology Co., Ltd., Shanghai, China) was used to measure the contact Angle of water. The DSC and TG of the composite fibers were measured using a synchronous thermal analyzer from setline, the heating rate was 10 °C/min, and the temperature curve was 35–800 °C. The ferroelectric energy storage performance of composite fiber was studied by TF analyzer

2000 (aixACCT Co., Eschweiler, Germany). The conductive tape was tightly attached to the upper and lower surfaces of the sample. The ferroelectric test was carried out at room temperature with a test frequency of 10 Hz and a test voltage range of 0.1–10 kV. GP6220 High temperature tensile testing machine (Gaopin Testing Instruments Co., LTD., Suzhou, China) was used to test the mechanical properties of composite fibers. The sample was cut into a size of 40 mm × 10 mm, the test temperature was 25 °C, and the test speed was 10 mm/min. The BDS40 broadband dielectric impedance spectrometer (NOVOCONTROL Co., Montabaur, Germany) was used to measure permittivity and dielectric loss. The test range was 10^1 to 10^6 Hz. The TH2828 LCR digital bridge from Changzhou Tonghui Electronics Co., Ltd. Changzhou, China was used to test the sensitivity of composite fibers. The sample was cut into a size of 20 mm × 20 mm, and the conductive tape was tightly attached to the upper and lower surfaces of the sample. Under the test conditions of 1.5 V and 100 Hz frequency, the composite fiber was subjected to a pressure test and the change in the capacitance value was recorded.

3. Results

3.1. SEM

It can be seen from Figure 2 that the fiber diameter was uniform, and the surface was smooth, indicating that the electrospinning process was not affected by the external environment and the spinning process was relatively stable. The composite fiber with a single addition of IL displayed a certain orientation. This may be due to the high charge density of the solution, which was formed by the increase in the ionic conductivity of the solution by ILs [28]. Under the action of the high electric field force, the extensional flow of the solution under the action of the electrostatic force was enhanced, which was why the composite fiber exhibited a certain orientation [29]. However, after the addition of Eu $(NO_3)_3 \cdot 6H_2O$ and ILs, the strong interaction between them had a certain effect on the orientation of the composite fibers, and the orientation was not as obvious as that of the PAN-IL composite fibers. As the IL concentration increased, the electrostatic force stretching increased, the dipoles became more polarized, and more PAN piezoelectric phases (planar zigzag conformation) were induced to form [15]. As shown in Figure 3, the bar graph represented the fiber diameter distribution, and the peak of the curve represented the average diameter. The average diameters of PAN@Eu-ILs composite fibers were as follows: 253 nm for PAN, 587 nm for PAN@Eu, 375 nm for PAN@ILs, 425 nm for PAN@Eu-3ILs, 475 nm for PAN@Eu-6ILs, and PAN@Eu-9ILs was 515 nm. The average diameter of the composite fibers added with Eu $(NO_3)_3 \cdot 6H_2O$ became obviously thicker, while that with the addition of ILs decreased slightly. First, after adding Eu $(NO_3)_3 \cdot 6H_2O$, the concentration of the solution increased. The higher solution concentration caused the thickest of the fiber diameter of the PAN@Eu. Meanwhile, due to the ionic conductivity of ILs, the spinning solution was stretched by stronger electrostatic force, and the diameter of the composite fibers decreased slightly. The EDS images in Figure 4 show that the composite fibers contain C, O, N, F, B, and Eu elements. It was confirmed that ILs and Eu $(NO_3)_3 \cdot 6H_2O$ were successfully added and uniformly distributed.

3.2. FT-IR

Figure 5a–d shows the FTIR spectra of the PAN@Eu-ILs composite fibers. Both PAN nanofibers and PAN composite fibers had similar infrared spectral curves. The characteristic peaks corresponding to PAN were as follows: The vibration bands at 2245 cm^{-1} was attributed to the stretching vibration of the -C≡N bond [30]. The vibrational peaks at 1450, 1360, and 1070 cm^{-1} were formed by the bending vibrations of the -CH_2-, -CH, and the -CN bonds, respectively [31]. Figure 5a shows that, compared with PAN, the characteristic peaks at 817 cm^{-1} and 741 cm^{-1} of PAN@Eu correspond to the vibrations of (C-N-C(O)) and (O-C-N) in the Eu^{3+}-PAN structure, respectively [32]. However, after the PAN@Eu composite fiber was doped with ILs, the vibration bands position shifted toward the lower wavenumber side. The vibration bands positions shifted more with the increase in IL concentration, which reflected the strong interaction between Eu $(NO_3)_3 \cdot 6H_2O$ and IL. From Figure 5b, we can see that a strong vibration band appeared at 1562 cm^{-1}, which was attributed to the stretching vibration of the imidazole ring of ILs [33]. With the increase in IL concentration, the vibration bands intensity gradually increased. Figure 5a,b confirms the successful doping of Eu $(NO_3)_3 \cdot 6H_2O$ and IL on PAN composite fibers. In order to further explore the effect of fluorescent fillers and ILs on the structure of PAN, we analyzed and calculated the two conformations of PAN composite fibers. The vibrational bands of 1250 cm^{-1} and 1230 cm^{-1} correspond to the planar zigzag and 3^1 helical conformation of PAN, respectively [34]. The planar zigzag conformation was more favorable for the generation of piezoelectric phase. The area ratio of the two peaks can be calculated by Equation (1) [30], thereby obtaining the changes of the two conformations in different PAN composite fibers, and further deducing the effect on the piezoelectric properties.

$$\Phi = \frac{S_{1250}}{S_{1250} + S_{1230}} \tag{1}$$

where S_{1230} and S_{1250} represent the peak areas at 1230 cm^{-1} and 1250 cm^{-1}, respectively. It can be seen from Figure 5d that after adding Eu $(NO_3)_3 \cdot 6H_2O$ and ILs, the intensity of the vibration bands at 1250 cm^{-1} was improved, indicating that the 3^1 helical conformation gradually changed to a planar zigzag conformation. In general, the higher the content of the planar zigzag conformation, the higher piezoelectric properties of PAN nanofibers [35]. The Φ of PAN@Eu reached 90.04%, the Φ of PAN-6ILs reached 92.01%, and the Φ of PAN@Eu-6ILs reached 94.74%. Compared with the PAN nanofibers (Φ reached 85.06%), the composite fillers with ILs and Eu $(NO_3)_3 \cdot 6H_2O$ simultaneously played a synergistic effect on the generation of the piezoelectric phase of PAN, which had the best piezoelectric performance. Therefore, when the amount of addition was appropriate, the best piezoelectric material can be obtained.

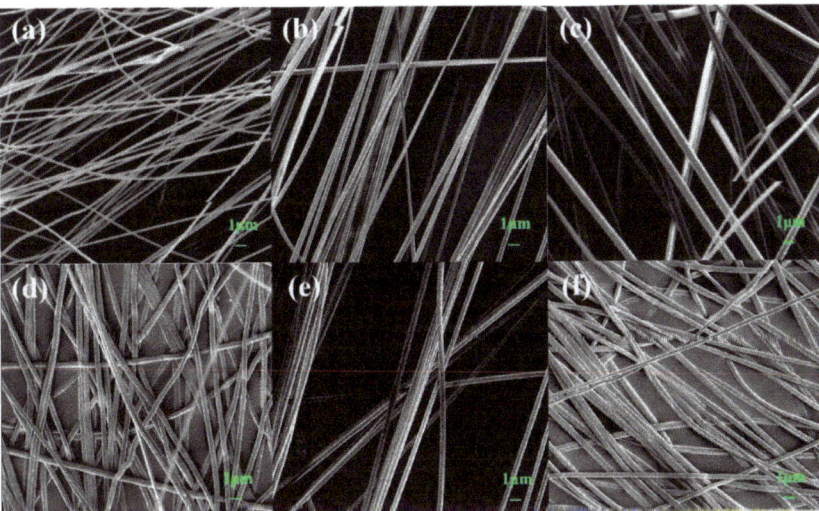

Figure 2. SEM of PAN@Eu/ILs composite fibers: (**a**) pure PAN; (**b**) PAN-6ILs; (**c**) PAN@Eu; (**d**) PAN@Eu-3ILs; (**e**) PAN@Eu-6ILs; (**f**) PAN@Eu-9ILs.

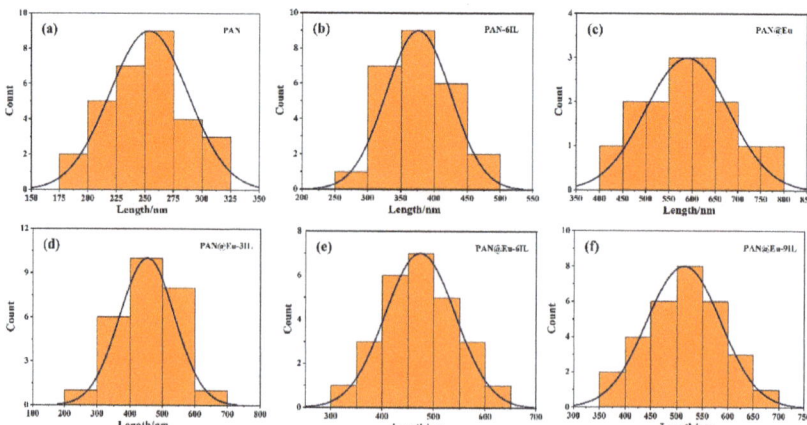

Figure 3. Diameter distribution of PAN@Eu/ILs composite fibers: (**a**) pure PAN; (**b**) PAN-6ILs; (**c**) PAN@Eu; (**d**) PAN@Eu-3ILs; (**e**) PAN@Eu-6ILs; (**f**) PAN@Eu-9ILs.

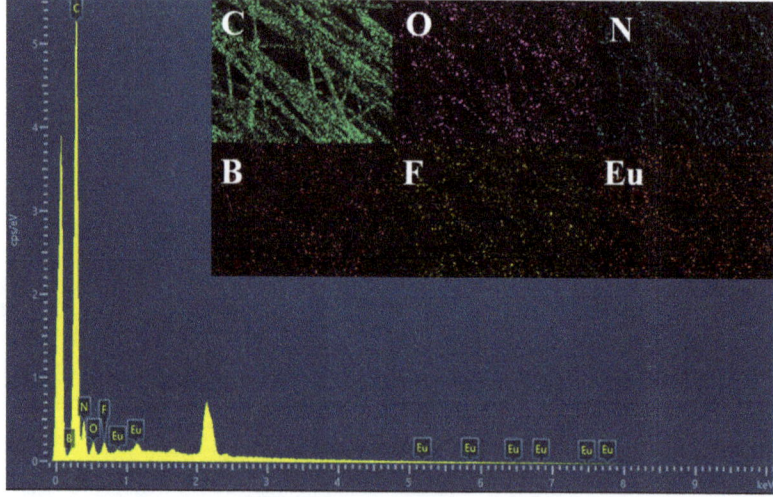

Figure 4. EDS images of PAN@Eu-6ILs composite fibers. The inset shows the EDS distribution image of each element in the PAN@Eu-6ILs composite fiber: C in PAN@Eu-6ILs, O in PAN@Eu-6ILs, N in PAN@Eu-6ILs, B in PAN@Eu-6ILs, F in PAN@Eu-6ILs, Eu in PAN@Eu-6ILs.

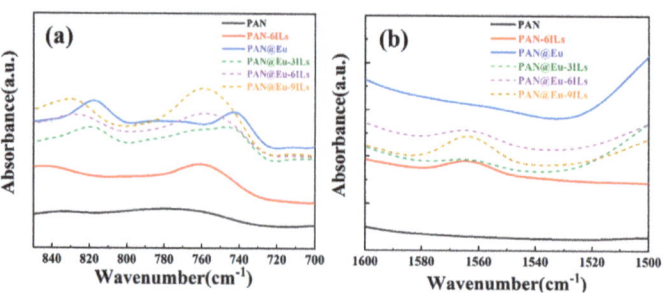

Figure 5. *Cont.*

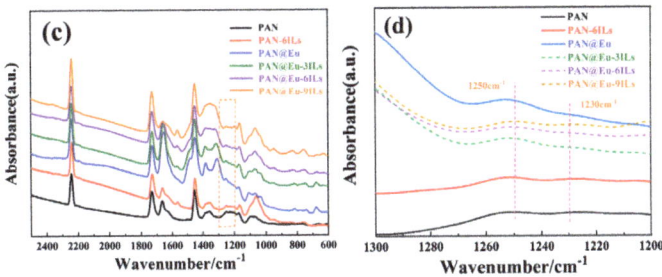

Figure 5. FT-IR spectra of PAN@Eu-ILs composite fibers in the frequency range of (**a**) 850 cm^{-1}–700 cm^{-1}; (**b**) 1600 cm^{-1}–1500 cm^{-1}; (**c**) 2500 cm^{-1}–600 cm^{-1}; (**d**) 1300 cm^{-1}–1200 cm^{-1}.

3.3. XRD

Figure 6a shows the XRD patterns of the PAN@Eu-ILs composite fibers. For pure PAN, the main characteristic peak at 2θ = 16.7° corresponded to the (100) facet peak of PAN [36], and the broad peak at approximately 2θ = 29° was assigned to the rotational disorder in the chain packing [37]. It was obvious that the crystal plane diffraction peak position of (100) was shifted to the right after adding ILs and Eu (NO$_3$)$_3$·6H$_2$O. In general, the right shift of the diffraction peak indicated that the addition of fillers promoted the transition from the 3^1 helical conformation to the planar zigzag conformation [38], which was beneficial to the improvement of the piezoelectric properties of the composite fibers. Among them, the (100) crystal plane diffraction peak of the PAN@Eu-6ILs composite fiber shifted to the right the most, which indicated that the plane zigzag conformation content was the most. The results corresponded to the FT-IR, and further verified the synergistic effect of composite fillers helped PAN composite fibers to achieve the best piezoelectric properties. The position statistics of the diffraction peaks of the (100) crystal plane is shown in Figure 6b. The (100) crystal plane positions of PAN@Eu and PAN-6ILs in the single-filler system were 16.91° and 17.02°, respectively, and the diffraction peak intensity of PAN@Eu was weak. In the dual-filler system, when the ILs contents were 3%, 6%, and 9%, the positions of the (100) crystal plane corresponded to 17.13°, 17.24° and 17.02°, respectively. The interplanar spacing (D) can be calculated by the Bragg Equation (2) [39]:

$$2d \sin \theta = n\lambda \qquad (2)$$

where d is the interplanar spacing of the corresponding crystal plane, and θ and λ are the diffraction angle and X-ray wavelength, respectively. The calculated results show that the interplanar spacings of PAN, PAN@Eu, and PAN-6ILs were 0.532 nm, 0.528 nm, and 0.524 nm, respectively, while in the dual filler system, the interplanar spacings of PAN@Eu-3ILs, PAN@Eu-6ILs and PAN@Eu-9ILs were 0.520 nm, 0.518 nm, and 0.524 nm, respectively. It was evident that the synergistic effect of the dual fillers further reduced the interplanar spacing. According to previous reports, the planar zigzag conformation has a smaller interplanar spacing [40]. This further confirmed the generation of the PAN piezoelectric phase.

3.4. DSC and TG

Figure 7 is the DSC and TG curves of PAN composite fibers. As shown in Figure 7a, the PAN composite fiber had a characteristic exothermic peak temperature (T$_p$) at 274 °C, corresponding to the cyclization reaction of PAN. The peak temperatures of PAN-6ILs, PAN@Eu, and PAN@Eu-6ILs composite fibers were 283.5 °C, 308 °C, and 282.3 °C, respectively. Because of ILs and Eu (NO$_3$)$_3$·6H$_2$O, the activation energy required for the cyclization of PAN composite fibers increased, and the T$_p$ of the composite fibers increased. This indicated that the PAN composite fibers obtained a uniform and stable structure with improved thermal stability [41]. Compared with the single-filler PAN composite fibers, the T_p of the dual-filler PAN@Eu-6ILs decreased, which may be the result of the strong interaction between ILs and Eu $(NO_3)_3 \cdot 6H_2O$. In addition, the cyclization reaction of different intermolecular cyano groups resulted in a slight elevation of the baseline of PAN composite fibers [42].

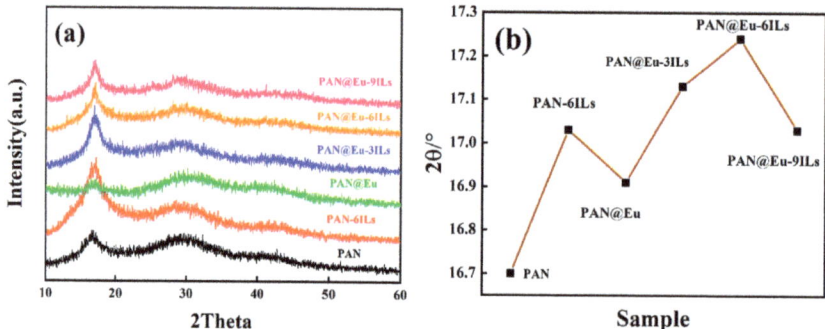

Figure 6. (a) XRD patterns of PAN@Eu-ILs composite fibers; (b) the position of the (100) crystal plane diffraction peak of PAN@Eu-ILs composite fibers.

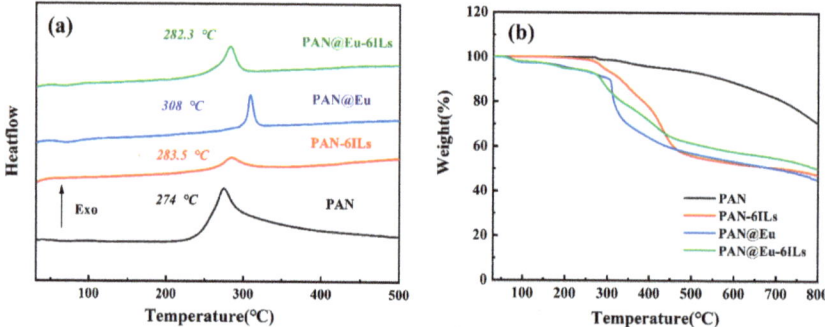

Figure 7. (a) DSC spectrum of PAN composite fibers; (b) TG curve of PAN composite fibers.

As shown in Figure 7b, PAN and PAN-6ILs began to lose a significant amount of weight at around 280 °C, which was mainly due to the decomposition of ILs and polymer chains. With the addition of fluorescent fillers, the PAN composite fibers began to dissociate rapidly at 308 °C, the thermal decomposition temperature shifted to high temperature, and the thermal stability was improved. However, the residual weight of PAN at 800 °C s was 71%, which was much higher than that of the PAN composite fibers with Eu $(NO_3)_3 \cdot 6H_2O$ and ILs added. For example, the residual weight of PAN@Eu-6ILs at 800 °C was only 51%. Although the residual weight decreased, the piezoelectric and ferroelectric energy storage properties in the PAN composite fibers were improved, which was beneficial to its better application.

3.5. Dielectric Properties

As shown in Figure 8a, the dielectric constant of PAN composite fibers increased significantly after the addition of ILs and Eu $(NO_3)_3 \cdot 6H_2O$. From Table 1, it can be concluded that the dielectric constant of PAN@Eu reached 3.68 at 10^3 Hz, which was 1.5 times that of the PAN nanofibers (2.48). In addition, the dielectric constant of the dual-filler PAN composite fiber was particularly improved. When the concentration of ILs was 9%, the dielectric constant of PAN@Eu-9ILs composite fibers at 10^3 Hz reached 6.6, which was 2.7 times that of the PAN nanofibers. The increase in the dielectric constant of PAN composite fibers in the low frequency region indicated the improvement of the interfacial polarization, which was due to the high polarity of the ILs [43,44]. In addition, the synergistic effect of the dual fillers had a positive effect on the enhanced dielectric constant. In terms of dielectric loss, the dielectric loss of PAN composite fibers increased slightly. Compared with other reported systems [45,46], this system remained at a low level (<0.14), which was beneficial for the long-term use of composite fibers as energy storage devices or piezoelectric sensors.

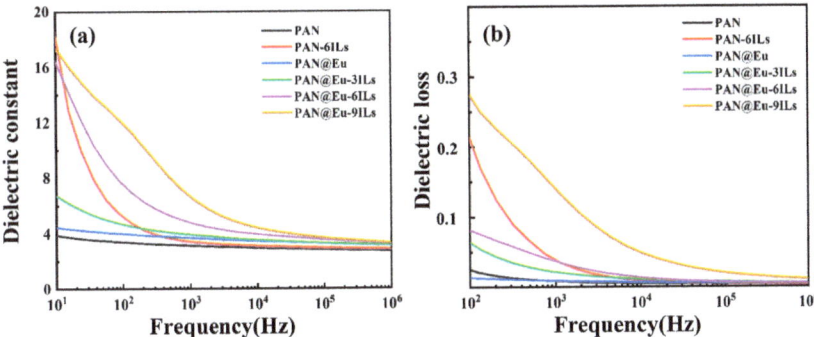

Figure 8. Dielectric properties of PAN@Eu-ILs composite fibers: (**a**) dielectric constant at different frequencies; (**b**) dielectric loss at different frequencies.

Table 1. Dielectric properties at 10^3 Hz and specific parameters of mechanical properties of PAN@Eu-ILs composite fibers.

Sample	Dielectric Constant/10^3 Hz	Dielectric Loss/10^3 Hz	The Maximum Stress/MPa	Breaking Elongation/%	The Elastic Modulus/MPa
PAN	2.48	0.019	4.49	92	4.88
PAN-6ILs	4.69	0.024	4.17	100	4.17
PAN@Eu	3.68	0.030	4.48	97	4.61
PAN@Eu-3ILs	3.91	0.042	3.38	114	2.96
PAN@Eu-6ILs	4.79	0.045	4.47	112	3.99
PAN@Eu-9ILs	6.60	0.139	5.35	102	5.25

3.6. Mechanical Properties and Ferroelectric Properties

Figure 9a is the stress–strain curve of PAN composite fibers, and Table 1 shows the specific parameters of the dielectric properties and mechanical properties. Polyacrylonitrile nanofibers certainly exhibited mechanical strength and stretchability [47,48]. The addition of ILs made the PAN nanofibers exhibit a certain orientation under the action of electrostatic force. This had a favorable effect on the mechanical properties of composite fibers. As shown in Figure 9a, the elongation at break of PAN composite fibers was improved, and the elongation was improved more obviously by ILs. In addition, the dual fillers had a better effect on the improvement of the tensile strength and the elastic modulus of the composite fibers, and the best was PAN@Eu-3ILs, whose tensile strength and the elastic modulus were significantly lower than PAN and single-filler PAN composite fibers. The more flexible PAN composite fibers were beneficial to its application as a sensor in human health monitoring. Figure 9b shows that a PAN composite fiber with a size of 40 mm × 40 mm could easily stand on the leaf of a green plant and the leaf was not bent. In addition, the composite fiber can also withstand bending and twisting without deformation. This demonstrates the light and soft characteristics of PAN composite fibers.

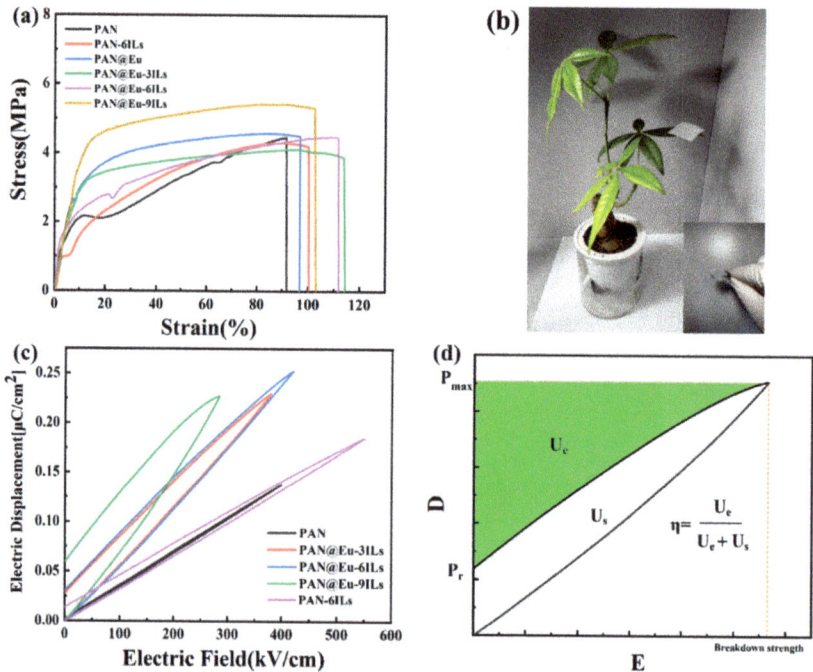

Figure 9. (**a**) Mechanical properties of PAN composite fibers; (**b**) DE hysteresis loop of PAN composite fiber; (**c**) photo of PAN composite fibers; (**d**) DE hysteresis loop diagram.

Figure 9c shows the single-phase hysteresis loop of the PAN composite fiber. In the ferroelectric test, we recorded the maximum test electric field before breakdown of the sample and defined it as the breakdown strength of the sample. In Figure 9d, E is the electric field strength, D is the electrical displacement, P_{max} is the maximum polarization value, P_r is the remanent polarization value, U_e is the energy density, U_s is the loss energy density, and η is the energy storage efficiency. As shown in Figure 9c, the maximum polarization value (P_{max}) of the dual-filler PAN@Eu-ILs composite fibers was slightly increased compared with that of the single-doped PAN-6ILs, which was the result of the synergistic effect of the dual fillers. Its effect on the breakdown strength was not obvious, but still slightly higher than that of pure PAN nanofibers. In addition, due to the influence of the dielectric loss of the PAN@Eu-ILs composite fibers, the remanent polarization value (P_r) kept increasing with the increase in the dielectric loss, and the corresponding energy loss also increased, which seriously reduced the energy storage efficiency. When the IL concentrations were 3%, 6%, and 9%, its P_r values were 0.027, 0.03, and 0.06 respectively, and the energy storage efficiency values were 81%, 80%, and 60%, respectively. However, under the effect of breakdown strength and enhanced P_{max}, a high energy storage density can still be maintained. Among them, PAN@Eu-6ILs had the best performance, its breakdown strength was 420 kV/cm^2, and its P_{max} was increased from 0.14 μC/cm^2 of PAN to 0.25 μC/cm^2. Under this condition, its energy density (U_e) reached 44.02 mJ/cm^3, while the U_e of PAN was 26.84 mJ/cm^3, and the energy density was increased by 1.64 times. It was obvious that the PAN@Eu-ILs system still maintained excellent energy storage performance. Compared with other reported systems [49,50], the PAN@Eu-ILs composite fibers had excellent energy conversion efficiency and good energy density. Therefore, it can be expected to be applied in the field of lightweight and high-efficiency energy devices in the future.

3.7. Hydrophobicity and Self-Cleaning Properties

In practical applications, the application of piezoelectric sensors may be disturbed by sweat and external moisture generated by the human body. Therefore, it was necessary to develop the hydrophobic function of composite fibers. Figure 10 shows the water contact angle of composite fibers. The water contact angle test showed that when the concentrations of IL were 0%, 3%, 6%, and 9%, the CAs of the PAN composite fibers reached 90.49°, 119.76°, 127.99°, and 109.51°, respectively. This indicated that the PAN-ILs composite fibers had excellent surface hydrophobicity. We also further explored the self-cleaning ability of PAN-ILs composite fibers. As we can see from the Figure 10e, the contaminants (sandy soil) attached to the surface of the composite fibers were quickly cleaned up under the scouring of the water flow, and the composite fiber recovered cleanly. Therefore, the PAN-ILs composite fiber had excellent performance in terms of hydrophobicity and self-cleaning. Unfortunately, the dual-filled PAN composite fiber was not hydrophobic, and it was penetrated rapidly when a droplet was dropped on the surface of the sample.

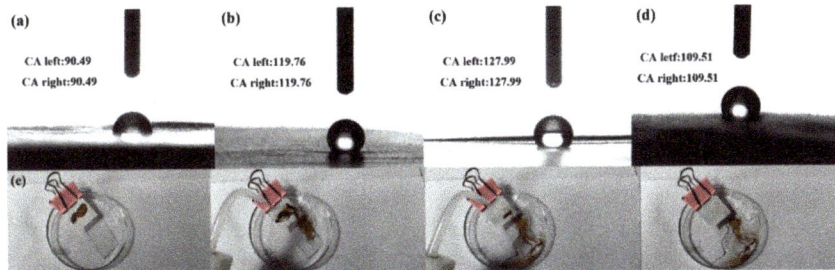

Figure 10. Water contact angles of (**a**) PAN; (**b**) PAN-3ILs; (**c**) PAN-6ILs; (**d**) PAN-9ILs; (**e**) self-cleaning process of PAN-ILs composite fibers.

3.8. Fluorescence Properties and Sensitivity

Figure 11 shows the working principle model of PAN composite fibers piezoelectric sensor. It can be seen that the -CH_2-CN- strand in the PAN was treated as a dipole, which exhibited a zigzag arrangement in the direction perpendicular to the fiber axis. In the electrospinning process, under the action of high electrostatic force, a large number of dipoles of double-filled PAN composite fibers rotated in the same direction, 3^1 helical conformation gradually changed to planar zigzag conformation, and the internal electric potential was enhanced. At the same time, the electrons in the electrode layer exhibited opposite potential and accumulated, finally forming an external current in the wire.

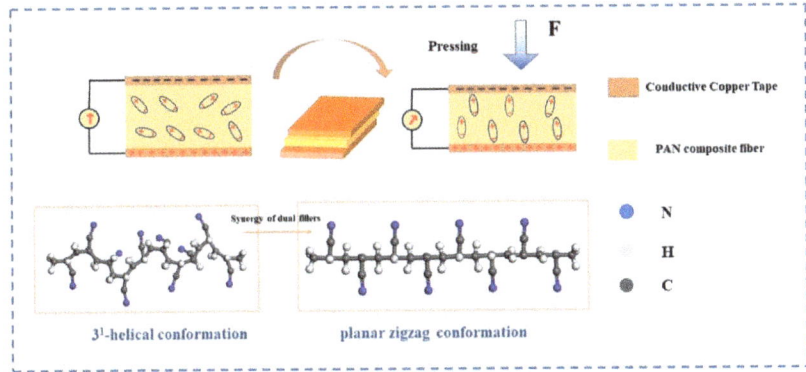

Figure 11. Schematic diagram of working principle of piezoelectric sensor.

Figure 12 shows the fluorescence spectra of PAN@Eu and PAN@Eu-ILs composite fibers. From the figure, it can be observed that all samples had diffraction peaks at 572, 590, 617, 652, and 689 nm, which correspond to the $^5D_0 \rightarrow {}^7F_0$, $^5D_0 \rightarrow {}^7F_1$, $^5D_0 \rightarrow {}^7F_2$, $^5D_0 \rightarrow {}^7F_3$, and $^5D_0 \rightarrow {}^7F_4$ energy level transitions of Eu^{3+} ions. This is consistent with the fluorescence spectrum of rare earth element europium [51]. Among them, the $^5D_0 \rightarrow {}^7F_2$ transition at 617 nm (red light) was the main emission peak. With the addition of ILs, the fluorescence intensity of composite fibers first increased and then decreased, and the fluorescence intensity of PAN@Eu-6ILs was the highest. However, excess ILs led to an increased probability of nonradiative transitions between Eu^{3+} ions, and in general, fluorescence properties were quenched when the probability of nonradiative transitions equaled the probability of radiative emission [52]. The corresponding band of the $^5D_0 \rightarrow {}^7F_2$ transition was weakened. Therefore, the fluorescence intensity of PAN@Eu-ILs composite fibers was weakened. Figure 12b presents the CIE color coordinate diagram; the CIE color model is a color space model created by the International Illumination Commission (ILC). The point corresponding to the CIE value of PAN@Eu-6ILs nanofibers (x = 0.601, y = 0.346) exhibited red emission. The red-light properties of the PAN@Eu-6ILs composite fiber enabled it to achieve certain anti-counterfeiting and lighting functions in multi-functional sensors.

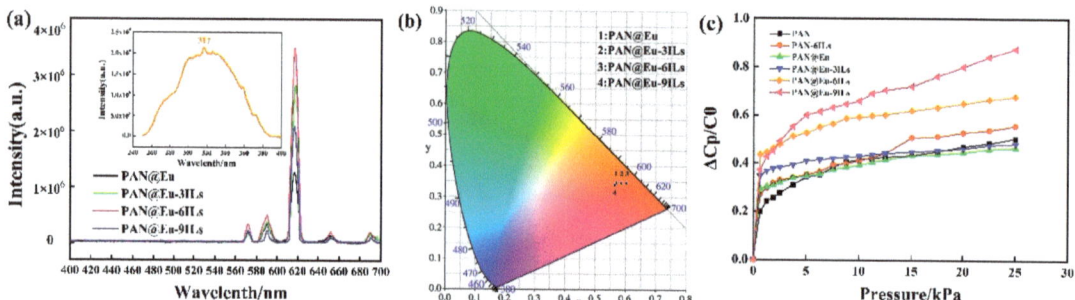

Figure 12. (a) Fluorescence properties of PAN@Eu-ILs composite fibers; (b) CIE diagram of PAN@Eu-ILs; (c) sensitivity of PAN composite fibers under different pressures.

Flexible pressure sensors with simple structure and low power consumption have attracted much attention due to their promising applications in wearable electronic products. However, assembling pressure sensors with high sensitivity, low detection limit, and wide dynamic range remains to be a great challenge [53]. In order to increase the service life of the sensor device, we designed the structure of the sensor. The composite fiber was cut into a size of 2 cm × 2 cm, and the upper and lower surfaces were attached with conductive tape, and finally packaged with PET. Figure 12c shows that after adding ILs or Eu $(NO_3)_3 \cdot 6H_2O$, the capacitance change of PAN nanofibers under the same pressure was improved to different degrees. In the low-pressure region (<1 kPa), the capacitance value of composite fibers changed suddenly, and the minimum monitoring limit was less than 0.625 kPa. In the high-pressure region (5–25 kPa), with the continuous increase in the applied external force, the change in the capacitance value tended to be gentle. The sudden change in the capacitance value can be calculated by the following formula analysis (3) [54]:

$$C = \varepsilon_0 \varepsilon_r \frac{A}{d} \quad (3)$$

where C is the capacitance, ε_0 is the vacuum dielectric constant, ε_r is the relative dielectric constant, A is the overlapping area between the two electrodes, and d is the distance between the two electrodes. It can be seen from the above formula that the capacitance of composite fibers is related to dielectric constant, contact area, and thickness [55]. The higher the dielectric constant, the greater the capacitance value. In addition, when composite fibers were subjected to pressure, their capacitance value increased with the decrease in distance.

When the composite fiber was subjected to a certain pressure, the movable positive and negative charges on its surface undergo a sudden change, which was the reason for the sudden change in the capacitance value. We further analyzed and calculated the sensitivity using formula (4) [56].

$$S = \frac{\sigma\left(\frac{\Delta(C_P - C_0)}{C_0}\right)}{\sigma P} \quad (4)$$

Here, S is the sensitivity of the composite fiber, C_P is the capacitance value when the pressure P is applied, and C_0 is the initial capacitance value. Table 2 shows the sensitivity values of PAN composite fiber. Among them, S is the sensitivity value of the low-pressure area (<0.6 kPa). It can be seen from Table 2 that the sensitivity of PAN was 0.30 kPa^{-1}, and the sensitivities of PAN-6ILs and PAN@Eu were 0.43 and 0.47 kPa^{-1}, respectively. In the dual-filler system, the sensitivity of PAN composite fibers first increased and then decreased with the increase in IL content. The sensitivity of PAN@Eu-6ILs reached 0.69 kPa^{-1}. The piezoelectric capacitance change of the composite fibers was influenced by the synergistic effect of the dual fillers and the ILs concentration. Another explanation can be that the certain orientation of nanofibers could improve piezoelectric sensitivity [57]. However, as the pressure increased, the sensitivity decreased rapidly. For example, the sensitivity of PAN@Eu-6ILs in the high-pressure region (0.007 kPa^{-1}) was much lower than that in the low-pressure region (0.69 kPa^{-1}). Therefore, PAN composite fibers are suitable for piezoelectric sensors for low pressure detection, and the sensitivity of this work has obvious advantages compared with other reported systems [58,59].

Table 2. Sensitivity values of PAN composite fibers under the action of low pressure and high pressure.

Sample	S/kPa^{-1}
PAN	0.30
PAN-6ILs	0.43
PAN@Eu	0.47
PAN@Eu3ILs	0.56
PAN@Eu-6ILs	0.69
PAN@Eu-9ILs	0.60

4. Discussion

In summary, we successfully prepared an IL and (Eu (NO$_3$)$_3$·6H$_2$O)-based PAN composite fiber. The results show that the PAN composite fibers could emit red fluorescence. Among them, the water contact angle (AC) of PAN-6ILs increased from 90.49° to 127.99° with excellent hydrophobic self-cleaning ability. Meanwhile, the PAN@Eu-6ILs of the dual-filler system exhibited good flexibility (elongation at break of 114% and elastic modulus of 2.98 MPa). In addition, the composite fibers exhibited efficient energy storage capacity. Under the action of an electric field of 420 kV/cm^3, the energy density of PAN@Eu-6ILs was 44.02 mJ/cm^3 and the energy storage efficiency was as high as 80%. More importantly, the composite fibers had a low detection limit (<0.625 kPa) and achieved a high sensitivity of 0.69 kPa^{-1}. Overall, the multifunctional PAN composite fibers are expected to be applied in energy storage devices, optoelectronic devices, flexible pressure piezo sensors, etc.

Author Contributions: Data curation, R.X., Y.H. and S.H.; formal analysis, R.X.; investigation, R.X.; methodology, Q.S.; supervision, Q.S.; validation, R.X.; visualization, R.X.; writing—original draft, R.X.; writing—review and editing, Q.S., Y.W. and Y.L. All authors have read and agreed to the published version of the manuscript.

Funding: This study was supported by the National Natural Science Foundation of China (No. 52073033), the fund of the Beijing Municipal Education Commission, China (No. 22019821001) and the Beijing excellent talents training fund (No. Z2019-042).

Institutional Review Board Statement: Not applicable.

Data Availability Statement: Not applicable.

Conflicts of Interest: There are no conflicts of interest among the authors.

References

1. Wu, Y.; Ma, Y.; Zheng, H.; Ramakrshna, S. Piezoelectric materials for flexible and wearable electronics: A review. *Mater. Des.* **2021**, *211*, 110164. [CrossRef]
2. Li, R.; Wei, X.; Xu, J.; Chen, J.; Li, B.; Wu, Z.; Wang, Z. Smart wearable sensors based on triboelectric nanogenerator for personal healthcare monitoring. *Micromachines* **2021**, *12*, 352. [CrossRef] [PubMed]
3. Liu, Y.; Khanbareh, H.; Halim, M.A.; Feeney, A.; Zhang, X.; Heidari, H.; Ghannam, R. Piezoelectric energy harvesting for self-powered wearab-le upper limb applications. *Nano. Select.* **2021**, *2*, 1459–1479. [CrossRef]
4. Han, S.; Chen, S.; Jiao, F. Insulating polymers for flexible thermoelectric composites: A multi-perspective review. *Compos Commun.* **2021**, *28*, 100914. [CrossRef]
5. Vallem, V.; Sargolzaeiaval, Y.; Ozturk, M.; Lai, Y.C.; Dickey, M.D. Energy harvesting and storage with soft and stretchable materials. *Adv. Mater.* **2021**, *33*, 2004832. [CrossRef] [PubMed]
6. Xiong, Y.; Shen, Y.; Tian, L.; Hu, Y.; Zhu, P.; Sun, R.; Wong, C.P. A flexible, ultra-highly sensitive and stable capacitive pressure sensor with convex microarrays for motion and health monitoring. *Nano Energy* **2020**, *70*, 104436. [CrossRef]
7. Guo, Y.; Zhong, M.; Fang, Z.; Wan, P.; Yu, G. A wearable transient pressure sensor made with MXene nanosheets for sensitive broad-range human–machine interfacing. *Nano Lett.* **2019**, *19*, 1143–1150. [CrossRef] [PubMed]
8. Qin, J.; Yin, L.J.; Hao, Y.N.; Zhong, S.L.; Zhang, D.L.; Bi, K.; Zhang, Y.X.; Zhao, Y.; Dang, Z.M. Flexible and stretchable capacitive sensors with different microstructures. *Adv. Mater.* **2021**, *33*, 2008267. [CrossRef] [PubMed]
9. Cao, X.; Xiong, Y.; Sun, J.; Zhu, X.; Sun, Q.; Wang, Z.L. Piezoelectric Nanogenerators Derived Self-Powered Sensors for Multifunctional Applications and Artificial Intelligence. *Adv. Funct. Mater.* **2021**, *31*, 2102983. [CrossRef]
10. Zhou, M.X.; Huang, Q.A.; Qin, M.; Zhou, W. A novel capacitive pressure sensor based on sandwich structures. *J. Microelectromech. Syst.* **2005**, *14*, 1272–1282. [CrossRef]
11. Yang, L.; Ma, J.; Zhong, W.; Liu, Q.; Li, M.; Wang, W.; Wu, Y.; Wang, Y.; Liu, X.; Wang, D. Highly accurate fabric piezoresistive sensor with anti-interference from both high humidity and sweat based on hydrophobic non-fluoride titanium dioxide nanoparticles. *J. Mater. Chem. C* **2021**, *9*, 5217–5226. [CrossRef]
12. Wang, S.; Shao, H.Q.; Liu, Y.; Tang, C.Y.; Zhao, X.; Ke, K.; Bao, R.Y.; Yang, M.B.; Yang, W. Boosting piezoelectric response of PVDF-TrFE via MXene for self-powered linear pressure sensor. *Compos. Sci. Technol.* **2021**, *202*, 108600. [CrossRef]
13. Khalifa, M.; Schoeffmann, E.; Lammer, H.; Mahendram, A.R.; Wuzella, G.; Anandhan, S. A study on electroactive PVDF/mica nanosheet composites with an enhanced γ-phase for capacitive and piezoelectric force sensing. *Soft Matter.* **2021**, *17*, 10891–10902. [CrossRef] [PubMed]
14. Wang, X.; Yang, Y. Effective energy storage from a hybridized electromagnetic-triboelectric nanogenerator. *Nano Energy* **2017**, *32*, 36–41. [CrossRef]
15. Yu, S.; Tai, Y.; Milam-Guerrero, J.; Nam, J.; Myung, N.V. Electrospun organic piezoelectric nanofibers and their energy and bio applications. *Nano Energy* **2022**, 107174. [CrossRef]
16. Gao, Y.; Meng, Q.B.; Wang, B.X.; Zhang, Y.; Mao, H.; Fang, D.W.; Song, X.M. Polyacrylonitrile Derived Robust and Flexible Poly (ionic liquid) s Nanofiber Membrane as Catalyst Supporter. *Catalysts* **2022**, *12*, 266. [CrossRef]
17. Hu, K.; Feng, J.; Hai, Q.; Jang, W.; Lyu, Z.; Lv, N. One-step construction of flexible conductive-piezoelectric nanoresistance network material for pressure sensing and positioning. *Colloid Surface A* **2022**, *641*, 128592. [CrossRef]
18. Shao, H.; Wang, H.; Cao, Y.; Ding, X.; Bai, R.; Chang, H.; Fang, J.; Jin, X.; Wang, W.; Lin, T. Single-layer piezoelectric nanofiber membrane with substantially enhanced noise-to-electricity conversion from endogenous triboelectricity. *Nano Energy* **2021**, *89*, 106427. [CrossRef]
19. Cai, T.; Bi, T.; Yang, Y.; Bi, E.; Xue, S. Preparation of high-performance polymer piezoelectric thin film by electric field induction technology. *Polymers* **2020**, *196*, 122486. [CrossRef]
20. Xu, P.; Luo, S.; Yu, J.; Cao, L.; Yu, S.; Sun, R.; Wong, C.P. Simultaneously enhanced permittivity and electric breakdown strength of polyacrylonitrile composites by introducing ultralow content BaSrTiO$_3$ nanofibers. *Adv. Eng. Mater.* **2019**, *21*, 1900817. [CrossRef]
21. Ohno, H. *Ionic Liquids. Functional Organic Liquids*; Wiley: Hoboken, NJ, USA, 2019; pp. 235–250. [CrossRef]
22. Barbosa, J.C.; Correia, D.M.; Goncalves, R.; Bermudez, V.Z.; Silva, M.M.; Mendez, S.L.; Costa, C.M. Enhanced ionic conductivity in poly (vinylidene fluoride) electrospun separator membranes blended with different ionic liquids for lithium ion batteries. *J. Colloid Interf. Sci.* **2021**, *582*, 376–386. [CrossRef] [PubMed]
23. Moni, G.; Mayeen, A.; Mohan, A.; George, J.J.; Thomas, S.; George, S.C. Ionic liquid functionalised reduced graphene oxide fluoroelastomer nanocomposites with enhanced mechanical, dielectric and viscoelastic properties. *Eur. Polym. J.* **2018**, *109*, 277–287. [CrossRef]
24. Jiang, N.; Li, H.; Hu, D.; Xu, Y.; Zhu, Y.; Han, X.; Zhao, G.; Chen, J.; Chang, X.; Xi, M.; et al. Stretchable strain and temperature sensor based on fibrous polyurethane film saturated with ionic liquid. *Compos. Commun.* **2021**, *27*, 100845. [CrossRef]
25. Zhang, H.; Chen, Y.; Xiao, J.; Song, F.; Wang, C.; Wang, H. Fabrication of enhanced dielectric PVDF nanocomposite based on the conjugated synergistic effect of ionic liquid and graphene. *Mater. Today Proc.* **2019**, *16*, 1512–1517. [CrossRef]

26. Yuan, Y.; Wang, X.; Liu, X.; Qian, J.; Zuo, P.; Zhuang, Q. Non-covalently modified graphene@poly(ionic liquid) nanocomposite with high-temperature resistance and enhanced dielectric properties. *Compos. Part A Appl. Sci. Manuf.* **2022**, *154*, 106800. [CrossRef]
27. Fu, G.; Shi, Q.; He, Y.; Xie, L.; Liang, Y. Electroactive and photoluminescence of electrospun P(VDF-HFP) composite nanofibers with Eu^{3+} complex and $BaTiO_3$ nanoparticles. *Polymer* **2022**, *240*, 124496. [CrossRef]
28. Zuo, W.; Zhu, M.; Yang, W.; Yu, H.; Chen, Y.; Zhang, Y. Experimental study on relationship between jet instability and formation of beaded fibers during electrospinning. *Polym. Eng. Sci.* **2005**, *45*, 704–709. [CrossRef]
29. Itoh, H.; Li, Y.; Chan, K.H.K.; Kotaki, M. Morphology and mechanical properties of PVA nanofibers spun by free surface electrospinning. *Polym. Bull.* **2016**, *73*, 2761–2777. [CrossRef]
30. Su, Y.; Zhang, W.; Lan, J.; Sui, G.; Zhang, H.; Yang, X. Flexible reduced graphene oxide/polyacrylonitrile dielectric nanocomposite films for high-temperature electronics applications. *ACS Appl. Nano Mater.* **2020**, *3*, 7005–7015. [CrossRef]
31. Zhang, H.; Quan, L.; Shi, F.; Li, C.; Liu, H.; Xu, L. Rheological behavior of amino-functionalized multi-walled carbon nanotube/polyacrylonitrile concentrated solutions and crystal structure of composite fibers. *Polymers* **2018**, *10*, 186. [CrossRef] [PubMed]
32. Adhikary, P.; Garain, S.; Ram, S.; Mandal, D. Flexible hybrid eu^{3+} doped P (VDF-HFP) nanocomposite film possess hypersensitive electronic transitions and piezoelectric throughput. *J. Polym. Sci. Pol. Phys.* **2016**, *54*, 2335–2345. [CrossRef]
33. Liu, Z.; Li, B.; Liu, Y.D.; Liang, Y. Integrated Bending Actuation and the Self-Sensing Capability of Poly (Vinyl Chloride) Gels with Ionic Liquids. *Adv. Funct. Mater.* **2022**, *32*, 2204259. [CrossRef]
34. Zhou, Y.; Sha, Y.; Liu, W.; Gao, T.; Yao, Z.; Zhang, Y.; Cao, W. Hierarchical radial structure of polyacrylonitrile precursor formed during the wet-spinning process. *RSC Adv.* **2019**, *9*, 17051–17056. [CrossRef]
35. Shao, H.; Wang, H.; Cao, Y.; Ding, X.; Fang, J.; Wang, W.; Jin, X.; Peng, L.; Zhang, D.; Lin, T. High-Performance Voice Recognition Based on Piezoelectric Polyacrylonitrile Nanofibers. *Adv. Electron. Mater.* **2021**, *7*, 2100206. [CrossRef]
36. Qiao, M.; Kong, H.; Ding, X.; Hu, Z.; Zhang, L.; Cao, Y.; Yu, M. Study on the changes of structures and properties of PAN fibers during the cyclic reaction in supercritical carbon dioxide. *Polymers* **2019**, *11*, 402. [CrossRef]
37. Han, C.; Liu, Q.; Xia, Q.; Wang, Y. Facilely cyclization-modified PAN nanofiber substrate of thin film composite membrane for ultrafast polar solvent separation. *J. Membrane Sci.* **2022**, *641*, 119911. [CrossRef]
38. Wang, W.; Zheng, Y.; Jin, X.; Sun, Y.; Lu, B.; Wang, H.; Fang, J.; Shao, H.; Lin, T. Unexpectedly high piezoelectricity of electrospun polyacrylonitrile nanofiber membranes. *Nano Energy* **2019**, *56*, 588–594. [CrossRef]
39. Koken, N.; Aksit, E.; Yilmaz, M. Nanofibers from chitosan/polyacrylonitrile/sepiolite nanocomposites. *Polym.-Plast. Tech. Mat.* **2021**, *60*, 1820–1832. [CrossRef]
40. Cai, T.; Yang, Y.; Bi, T.; Bi, E.; Li, Y. $BaTiO_3$ assisted PAN fiber preparation of high performance flexible nanogenerator. *Nanotechnology* **2020**, *31*, 24LT01. [CrossRef]
41. Samimi-Sohrforozani, E.; Azimi, S.; Abolhasani, A.; Malekian, S.; Arbab, S.; Zendehdel, M.; Abolhasani, M.M. Development of Porous Polyacrylonitrile Composite Fibers: New Precursor Fibers with High Thermal Stability. *Electron. Mater.* **2021**, *2*, 454–465. [CrossRef]
42. Wu, M.; Wang, Q.; Li, K.; Wu, Y.; Liu, H. Optimization of stabilization conditions for electrospun polyacrylonitrile nanofibers. *Polym. Degrad. Stabil.* **2012**, *97*, 1511–1519. [CrossRef]
43. Feng, Y.; Chen, P.; Zhu, Q.; Qin, B.; Li, X.; Deng, Q.; Li, X.; Peng, C. Boron nitride nanosheet-induced low dielectric loss and conductivity in PVDF-based high-k ternary composites bearing ionic liquid. *Mater Today Commun.* **2021**, *26*, 101896. [CrossRef]
44. Xu, P.; Gui, H.; Hu, Y.; Bahader, A.; Ding, Y. Dielectric properties of polypropylene-based nanocomposites with ionic liquid-functionalized multiwalled carbon nanotubes. *J. Electron. Mater.* **2014**, *43*, 2754–2758. [CrossRef]
45. Parangusan, H.; Ponnamma, D.; Almaadeed, M.A.A. Investigation on the effect of γ-irradiation on the dielectric and piezoelectric properties of stretchable PVDF/Fe–ZnO nanocomposites for self-powering devices. *Soft Matter.* **2018**, *14*, 8803–8813. [CrossRef]
46. Sahoo, R.; Mishra, S.; Unnikrishnan, L.; Mohapatra, S.; Nayak, S.K.; Anwar, S.; Ramadoss, A. Enhanced dielectric and piezoelectric properties of Fe-doped ZnO/PVDF-TrFE composite films. *Mat. Sci. Semicon. Proc.* **2020**, *117*, 105173. [CrossRef]
47. Malik, S.; Sundarraian, S.; Hussain, T.; Nazir, A.; Ramakrishna, S. Fabrication of Highly Oriented Cylindrical Polyacrylonitrile, Poly (lactide-co-glycolide), Polycaprolactone and Poly (vinyl acetate) Nanofibers for Vascular Graft Applications. *Polymers* **2021**, *13*, 2075. [CrossRef]
48. Yu, Y.; Zhang, F.; Liu, Y.; Zheng, Y.; Xin, B.; Jiang, Z.; Peng, X.; Jin, S. Waterproof and breathable polyacrylonitrile/(polyurethane/fluorinated-silica) composite nanofiber membrane via side-by-side electrospinning. *J. Mater. Res.* **2020**, *35*, 1173–1181. [CrossRef]
49. Yang, Y.; Gao, Z.S.; Yang, M.; Zheng, M.S.; Wang, D.R.; Zha, J.W.; Wen, Y.Q.; Dang, Z.M. Enhanced energy conversion efficiency in the surface modified $BaTiO_3$ nanoparticles/polyurethane nanocomposites for potential dielectric elastomer generators. *Nano Energy* **2019**, *59*, 363–371. [CrossRef]
50. Yin, G.; Yang, Y.; Song, F.; Renard, C.; Dang, Z.M.; Shi, C.Y.; Wang, D. Dielectric elastomer generator with improved energy density and conversion efficiency based on polyurethane composites. *ACS Appl. Mater. Inter.* **2017**, *9*, 5237–5243. [CrossRef]
51. He, Y.; Fu, G.; He, D.; Shi, Q.; Chen, Y. Fabrication and characterizations of Eu^{3+} doped PAN/$BaTiO_3$ electrospun piezoelectric composite fibers. *Mater. Lett.* **2022**, *314*, 131888. [CrossRef]

52. Yang, Y.G.; Wu, F.N.; Lv, X.S.; Yu, H.J.; Zhang, H.D.; Li, J.; Hu, Y.Y.; Liu, B.; Wang, X.P.; Wei, L. Luminescence investigation of lanthanum ions (Eu^{3+} or Tb^{3+}) doped $SrLaGa_3O_7$ fluorescent powders. *Opt. Mater.* **2020**, *107*, 110010. [CrossRef]
53. Chen, M.; Li, K.; Cheng, G.; He, K.; Li, W.; Zhang, D.; Li, W.; Feng, Y.; Wei, L.; Li, W.; et al. Touchpoint-tailored ultrasensitive piezoresistive pressure sensors with a broad dynamic response range and low detection limit. *ACS Appl. Mater. Inter.* **2018**, *11*, 2551–2558. [CrossRef]
54. Zhou, Q.; Ji, B.; Wei, Y.; Hu, B.; Gao, Y.; Xu, Q.; Zhou, J.; Zhou, B. A bio-inspired cilia array as the dielectric layer for flexible capacitive pressure sensors with high sensitivity and a broad detection range. *J. Mater. Chem. A* **2019**, *7*, 27334–27346. [CrossRef]
55. Wang, Z.; Si, Y.; Zhao, C.; Yu, D.; Wang, W.; Sun, G. Flexible and washable poly (ionic liquid) nanofibrous membrane with moisture proof pressure sensing for real-life wearable electronics. *ACS Appl. Mater. Inter.* **2019**, *11*, 27200–27209. [CrossRef]
56. Fu, G.; Shi, Q.; Liang, Y.; He, Y.; Xue, R.; He, S.; Chen, Y. Fluorescent markable multi-mode pressure sensors achieved by sandwich-structured electrospun P (VDF-HFP) nanocomposite films. *Polymer* **2022**, *254*, 125087. [CrossRef]
57. Shehata, N.; Elnabawy, E.; Abdelkader, M.; Hassanin, A.H.; Salah, M.; Nair, R.; Bhat, S. Static-aligned piezoelectric poly (vinylidene fluoride) electrospun nanofibers/MWCNT composite membrane: Facile method. *Polymers* **2018**, *10*, 965. [CrossRef]
58. Sharma, S.; Chhetry, A.; Sharifuzzaman, M.; Yoon, H.; Park, J.Y. Wearable capacitive pressure sensor based on MXene composite nanofibrous scaffolds for reliable human physiological signal acquisition. *ACS Appl. Mater. Inter.* **2020**, *12*, 22212–22224. [CrossRef]
59. Fu, G.; He, Y.; Liang, Y.; He, S.; Xue, R.; Wu, Y.; Yu, W.; Shi, Q. Enhanced piezoelectric performance of rare earth complex-doped sandwich-structured electrospun P(VDF-HFP) multifunctional composite nanofiber membranes. *J. Mater. Sci.-Mater.* **2022**, *33*, 22183–22195. [CrossRef]

Article

Insights into the Influence of Different Pre-Treatments on Physicochemical Properties of Nafion XL Membrane and Fuel Cell Performance

Asmaa Selim [1,2,*], **Gábor Pál Szijjártó** [1] **and András Tompos** [1]

1. Institute of Materials and Environmental Chemistry, Excellence Centre of the Hungarian Academy of Sciences, Research Centre for Natural Sciences, Magyar Tudósok Körútja 2, 1117 Budapest, Hungary
2. Chemical Engineering and Pilot Plat Department, Engineering and Renewable Energy Research Institute, National Research Centre, 33 El Bohouth Street, Giza 12622, Egypt
* Correspondence: asmaa.selim@ttk.hu

Abstract: Perfluorosulfonic acid (PFSA) polymers such as Nafion are the most frequently used Proton Exchange Membrane (PEM) in PEM fuel cells. Nafion XL is one of the most recently developed membranes designed to enhance performance by employing a mechanically reinforced layer in the architecture and a chemical stabilizer. The influence of the water and acid pre-treatment process on the physicochemical properties of Nafion XL membrane and Membrane Electrode Assembly (MEA) was investigated. The obtained results indicate that the pre-treated membranes have higher water uptake and dimensional swelling ratios, i.e., higher hydrophilicity, while the untreated membrane demonstrated a higher ionic exchange capacity. Furthermore, the conductivity of the acid pre-treated Nafion XL membrane was ~ 9.7% higher compared to the untreated membrane. Additionally, the maximum power densities obtained at 80 °C using acid pre-treatment were ~ 0.8 and 0.93 W/cm^2 for re-cast Nafion and Nafion XL, respectively. However, the maximum generated powers for untreated membranes at the same condition were 0.36 and 0.66 W/cm^2 for re-cast Nafion and Nafion XL, respectively. The overall results indicated that the PEM's pre-treatment process is essential to enhance performance.

Keywords: proton exchange membranes; Nafion XL; chemical treatment; hydration degree; dimensional swelling; water uptake; perfluorosulfonic acid/PFSA; fuel cell; chemical-physical properties

Citation: Selim, A.; Szijjártó, G.P.; Tompos, A. Insights into the Influence of Different Pre-Treatments on Physicochemical Properties of Nafion XL Membrane and Fuel Cell Performance. *Polymers* **2022**, *14*, 3385. https://doi.org/10.3390/polym14163385

Academic Editor: Jung-Chang Wang

Received: 28 July 2022
Accepted: 15 August 2022
Published: 18 August 2022

Publisher's Note: MDPI stays neutral with regard to jurisdictional claims in published maps and institutional affiliations.

Copyright: © 2022 by the authors. Licensee MDPI, Basel, Switzerland. This article is an open access article distributed under the terms and conditions of the Creative Commons Attribution (CC BY) license (https://creativecommons.org/licenses/by/4.0/).

1. Introduction

It is challenging to reliably and consistently apply the valuable electric energies provided by renewable sources such as solar and wind because of their instability and intermittency during generation. Using energy storage devices based on hydrogen and fuel cell technologies might significantly boost the utilization rate and stability of renewable energy sources, which would be a significant step toward solving this problem [1]. Proton Exchange Membrane Fuel Cells (PEMFCs) have made great strides toward becoming the next generation of green energy technologies [2–4].

Moreover, it is becoming more common for PEMFCs to be used in a variety of industrial applications, moving up to several hundreds of kW in stationary and transport systems from milliwatts in mobile devices [5–7]. Membrane Electrode Assemblies (MEAs) have three essential components; the porous gas diffusion layer responsible for the even distribution of reactants over the surface, electrocatalysts for the anode and a cathode side implementing the electrochemical reactions, and, between them there is the polymer electrolyte membrane or the proton exchange membrane (PEM). PEM acts as an electrolyte that transfers the protons from the anode to the cathode while preventing the conduction of electrons. The electrodes usually consists of carbon-based papers/felts and the electrocatalyst, mainly a Pt-based material [3,8–10]. In the operation, H_2 and O_2 or air is fed to the anode and cathode, respectively. H_2 is oxidized over the anode catalyst producing protons (H^+) and electrons (e^-). PEM only allows the H^+ to pass through the membrane while the electric current flows through an external electric circuit towards the cathode side. On the cathode side, water forms as a result of the reduction of O_2 [11–14].

Hence, PEM has several functions during the process, such as separating the reactants to prevent the mixing of gases on both electrode sides, conducting the proton through it, and deflecting the electrons towards the external circuit. The proton exchange membrane is a critical component PEMFCs. The membrane has to have high ionic conductivity, low H_2 crossover, and good thermal, electrochemical, and mechanical properties in dry and hydrated states [3,12,14–16]. Despite the advancements made at several levels in reducing its cost and improving its performance and durability, the need for ever-increasing longevity and lower costs necessitates the use of more robust components.

One of the most frequently used polymers for PEM fuel cells is perfluorosulfonic acid (PFSA) polymers. PFSA usually consists of the following two parts linked through an ether group: hydrophilic side chains with sulfonic acid groups and the hydrophobic backbone based on perfluorinated polytetrafluoroethylene (PTFE). The hydrophilic part is responsible for proton conduction in the hydrated state, whereas the morphological and mechanical stability are attributed to the hydrophobic backbone [17,18].

Nafion is the most studied and commercialized PFSA based polymer, and it was developed by DuPont de Nemours & Co in the 1960s. Nafion has high proton conductivity and acceptable chemical and mechanical stability at low and intermediate temperatures. The reason behind the high conductivity of Nafion is the existence of SO_3H, which facilitates proton conduction through the hydrated membrane, as described before. In addition, the hydrophobic and hydrophilic domains which have been introduced result in significant phase separation/segregation between the two domains that form relatively wide microchannels, which is beneficial to the transportation of protons [19]. However, as a result of the plasticizing effect of water on the Nafion chains, it has a significant hydrogen permeability in the hydrated state compared to the dry state [20,21]. Moreover, Nafion membranes are sensitive to both chemical and mechanical degradation. In the chemical process, hydroxide radicals formed as a byproduct of the electrochemical reaction attack the ionomer, resulting in its decomposition, whereas the swelling-induced mechanical stresses during hydration result in pinholes growth, delamination, and creep [22–24].

Recently, DuPont designed a membrane that shows advantageous mechanical-chemical stability called Nafion XL. Nafion XL is considered as a sandwich composite membrane consisting of two layers of Nafion impregnated with radical scavengers and, between them a microporous PTFE-rich support layer exists. The former layer can alleviate chemical attack through the radical scavengers, which can neutralize free radicals before they attack the ionomer. The latter is responsible not only for improving mechanical stability but also for reducing the dimensional swelling, which provides resistance to mechanical stress in the hydrated state. Nafion XL membranes were reported to have higher strength and toughness, less shrinkage stress, enhanced durability to creep and fatigue, as well as superior resistance during accelerated stress tests (AST) [23–28]. It is generally recommended in the related work to pre-treat such commercial membranes to improve performance [3]. The multi-step pre-treatment procedure usually consists of treatment with hydrogen peroxide in order to remove organic impurities, with deionized water used for rinsing, and protonation performed with an acid.

Before starting the fabrication of the MEA, the membrane was subjected to a treatment involving both water and acid. The goal of this study was to achieve a reference state for a commercial Nafion XL membrane. The outcomes were compared with those of a self-synthesized, "re-cast" Nafion membrane both before and after the pre-treatment, as well as those of an untreated Nafion XL membrane. The effect of the pre-treatment technique on the water uptake, dimensional swelling degree, ion exchange capacity (IEC), and hydration degree of the treated membranes was investigated. Conductivity and single fuel cell test results employing treated and untreated Nafion XL, as well as re-cast Nafion, are provided in order to verify the influence of the pre-treatment process on the electrochemical properties of the membranes. Results of re-cast Nafion tests were also reported.

2. Materials and Methods

2.1. Materials

DuPont Nafion solution (D520–1000 EW) containing 5 wt% copolymer resin was purchased from the fuel cell store. A commercial Nafion XL membrane with a thickness of ca. 27.5 μm was purchased from Ion Power GmbH, München, Germany. Dimethyl acetamide (DMAc) and sulfuric acid were obtained from VWR Chemicals, Budapest, Hungary. Carbon paper type H23C6 for the GDE preparation was obtained from Freudenberg FCCTSE&CO, Weinheim, Germany. Catalyst ink C-40-PT containing 40% Pt loading was purchased from QuinTech, Göppingen, Germany. Millipore water was used for all the membrane preparations was obtained in-house.

2.2. Membrane Preparation and Pre-Treatment

In this study, two PFSA membranes were used, namely commercial Nafion XL and laboratory prepared re-cast Nafion. For the re-cast Nafion, the solution casting process was used. Briefly, after evaporating the solvent from the Nafion solution, a suitable amount of the dissolved resin in DMAC was cast on a glass Petri dish. The produced membrane was dried at 80 °C for 24 h, followed by annealing at 120 for 4 h. The re-cast membrane was then obtained after immersing it in DI water for a few minutes. Both the re-cast and Nafion XL membrane were treated with deionized water and 0.5 M H_2SO_4. The pre-treatment process details are listed in Table 1. Considering the as-received and the re-cast membrane, deionized, acid-treated membranes are referred to as AsR, DI, and Acid throughout this paper, respectively.

Table 1. Nomenclature, a brief description of pre-treatment processes.

Nomenclature/Membrane	Pre-Treatment
AsR	Used as received or cast without any treatment
DI	Heated in DI water at 80 °C for 1 h
Acid	Heated in 0.5 M H_2SO_4 at 80 °C for 1 h

2.3. Membrane Characterization

2.3.1. Physicochemical Properties

Water uptake and the dimensional swelling ratio in terms of area (In-plane) and thickness (Through-plane) were calculated using Equations (1)–(3). Square membranes with a side length of 15 mm were used for these measurements, three measurements were performed at different locations and the average was obtained. First, the membrane samples were immersed in DI for 24 h at RT, then the surface water was removed with tissue paper and the wet mass and the wet dimensions of the membranes were noted. Finally, adsorbed water was totally evaporated by heating all the samples in a vacuum oven at 50 °C overnight, and the dry weight and size were recorded [29–31]. The water uptake (WU) was determined as follows:

$$WU(\%) = [(W_w - W_d)/W_d] \times 100 \quad (1)$$

where W_w and W_d are the weights of wet and dry membrane samples, respectively. In-plane and through-plane swelling ratios were determined as follows:

$$In-plane\ SR\ (area) = [(A_w - A_d)/A_d] \times 100 \quad (2)$$

$$Through-plane\ SR\ (Thickness) = [(T_w - T_d)/T_d] \times 100 \quad (3)$$

where A_w, A_d, T_w, T_d are the area and the thickness of the wet and dry membrane samples (cm), respectively.

Ion Exchange Capacity (IEC) was determined as the ratio of number of moles of sulfonic acid group per weight of dried membrane in gram. Acid-base titration was used to investigate the experimental values of IEC. First, all membranes were fully dried in an oven at 80 °C overnight, and weighed.

Secondly, the samples were soaked in 1 M NaCl solution for 24 h under continuous stirring after being cut into small pieces. Finally, the solutions were titrated with 0.01 M NaOH solution using methyl orange as an indicator. The IEC was calculated with the following equation [32]:

$$IEC = [(C_{NaOH} * V_{NaOH})/W_d] \quad (4)$$

where C_{NaOH} = 0.01 M, V_{NaOH} is the volume of the NaOH solution used for titration and W_d is the initial dry weight of the membrane.

Hydration degree, λ, is expressed as the number of water molecules available per SO_3H group and considered as one of the essential characteristics for proton exchange membranes. The λ was determined for all the membranes from water uptake and ion exchange capacity from Equation (5) [33].

$$\lambda = [(10 * WU)/(IEC * 18)] \quad (5)$$

2.3.2. Electrochemical Characterization and MEA Performance

MEAs were prepared as follows. Gas Diffusion Electrodes (GDE) with 0.15 mg/cm^2 Pt content were obtained by spray painting the catalyst ink onto the surface of the carbon paper; the catalyst ink consists of C-40-PT with 40 m/m% Pt content, Nafion solution (5 m/m%) and 2-propanol. For the negative electrode, anode, and for the positive electrode, cathode the Pt content was 0.15 mg/cm^2. GDEs were heated at 80 °C before annealing at 120 °C for 30 min each. Membrane Electrode Assembly was prepared by pressing the membranes of 16 cm^2 between two GDEs for 3 min under 59.4 kg/cm^2 and 120 °C.

MEAs were characterized in single cells using VMP-300 multichannel potentiostat (BioLogic). Polarization curves were obtained after activating the membranes at 80 °C for 4 h at 0.4 V. The measurements were performed in a temperature range of 25–95 °C and under 50% and 30% relative humidity for H_2 and O_2, respectively. Back pressures were fixed at 250 kPa and 230 kPa for H_2 and O_2, respectively [34].

Membrane proton conductivity measurements were carried out at room temperature using potentiostatic electrochemical impedance spectroscopy (PEIS), and the frequency range was from 100 kHz to 10 mHz while the amplitude was 10 mV of oscillating voltage. Nitrogen and hydrogen flow was equal to 200 mL/min on both the cathode and the anode. Impedance spectra were recorded where the cathode was acting as the working electrode while the anode was considered as the unified reference and the counter electrode [35]. PEIS measurements were evaluated using the EC-lab program of BioLogic. Membrane resistance was calculated from the low intersect of the Nyquist plot with the z-axis. Membrane conductivity (S.cm^{-1}) was obtained using the following equation:

$$\sigma = L/(R \times A) \quad (6)$$

where L and A are the thickness and area of the membrane in cm and cm^2, respectively, while R is the membrane resistance in Ω.

3. Results and Discussions

The thickness of the membrane was measured using a Mitutoyo micrometer and the dimensions of the membrane were measured with caliper feet.

Both re-cast and XL membranes had almost the same initial thickness of 27.8 and 27.5 μm, respectively. As shown in Figure 1, pre-treatment processes can result in a considerable increase in the thickness of the membranes. Specifically, the re-cast membrane shows approximately a 12 and 28% increase in its thickness after DI and Acid pre-treatment, respectively, whereas Nafion XL had a lower expansion in the thickness of approximately 5 and 15% for the same pre-treatments. The thickness increment may be due to the swelling of the ionic micelle nanostructure after the pre-treatment process. A similar trend was reported by Jiang et al. [36]. The smaller increase in thickness upon different treatments of Nafion XL may be due to the existence of the reinforcement layer and the radical scavengers impregnated into the Nafion XL structure hindering dimension change. Nevertheless, a moderate thickness growth may be beneficial for the chemical and physical properties of both the membranes, which is illustrated in the following chapter.

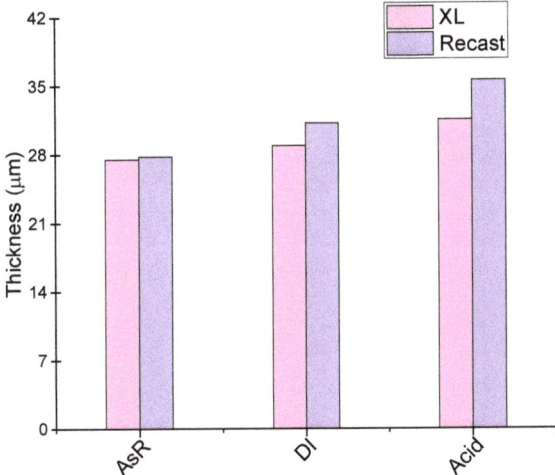

Figure 1. Influence of the different pre-treatments on the thickness of the re-cast and XL membrane.

Generally, the performance of the polymer electrolyte membranes relies on its hydration and ability to retain water. The hydration properties of the membranes are usually investigated from the water uptake and dimensional swelling ratio and hydration degree. The influence of the pre-treatment processes on the swelling ratio both in-plane and through-plane, the water uptake of the membranes as well as the hydration degree results obtained in DI water at room temperature are presented in Figure 2a–c.

Based on the results reported in the literature, the different types of pre-treatments eventually result in higher water uptake in comparison to that obtained for the received samples due to the increase in the free volume available in the Nafion matrix. Upon pre-treatment, the membrane nanostructure was changed thereby allowing more water molecules to be absorbed [37]. The water uptake for the re-cast membranes is consistent with the previous reports [38]. The pre-treated membrane water uptake was very high compared to the AsR membrane (Figure 2a), and consequently the hydration degree was also higher (Figure 2c). Nevertheless, Nafion XL demonstrated only a slight increase in the water uptake (Figure 2a) as well as in the hydration degree (Figure 2c) after the pre-treatments. Most probably, the presence of the additives and the reinforcing layer affect the quasi-equilibrium state of the membrane during hydration.

On the other hand, to establish the influence of water uptake on the dimensional swelling of the membrane, the membrane swelling ratio in the in-plane and through-plane directions were obtained and the results are presented in Figure 2b. It is seen that both membranes possessed higher swelling ratios in both directions after pre-treatment. This is due to the fact that pre-treatment both with water or acid can erase the membrane thermal history and overwrite the hydrophilic passageways by connection previously isolated free volume voids via microscopic channels and allowing for the formation of large hydrophilic routes [39].

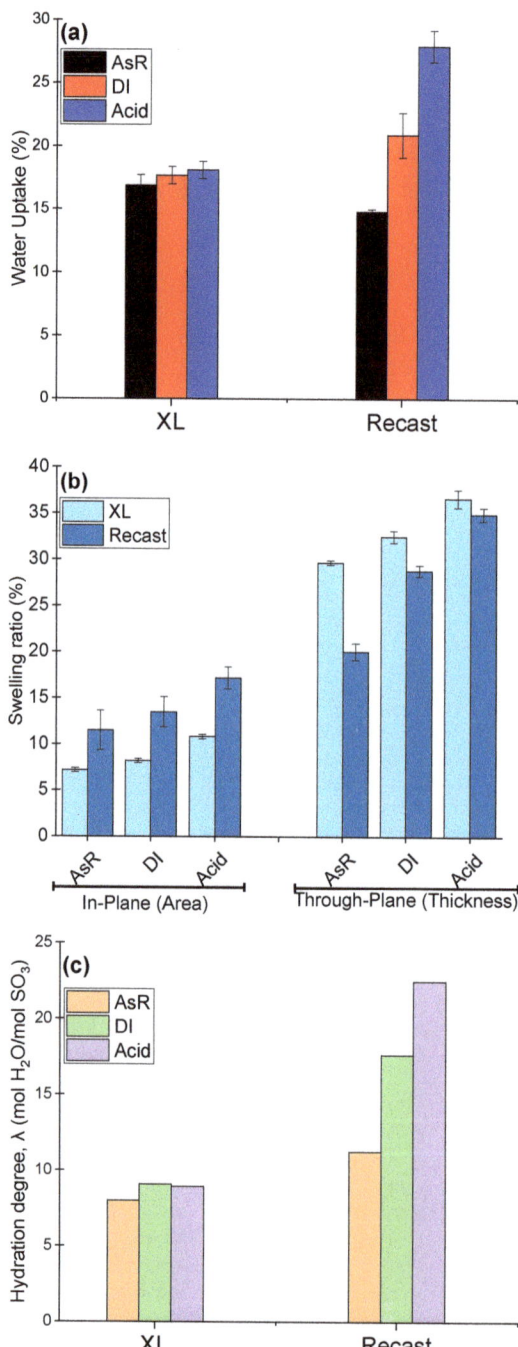

Figure 2. Hydration properties for re-cast Nafion and Nafion XL at different treatment, (**a**) Water uptake, (**b**) Dimensional swelling ratio, (**c**) Hydration degree.

It is not surprising that Nafion XL exhibited larger expansion in thickness due to the presence of the reinforcement layer, which restricts the expansion in the in-plane direction. Additionally, the possibility of the anisotropic structure of the reinforcement layer itself in its plane can lead to an even greater through-plane swelling. Considering the fact that the in-plane swelling can result in cell failure due to the possible separation of the MEA components, Nafion XL has advantageous properties. However, the through-plane swelling can also lead to increased pressure between the MEA components, which can be treated by the extra compression of the MEA. Essentially, the through-plane swelling has a lower impact on the cell failure [40]. In general, from Figures 1 and 2, the crucial role of the reinforcement layer and the additives in the XL structure can be concluded for the membrane stability even after the pre-treatment processes.

Ion Exchange Capacity is one of the most decisive factors influencing the resistance of an ion exchange membrane [36]. In Table 2, the IEC values for both re-cast and XL membranes before and after treatments can be seen. The values for a given membrane are in the same order of magnitude regardless of the different treatments applied. It is recognized that after DI treatment, the IEC values for both re-cast and XL membranes were somewhat lower than for untreated membranes, whereas the IEC values of acid-treated membranes were similar to that of AsR membranes due to the close pKa of the sulfuric acid and the sulfonic acid groups in the Nafion chain, which led to masking the effect of treatment [41]. The lower IEC values obtained after DI-treatment are a consequence of increased hydrophilicity and water uptake upon the pre-treatment process, which eventually can prevent the total acid capacity of the membrane being utilized, resulting in incomplete protonation and consequently lower IEC values [41–43].

Table 2. IEC and conductivity values for untreated and treated re-cast and Nafion XL membranes.

Nomenclature/Membrane	IEC (mmol/g)		Conductivity (mS/cm)	
	Re-Cast	XL	Re-Cast	XL
AsR	0.73	1.18	4.89	50.55
DI	0.66	1.09	12.17	50.92
Acid	0.69	1.13	22.76	55.44

The proton conductivity values of the as received and treated membranes at 25 °C were calculated from the impedance spectra and the Nyquist plot. The total resistance of the MEA was calculated from Equation (6) using the electrode resistance of 0.0228 Ω derived from the reported conductivity of the Nafion XL membrane by the supplier. Table 2 presents the obtained conductivity values for the treated and as-received re-cast and XL Nafion.

The conductivity of the PFSA membranes was reported to be improved by pre-treatment in water or acid. From the results, it can be seen that treating membranes in water and acid results in enhanced conductivity due to a greater water uptake, swelling ratio and hydration degree as discussed before. Eventually, a higher concentration of protons was obtained in the polymer matrix, which was most probably complemented by the formation of new conduction channels with different tortuosity and water networks. It is also recognized that the increase in the conductivity followed the same trend as the water uptake (Figure 2a).

Membrane electrode assemblies prepared from all treated and un-treated membranes were electrochemically characterized. The fuel cell performance was investigated for all the membranes using an in situ single fuel cell at the conventional temperature (80 °C) and relatively low humidity of 50% and 30% for H_2 and O_2, respectively. In Figure 3, the U-I polarization curves (a) and the power density curves (b) are plotted. Despite the lower IEC values, compared to the AsR membranes both water and acid treated re-cast and XL Nafion possess higher performance, which can be attributed to the higher affinity of the treated membranes to retain water leading to higher proton conductivity through the membranes.

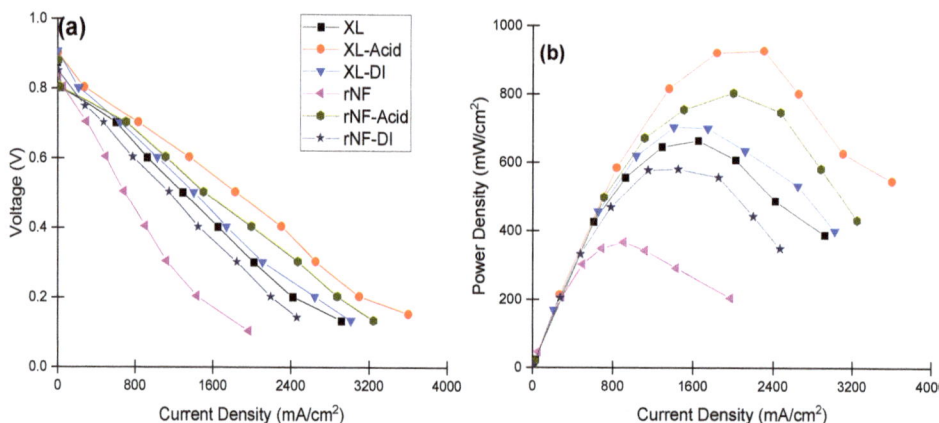

Figure 3. I-V (**a**) and power density (**b**) curves for all the membranes at 80 °C and relative humidity of 50%, 30% for H_2 and O_2, respectively.

Single cell efficiency was investigated by varying the operating temperature between 25 and 60 °C. Maximum power densities were obtained from the power density curves while the current densities were evaluated at 0.4 V from the polarization curves recorded according to the New European Driving Cycle protocol [43] and results are plotted in Figure 4.

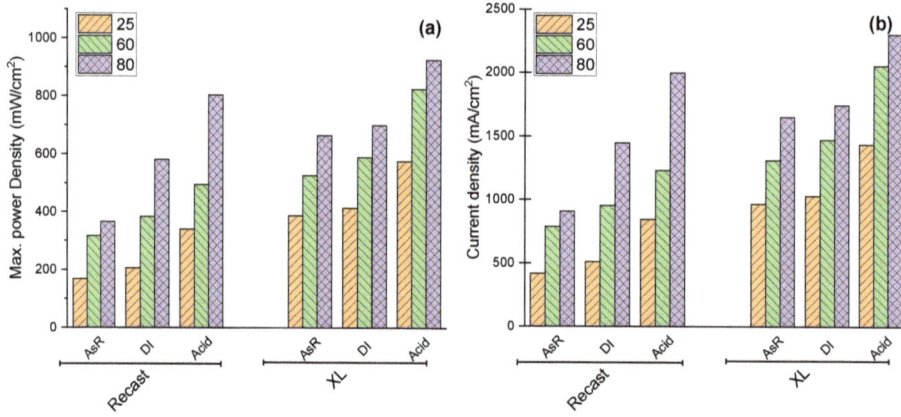

Figure 4. Influence of temperature on the single cell performance of treated and un-treated membranes at 25, 60 and 80 °C Maximum power density (**a**) and current density at 0.4 V (**b**).

As can be seen, the acid-treated membrane showed improved performance with increasing temperature. For all membranes, the highest power density was achieved at 80 °C. Additionally, the trends in current density change in Figure 4 (b) follow those for the maximum power density.

4. Conclusions

Commercial membrane Nafion XL and re-cast membrane (from Nafion solution) were treated in DI and sulfuric acid, followed by characterization and testing in a single fuel cell. Treatment in water is usually applied for cleaning membranes and improving their water affinity while treatment in sulfuric acid is usually used for enhancing the membrane hydrophilicity. Chemical-physical characterization was performed and the electrochemical properties were studied for all treated and untreated membranes. Acid-treated membranes showed a higher water uptake, in-plane, and through-plane swelling ratio compared to the as-received and prepared membranes. Nevertheless, the water uptake as well as the hydration degree of Nafion XL after the different pre-treatments was lower compared to those of the re-cast Nafion membrane as a result of the presence of a reinforcement layer in Nafion XL. The ion exchange capacity for the treated membranes was slightly lower than that for the as-received membranes, which however did not lead to the deterioration of proton conductivity. Finally, increased water uptake, degree of hydration, and the formation of new conduction channels resulted in better proton conductivity of the acid-treated membranes as compared to the as-received samples. As a result of the improved proton conductivity achieved after acid treatment, better performance of the commercial and the re-cast membranes was observed for the fuel cell tests compared to the results obtained on the as-received membranes.

Author Contributions: Conceptualization, A.S. and G.P.S.; Data curation, A.S.; Formal analysis, A.S.; Funding acquisition, A.T.; Investigation, A.S.; Methodology, A.S., G.P.S.; Project administration, A.T.; Resources, A.T.; Supervision, A.T.; Validation, A.S.; Visualization, A.S.; Writing original draft, A.S.; Writing—Review and editing, A.S. All authors have read and agreed to the published version of the manuscript.

Funding: This research was funded by Project no. RRF-2.3.1-21-2022-00009, titled National Laboratory for Renewable Energy has been implemented with the support provided by the Recovery and Resilience Facility of the European Union within the framework of Programme Széchenyi Plan Plus.

Institutional Review Board Statement: Not applicable.

Informed Consent Statement: Not applicable.

Data Availability Statement: The data presented in this study are available on request from the corresponding author.

Conflicts of Interest: The authors declare no conflict of interest.

References

1. Wong, C.Y.; Wong, W.Y.; Ramya, K.; Khalid, M.; Loh, K.S.; Daud, W.R.W.; Lim, K.L.; Walvekar, R.; Kadhum, A.A.H. Additives in Proton Exchange Membranes for Low- and High-Temperature Fuel Cell Applications: A Review. *Int. J. Hydrog. Energy* **2019**, *44*, 6116–6135. [CrossRef]
2. Prykhodko, Y.; Fatyeyeva, K.; Hespel, L.; Marais, S. Progress in Hybrid Composite Nafion®-Based Membranes for Proton Exchange Fuel Cell Application. *Chem. Eng. J.* **2021**, *409*, 127329. [CrossRef]
3. Mahato, N.; Jang, H.; Dhyani, A.; Cho, S. Recent Progress in Conducting Polymers for Hydrogen Storage and Fuel Cell Applications. *Polymers.* **2020**, *12*, 2480. [CrossRef] [PubMed]
4. Paddison, S.J.; Gasteiger, H.A. PEM Fuel Cells: Materials and Design Development Challenges. In *Fuel Cells and Hydrogen Production*; Springer New York: New York, NY, USA, 2012; pp. 173–193.
5. Scott, K.; Shukla, A.K. Polymer Electrolyte Membrane Fuel Cells: Principles and Advances. *Rev. Environ. Sci. Biotechnol.* **2004**, *3*, 273–280. [CrossRef]
6. Kocha, S.S. Polymer electrolyte membrane (PEM) fuel cells, automotive applications. In *Fuel Cells and Hydrogen Production*; Springer New York: New York, NY, USA, 2019; pp. 135–171.
7. Karimi, M.B.; Mohammadi, F.; Hooshyari, K. Effect of Deep Eutectic Solvents Hydrogen Bond Acceptor on the Anhydrous Proton Conductivity of Nafion Membrane for Fuel Cell Applications. *J. Memb. Sci.* **2020**, *605*, 118116. [CrossRef]
8. Amjadi, M.; Rowshanzamir, S.; Peighambardoust, S.J.; Sedghi, S. Preparation, Characterization and Cell Performance of Durable Nafion/SiO$_2$ Hybrid Membrane for High-Temperature Polymeric Fuel Cells. *J. Power Sources.* **2012**, *210*, 350–357. [CrossRef]
9. Amjadi, M.; Rowshanzamir, S.; Peighambardoust, S.J.; Hosseini, M.G.; Eikani, M.H. Investigation of Physical Properties and Cell Performance of Nafion/TiO$_2$ Nanocomposite Membranes for High Temperature PEM Fuel Cells. *Int. J. Hydrog. Energy* **2010**, *35*, 9252–9260. [CrossRef]
10. Alanazi, A.; Ogungbemi, E.; Wilberforce, A.; Ijaodola, O.S.; Vichare, P.; Olabi, A.-G. State-of-the-Art Manufacturing Technologies of PEMFC Components. In Proceedings of the 10th International Conference on Sustainable Energy and Environmental Protection (SEEP 2017), Bled, Slovenia, 27–30 June 2017; pp. 189–198. [CrossRef]
11. Goh, J.T.E.; Rahim, A.R.A.; Masdar, M.S.; Shyuan, L.K. Enhanced Performance of Polymer Electrolyte Membranes via Modification with Ionic Liquids for Fuel Cell Applications. *Membranes* **2021**, *11*, 395. [CrossRef]

12. de Bruijn, F.A.; Makkus, R.C.; Mallant, R.K.A.M.; Janssen, G.J.M. Chapter Five: Materials for state-of-the-art PEM fuel cells, and their suitability for operation above 100 °C. In *Advances in Fuel Cells*; Elsevier Inc: Amsterdam, The Netherlands, 2007; pp. 235–336.
13. Rhee, H.W.; Ghil, L.-J. Chapter fourteen: Polymer nanocomposites in fuel cells. In *Advances in Polymer Nanocomposites*; Woodhead Publishing Limited: Shaston, UK, 2012; pp. 433–471.
14. Chia, M.Y.; Thiam, H.S.; Leong, L.K.; Koo, C.H.; Saw, L.H. Study on Improvement of the Selectivity of Proton Exchange Membrane via Incorporation of Silicotungstic Acid-Doped Silica into SPEEK. *Int. J. Hydrog. Energy* **2020**, *45*, 22315–22323. [CrossRef]
15. Wycisk, R.; Pintauro, P.N.; Park, J.W. New Developments in Proton Conducting Membranes for Fuel Cells. *Curr. Opin. Chem. Eng.* **2014**, *4*, 71–78. [CrossRef]
16. Kim, J.; Yamasaki, K.; Ishimoto, H.; Takata, Y. Ultrathin Electrolyte Membranes with PFSA-Vinylon Intermediate Layers for PEM Fuel Cells. *Polymers* **2020**, *12*, 1730. [CrossRef] [PubMed]
17. Yandrasits, M.A.; Lindell, M.J.; Hamrock, S.J. New Directions in Perfluoroalkyl Sulfonic Acid–Based Proton-Exchange Membranes. *Curr. Opin. Electrochem.* **2019**, *18*, 90–98. [CrossRef]
18. Schalenbach, M.; Hoefner, T.; Paciok, P.; Carmo, M.; Lueke, W.; Stolten, D. Gas Permeation through Nafion. Part 1: Measurements. *J. Phys. Chem. C* **2015**, *119*, 25145–25155. [CrossRef]
19. Schalenbach, M.; Hoeh, M.A.; Gostick, J.T.; Lueke, W.; Stolten, D. Gas Permeation through Nafion. Part 2: Resistor Network Model. *J. Phys. Chem. C* **2015**, *119*, 25156–25169. [CrossRef]
20. Kusoglu, A.; Karlsson, A.M.; Santare, M.H.; Cleghorn, S.; Johnson, W.B. Mechanical Behavior of Fuel Cell Membranes under Humidity Cycles and Effect of Swelling Anisotropy on the Fatigue Stresses. *J. Power Sources* **2007**, *170*, 345–358. [CrossRef]
21. Mukundan, R.; Baker, A.M.; Kusoglu, A.; Beattie, P.; Knights, S.; Weber, A.Z.; Borup, R.L. Membrane Accelerated Stress Test Development for Polymer Electrolyte Fuel Cell Durability Validated Using Field and Drive Cycle Testing. *J. Electrochem. Soc.* **2018**, *165*, F3085–F3093. [CrossRef]
22. Kusoglu, A.; Santare, M.H.; Karlsson, A.M. Aspects of Fatigue Failure Mechanisms in Polymer Fuel Cell Membranes. *J. Polym. Sci. Part B Polym. Phys.* **2011**, *49*, 1506–1517. [CrossRef]
23. Prabhakaran, V.; Arges, C.G.; Ramani, V. Investigation of Polymer Electrolyte Membrane Chemical Degradation and Degradation Mitigation Using in Situ Fluorescence Spectroscopy. *Proc. Natl. Acad. Sci. USA* **2012**, *109*, 1029–1034. [CrossRef]
24. Baker, A.M.; Torraco, D.; Judge, E.J.; Spernjak, D.; Mukundan, R.; Borup, R.L.; Advani, S.G.; Prasad, A.K. Cerium Migration during PEM Fuel Cell Assembly and Operation. *ECS Trans.* **2015**, *69*, 1009–1015. [CrossRef]
25. Khattra, N.S.; Lu, Z.; Karlsson, A.M.; Santare, M.H.; Busby, F.C.; Schmiedel, T. Time-Dependent Mechanical Response of a Composite PFSA Membrane. *J. Power Sources* **2013**, *228*, 256–269. [CrossRef]
26. Xing, Y.; Li, H.; Avgouropoulos, G. Research Progress of Proton Exchange Membrane Failure and Mitigation Strategies. *Materials* **2021**, *14*, 2591. [CrossRef] [PubMed]
27. Zhang, D.; Xin, L.; Xia, Y.; Dai, L.; Qu, K.; Huang, K.; Fan, Y.; Xu, Z. Advanced Nafion Hybrid Membranes with Fast Proton Transport Channels toward High-Performance Vanadium Redox Flow Battery. *J. Memb. Sci.* **2021**, *624*, 119047. [CrossRef]
28. Passalacqua, E.; Pedicini, R.; Carbone, A.; Gatto, I.; Matera, F.; Patti, A.; Saccà, A. Effects of the Chemical Treatment on the Physical-Chemical and Electrochemical Properties of the Commercial NafionTM Nr212 Membrane. *Materials* **2020**, *13*, 1–16. [CrossRef]
29. Yu, L.; Lin, F.; Xu, L.; Xi, J. Structure-Property Relationship Study of Nafion XL Membrane for High-Rate, Long-Lifespan, and All-Climate Vanadium Flow Batteries. *RSC Adv.* **2017**, *7*, 31164–31172. [CrossRef]
30. Ye, J.; Cheng, Y.; Sun, L.; Ding, M.; Wu, C.; Yuan, D.; Zhao, X.; Xiang, C.; Jia, C. A Green SPEEK/Lignin Composite Membrane with High Ion Selectivity for Vanadium Redox Flow Battery. *J. Memb. Sci.* **2019**, *572*, 110–118. [CrossRef]
31. Teixeira, F.C.; de Sá, A.I.; Teixeira, A.P.S.; Ortiz-Martínez, V.M.; Ortiz, A.; Ortiz, I.; Rangel, C.M. New Modified Nafion-Bisphosphonic Acid Composite Membranes for Enhanced Proton Conductivity and PEMFC Performance. *Int. J. Hydrogen. Energy* **2021**, *46*, 17562–17571. [CrossRef]
32. Yazici, M.S.; Dursun, S.; Borbáth, I.; Tompos, A. Reformate Gas Composition and Pressure Effect on CO Tolerant Pt/Ti0.8Mo0.2O2–C Electrocatalyst for PEM Fuel Cells. *Int. J. Hydrogen. Energy* **2021**, *46*, 13524–13533. [CrossRef]
33. Yang, J.; Li, Y.; Huang, Y.; Liang, J.; Shen, P.K. Dynamic Conducting Effect of WO3/PFSA Membranes on the Performance of Proton Exchange Membrane Fuel Cells. *J. Power Sources* **2008**, *177*, 56–60. [CrossRef]
34. Jiang, B.; Yu, L.; Wu, L.; Mu, D.; Liu, L.; Xi, J.; Qiu, X. Insights into the Impact of the Nafion Membrane Pretreatment Process on Vanadium Flow Battery Performance. *ACS Appl. Mater. Interfaces* **2016**, *8*, 12228–12238. [CrossRef]
35. Di Noto, V.; Fontanella, J.J.; Wintersgill, M.C.; Giffin, G.A.; Vezzù, K.; Piga, M.; Negro, E. Pressure, Temperature, and Dew Point Broadband Electrical Spectroscopy (PTD-BES) for the Investigation of Membranes for PEMFCs. *Fuel Cells* **2013**, *13*, 48–57. [CrossRef]
36. Alberti, G.; Narducci, R.; Sganappa, M. Effects of Hydrothermal/Thermal Treatments on the Water-Uptake of Nafion Membranes and Relations with Changes of Conformation, Counter-Elastic Force and Tensile Modulus of the Matrix. *J. Power Sources* **2008**, *178*, 575–583. [CrossRef]
37. Baker, A.M.; Wang, L.; Johnson, W.B.; Prasad, A.K.; Advani, S.G. Nafion Membranes Reinforced with Ceria-Coated Multiwall Carbon Nanotubes for Improved Mechanical and Chemical Durability in Polymer Electrolyte Membrane Fuel Cells. *J. Phys. Chem. C* **2014**, *118*, 26796–26802. [CrossRef]

38. Kuwertz, R.; Kirstein, C.; Turek, T.; Kunz, U. Influence of Acid Pretreatment on Ionic Conductivity of Nafion® Membranes. *J. Memb. Sci.* **2016**, *500*, 225–235. [CrossRef]
39. Kinumoto, T.; Inaba, M.; Nakayama, Y.; Ogata, K.; Umebayashi, R.; Tasaka, A.; Iriyama, Y.; Abe, T.; Ogumi, Z. Durability of Perfluorinated Ionomer Membrane against Hydrogen Peroxide. *J. Power Sources* **2006**, *158*, 1222–1228. [CrossRef]
40. Lin, Q.; Sun, X.; Chen, X.; Shi, S. Effect of Pretreatment on Microstructure and Mechanical Properties of NafionTM XL Composite Membrane. *Fuel Cells* **2019**, *19*, 530–538. [CrossRef]
41. Young, S.K.; Mauritz, K.A. Nafion®/(Organically Modified Silicate) Nanocomposites via Polymer in Situ Sol-Gel Reactions: Mechanical Tensile Properties. *J. Polym. Sci. Part B Polym. Phys.* **2002**, *40*, 2237–2247. [CrossRef]
42. Di Noto, V.; Gliubizzi, R.; Negro, E.; Pace, G. Effect of SiO_2 on Relaxation Phenomena and Mechanism of Ion Conductivity of [Nafion/$(SiO_2)_x$] Composite Membranes. *J. Phys. Chem. B* **2006**, *110*, 24972–24986. [CrossRef] [PubMed]
43. Tsotridis, G.; Pilenga, A.; Marco, G.; Malkow, T. *EU Harmonised Test Protocols for PEMFC MEA Testing in Single Cell Configuration for Automotive Applications*; Publications Office of the European Union: Luxembourg, 2015.

A Novel Vibration Piezoelectric Generator Based on Flexible Piezoelectric Film Composed of PZT and PI Layer

Jia Wang, Yujian Tong, Chong Li *, Zhiguang Zhang and Jiang Shao *

School of Mechanical Engineering, Jiangsu University of Science and Technology, Zhenjiang 212100, China; wjjzhb@just.edu.cn (J.W.); 202020030@stu.just.edu.cn (Y.T.); 172210202331@stu.just.edu.cn (Z.Z.)
* Correspondence: lichong@just.edu.cn (C.L.); jiang-shao@hotmail.com (J.S.); Tel.: +86-511-8444-5385 (C.L.)

Abstract: A novel piezoelectric generator based on soft piezoelectric film consisting of a polyimide (PI) sheet and lead zirconate titanate (PZT) is proposed to generate electric energy under the operating conditions of low-frequency and small-amplitude vibration. The theoretical model and working principle of the piezoelectric generator are discussed in detail. Using ANSYS software, a finite element analysis of the static and modal characteristics of the piezoelectric generator is carried out. Further, the output of the prepared piezoelectric generator is investigated by a home-made experimental platform. Results show that the transient excitation voltage of the generator increases with the increase in load resistance, and the continuous excitation voltage increases first and then remains almost stable. The maximum continuous power produced by the piezoelectric generator is about 4.82 mW. Furthermore, the continuous excitation voltage and power are in accordance with the simulation values when the load resistances are 20 kΩ and 25 kΩ, respectively.

Keywords: piezoelectric generator; PI film; experimental analysis; simulation

1. Introduction

Recently, renewable energies have gained considerable attention due to the negative effects of fossil fuel consumption, such as global warming [1,2]. As a result, significant efforts have been devoted to developing advanced energy-harvesting technologies to generate electric power from renewable energies, such as solar, ocean wave, wind, mechanical vibration and so on [3,4].

Among these renewable energies, mechanical vibration is an attractive option because of its abundance in the natural environment [5]. Moreover, mechanical-vibration-based generator has the advantages of relative high-power density, higher potential and longer lifespan [6–9]. Consequently, several strategies, such as piezoelectric mechanism, electromagnetic mechanism and electrostatic mechanism, have been introduced to develop a mechanical-vibration-based energy harvester [10–13]. The piezoelectric generator is based on the piezoelectric mechanism, which has been extensively investigated and is understood to be related to mechanical–electrical energy conversion. For instance, using the longitudinal vibration of the drill pipes, Zheng et al. [14] designed a while-drilling energy-harvesting device. The designed energy-harvesting device was utilized as a continuous power supply for downhole instruments during the drilling procedure. When the thickness of the piezoelectric patches was 1.2–1.4 mm, the designed device presented the best energy harvest performance, with a peak voltage of 15–40 V. To monitor the random vibration of rails, an efficient rail-borne piezoelectric energy harvester was used to collect energy from the random railway vibrations by Yang et al. [15]. The output power peaks at the first two resonance frequencies were 1036.9 and 8.01 mW/Hz. To improve the working frequency bandwidth and environmental robustness of the piezoelectric vibration energy harvester, a multi-frequency response piecewise-linear piezoelectric vibration energy harvester was developed [16]. Based on the electromechanical coupling and the dynamic response, a theoretical model of the energy harvester was established. Results showed that the energy generated by the multi-frequency response piecewise-linear piezoelectric energy harvester was 194% of the energy generated by its linear counterpart under the same excitation conditions. Vibration energy harvesting from backpacks has the potential to generate electrical power, and various energy-harvesting backpacks have been designed. However, dynamics between the human body and the backpack have an important influence on the dynamic performance of the backpacks. Therefore, to improve human comfort, Liu et al. [17] investigated the dynamic interaction between the human body and the energy-harvesting backpacks. It was found that tuning the backpack parameters can reduce the ground reaction forces in the push-off phase and potentially improve human comfort. To overcome the low energy utilization of a traditional piezoelectric energy harvester, a uniform stress distribution of bimorph has been designed for piezoelectric energy harvesting [18]. Further, Morel et al. [19] proposes a general, normalized, and unified performance evaluation of the various electrical strategies to tune the harvester's frequency response. With a thorough analysis, the influence of the tunable electrical interfaces on the electromechanical generator response was investigated.

As the key components of the piezoelectric generator, piezoelectric smart materials, typically PZT, possessing advantages of small size, fast response and high energy density are widely utilized in energy-harvesting applications [20–22], and a number of PZT-based generators have been developed. A piezoelectric generator based on vortex-induced vibration was proposed to convert flow energy underwater to electrical energy [23]. This generator consists of a piezoelectric cantilever beam, connecting device, springs, bluff body and displacement sensor. The output voltage is derived from the flow–solid-electric coupling equations. The experimental test of the piezoelectric generator performance at different water speeds shows good agreement with the theoretical results. Due to the inefficiency of harvesting energy from low-frequency vibrations of traditional piezoelectric cantilever structures, a novel piezoelectric generator used a cantilevered bimorph with a tip mass and a pair of preloading springs was designed [24]. The harvester was modeled as a Euler–Bernoulli beam, and the piezoelectric material is assumed to be linear. It was found that changing the preloading of the spring helped reduce the natural frequency of the cantilever. Further, in order to improve the adaptive range of airflow velocity, an airflow piezoelectric generator based on multi-harmonic excitation was proposed [25]. The flow field characteristics were obtained by the computational fluid dynamics (CFD) method. The result showed that periodic compression and expansion of air were presented and a standing wave was formed inside the resonator. The results verified the viability of the multi-harmonic excitation energy-exchange method. An eye-shaped generator consisting of a rectangular ceramic and two elastic body plates was also developed [26]. Once tension is applied to both ends of the elastic body, it causes the elastic body to deform, resulting in a positive piezoelectric effect to generate electric energy. The proposed generator is relatively durable because the forces are not applied directly to the ceramic. However, despite their outstanding piezoelectric properties, ceramic materials are limited in the field of energy collection owing to their inherent disadvantages, such as rigidity, brittleness and low voltage coefficient [27–29].

Therefore, piezoelectric polymers were introduced to overcome the abovementioned drawbacks of ceramic materials. Piezoelectric polymers demonstrate better dielectric behavior and field strength, being able to tolerate high driving voltage [30]. The popular piezoelectric polymers are polyvinylidene fluoride (PVDF), polylactic acids (PLA), polyurethanes (PU) and PI [31–34]. As a promising piezoelectric material, PI has been widely used, owing to its outstanding thermal stability, its excellent mechanical durability and the exceptional designability of its piezoelectric properties. For instance, Dagdeviren et al. [35] reported a piezoelectric generator to collect mechanical energy generated from the movement of heart, lung, and diaphragm. The developed piezoelectric generator was based on PZT and PI film, where PI worked as flexible matrix and encapsulation layer, while PZT was in a configuration of ribbons sandwiched between gold and platinum electrodes. The averaged power density of the proposed piezoelectric generator reached 1.2 µW/cm^2.

Although significant progress has been made, it still remains a challenge to gather electric energy under small-vibration amplitude. In this paper, a novel vibration piezoelectric generator capable of amplifying the vibration amplitude and generating electrical energy was developed based on soft piezoelectric PZT/PI film. The performance of prepared generator was evaluated by a home-made experimental platform. The results indicated that the maximum output of the developed piezoelectric generator could reach 4.82 mW. Moreover, simulation studies with respect to continuous excitation voltage and power were also conducted, and showed good agreement with experimental results, while the load resistances were chosen to be 20 kΩ and 25 kΩ, respectively.

2. Operating Principle

In a typical cantilever beam structure piezoelectric generator, one end of the cantilever beam is fixed, and the other end mechanically vibrates [36–39], transforming mechanical energy into elastic potential energy. The alternating voltage is generated by releasing the strained piezoelectric layer.

In this study, the proposed piezoelectric generator 2qs based on a two-stage amplitude amplification mechanism, as shown in Figure 1. In the piezoelectric power generator, the amplification mechanism 2qs added to the force transfer mechanism, which was used to evenly transfer the force to both ends of the elastic mechanism, so that both ends of the elastic mechanism could achieve the same bending, and the bending degree was the same. The piezoelectric generator consisted of piezoelectric macro fiber composite (MFC), primary amplifier, second stage amplifier, restorative spring and support. The primary amplifier was responsible for the first step amplification of vibration amplitude by means of lever amplification principle, whereas the second amplifier achieved the second amplification of the amplitude by using an X–type mechanism. The piezoelectric MFC was attached to an elastic steel plate and deformed simultaneously with the elastic steel plate. The restorative spring was employed to restore the initial state under the restorative force when the lever of the primary amplifier was pressed.

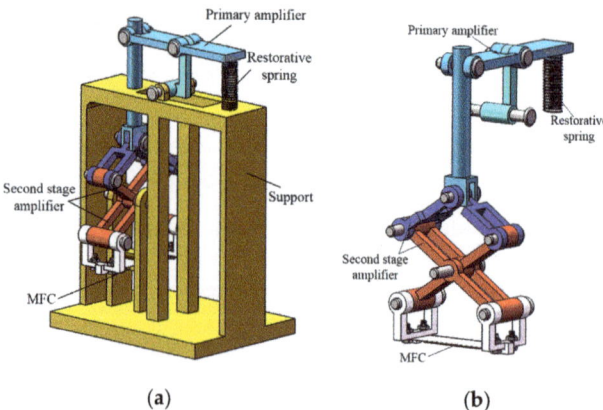

Figure 1. Structure of the vibration piezoelectric generator. (**a**) Overall structure; (**b**) amplitude amplification mechanism.

The operating principle of the proposed vibration piezoelectric generator is shown in Figure 2. At the initial moment, the primary amplifier is in the horizontal state and the second stage amplifier is in a relaxed state with undeformed MFC. In this state, the restorative spring is free. When the vibration acts on the end of the primary amplifier, the restorative force spring at the end of the lever is compressed. Under the action of the primary lever, the secondary amplifying mechanism is lifted to drive the bending deformation of both ends of the MFC. Due to the positive piezoelectric effect, the bending deformation of MFC produces voltage. Under the cyclic vibration, continuous voltage output is obtained at both ends of the MFC.

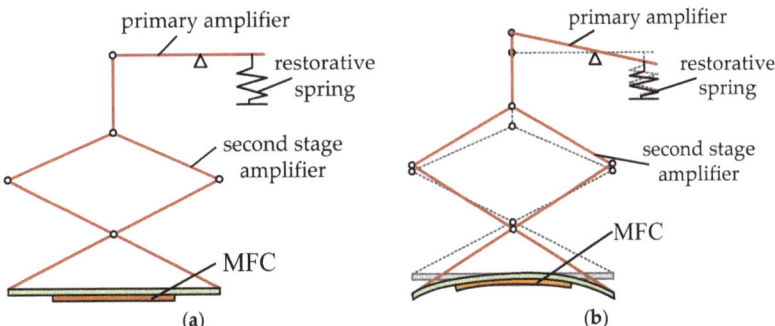

Figure 2. Operating principle of the vibration piezoelectric generator. (**a**) The MFC is not deformed; (**b**) the MFC is deformed.

Compared with other kinds of piezoelectric generators [36–39], the proposed piezoelectric generator has the function of multistage displacement amplification. The primary amplification mechanism realizes the first amplification of vibration displacement through the lever principle, and the second-stage amplification mechanism can realize multistage amplification of vibration displacement. Therefore, even if the external input vibration displacement is very small, the piezoelectric generator can also generate large electric energy through the displacement amplification function.

Here, the fixing mode at both ends of the elastic steel plate can be simplified into simple support at both ends, as shown in Figure 3. When the piezoelectric generator is in operation, the elastic steel plate can be regarded as consisting of two cantilever plates. When external vibration is applied to the input end of the piezoelectric generator, the vibration displacement is amplified by multistage displacement amplification. Multistage displacement amplification mechanism drives the reciprocating motion at both ends of the elastic steel plate. Under multiple actions, the piezoelectric MFC produces bending deformation, which in turn generates electrical energy under the piezoelectric positive effect.

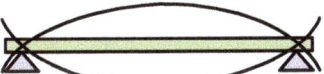

Figure 3. Fixed mode of the elastic steel plate.

3. Electromechanical Model

The proposed vibration piezoelectric generator amplifies the amplitude through two-stage amplification, and the efficiency of energy collection is improved. Figure 4 presents the displacement calculation diagram of each part under external excitation, where δ_1 is excitation amplitude value.

For the proposed piezoelectric generator, a sinusoidal excitation of amplitude δ_1 is used, and it is expressed as:

$$e(t) = \delta_1 \sin(2\pi f t), \qquad (1)$$

where f is exciting frequency. Under the action of primary amplifier, the displacement amplitude of the upper end of the second-stage amplification is calculated as:

$$\delta_2(t) = \frac{l_2 \delta_1}{l_1} \sin(2\pi f t), \qquad (2)$$

where l_1 and l_2 are the lengths of the two sections of the lever. The second magnification consists of three isosceles triangles, and the angle between the two equal sides is θ. With the first lever amplification, the triangle CDE is elevated by δ_2. Then the angle between equal sides is expressed as:

$$\theta_1(t) = 2\arccos\left(\cos\frac{\theta}{2} + \frac{\delta_2(t)}{l_3}\right), \quad (3)$$

where l_3 is equilateral length of a triangle. The height increment of ΔDFE and ΔGFH can be written as:

$$\delta_3(t) = l_3\left(\cos\frac{\theta_1(t)}{2} - \cos\frac{\theta}{2}\right) \quad (4)$$

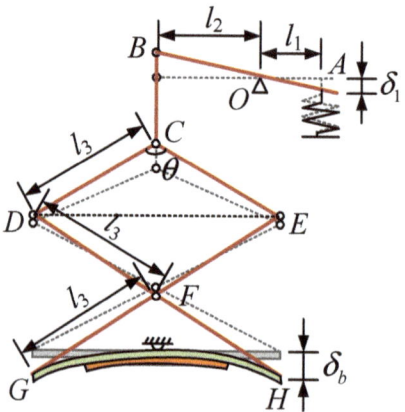

Figure 4. Displacement calculation diagram of each part under external excitation.

The displacement at the end of the MFC base is denoted as:

$$\delta_b(t) = 2\delta_3(t) + \delta_2(t) - \delta_2(t) = 2\delta_3(t), \quad (5)$$

The length of the MFC base and MFC are l_b and l_p, respectively. Therefore, the displacement amplitude at the end of the piezoelectric MFC is calculated as:

$$\delta_p(t) = \delta_b(t)\frac{l_p}{l_b}, \quad (6)$$

Figure 5 shows the deformed MFC, where r and α are the radius and angle of the deformed MFC, respectively. The relationship between r and α can be derived as:

$$\begin{cases} \alpha r = l_p \\ r(1 - \cos\frac{\alpha}{2}) = \delta_p \end{cases}, \quad (7)$$

From Equation (7), it can be concluded that

$$\alpha = \frac{l_p}{\delta_p}\left(1 - \cos\frac{\alpha}{2}\right), \quad (8)$$

Therefore, the shortened length of MFC is expressed as:

$$\begin{aligned}\delta_M(t) &= \alpha(r + h_b) - \alpha r \\ &= \frac{l_p h_b}{\delta_p(t)}\left(1 - \cos\frac{\alpha}{2}\right)\end{aligned}, \quad (9)$$

where h_b is the thickness of the MFC base.

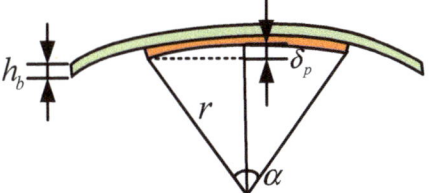

Figure 5. MFC deformation calculation diagram.

Figure 6 presents the equivalent circuit of piezoelectric MFC, where R is load resistance. When the piezoelectric MFC is subjected to bending deformation, assuming a linear change between stress and strain, the output charge is [40]

$$Q(t) = d_{31} E S_p \frac{\delta_M(t)}{l_p}, \qquad (10)$$

where d_{31}, E, S_p are piezoelectric strain constant, Young modulus of MFC and surface area of the piezoelectric material, respectively.

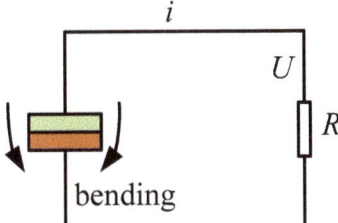

Figure 6. Equivalent circuit of piezoelectric MFC.

The output current of load is calculated by:

$$i(t) = \frac{\partial Q(t)}{\partial t} = d_{31} E S_p \frac{\partial \delta_M(t)}{l_p \partial t}, \qquad (11)$$

The output power of the proposed piezoelectric generator is expressed as:

$$P(t) = \frac{Q^2(t)}{2C} = d_{31}^2 E^2 S_p^2 \frac{\delta_M^2(t)}{2Cl_p^2}, \qquad (12)$$

where C is the capacitance of piezoelectric MFC. The effective value of the output voltage is calculated by the following expression:

$$U_{eff}(t) = \sqrt{P(t)R}, \qquad (13)$$

Thus, the peak output voltage of the generator can be derived as:

$$U_{p-p}(t) = \sqrt{2} U_{eff}(t), \qquad (14)$$

4. Structural Simulation Analysis

In order to verify the rationality of the designed structure, the finite element simulation software ANSYS was used to simulate the structural performance of the proposed piezoelectric generator. The elastic steel plate of the piezoelectric vibrator is 65 Mn-high elastic quenched spring steel plate with a thickness of 0.1 mm. Due to its excellent production process, the 65 Mn-high elastic quenched steel plate has the characteristics of high elasticity, high wear resistance, high toughness and good flatness. Its mechanical properties were good, with high tensile strength and elongation. Therefore, it is suitable as an elastic plate for the piezoelectric vibrator in a vibratory piezoelectric energy-collecting device.

The support structure and the force transfer mechanism in the device are made of PLA material and 3D-printed in a single form. The printed structure has the characteristics of high cleanliness and strength, fine appearance, etc. In addition, the accuracy also meets the requirements, being suitable for the assembly verification of the experimental device and for the appearance and structural components of this design. The printed device met the requirements after assembly.

Before the ANSYS simulation, it was necessary to set the basic parameters of the material of the piezoelectric generator, including the material density, elastic modulus and Poisson's ratio. Only by setting these parameters accurately could the desired effect be achieved in the simulation and the correct results obtained. Table 1 shows the basic parameters of materials, while Table 2 gives the parameter information of the MFC piezoelectric film. The piezoelectric MFC film was composed of a piezoelectric material with two sides, and an adhesive backing sheet attached on one side. The slicing of the piezoelectric material constituted a plurality of piezoelectric fibers in juxtaposition. A conductive film was then adhesively bonded to the other side of the piezoelectric material, and the adhesive backing sheet was removed. The conductive film had first and second conductive patterns formed thereon, which were electrically isolated from one another and in electrical contact with the piezoelectric material. The first and second conductive patterns of the conductive film each had a plurality of electrodes forming a pattern of interdigitated electrodes. A second film was then bonded to the other side of the piezoelectric material. The second film may have had a pair of conductive patterns similar to the conductive patterns of the first film [41].

Table 1. Basic parameters of materials.

Materials	PLA	Spring Steels
Use	Support structure	Elastic vibrator
Density (kg/m^3)	1.12	7.81
Young's modius (GPa)	2.7	197
Poisson's ratio	0.39	0.25
Buk modulus (GPa)	4.1	131
Shear modulus (GPa)	0.97	78.8

Table 2. Parameter information of MFC piezoelectric film.

Parameter	Values	Parameter	Values
Working mode	d_{31}	Effective working length	28 mm
Thickness	300 μm	Effective working width	14 mm
Electrode	Standard lead-free solder S-Sn99Cu1	Total length	37 mm
Capacitance	48 nF	Total width	18 mm
Upper limit of operating frequency	<1 MHz		

Finite element analysis can simplify complex problems and obtain an appropriate and similar solution for each small problem. In the working process of the proposed piezoelectric generator, the components of the generator re subjected to more or less tension, pressure or bending deformation, so the statics analysis of the device could be carried out using ANSYS finite element analysis software to analyze its overall deformation and stress.

The static analysis of vibratory piezoelectric generator was carried out using ANSYS software. In order to verify the performance of the piezoelectric generator, the structural deformation of the piezoelectric generator under low amplitude excitation was simulated, and the amplitude of external excitation was set as 0.5 mm to solve the deformation of each part of the piezoelectric generator. At the same time, in order to study the characteristic of piezoelectric generator under external excitation, the stress of the piezoelectric generator was analyzed. The deformation and stress analysis results are shown in Figure 7.

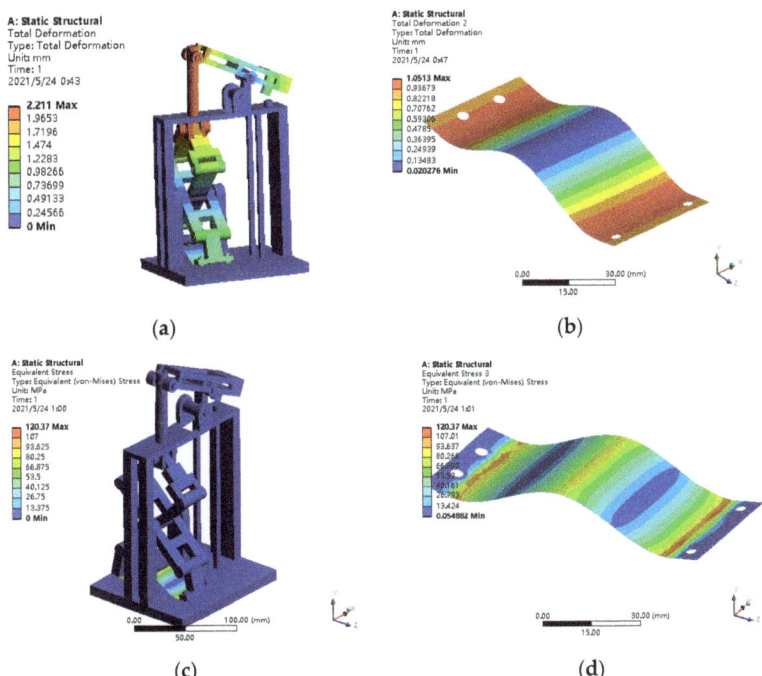

Figure 7. Deformation and stress of piezoelectric generator under external excitation. (**a**) Overall deformation; (**b**) elastic steel plate deformation; (**c**) overall stress; (**d**) elastic steel plate stress.

As shown in Figure 7a,b, the overall maximum deformation of the piezoelectric generator was 2 mm, and the maximum deformation was located at the vertical rod of the piezoelectric generator, which met the allowable requirements. Other connectors of the generator also had more or less deformation under the action of external forces, but they were less than 2 mm, meeting the use requirements for PLA materials. Moreover, the maximum deformation of the elastic steel plate, the key component of the piezoelectric generator, was only 1 mm.

According to the mechanical properties of the material, the tensile and compression yield strength of PLA was 40–51 MPa, and the yield limit of the elastic steel plate was 825–925 MPa. In the finite element analysis results of Figure 7c,d, the maximum stress of the elastic steel plate was 120.37 MPa, less than the yield strength of the spring steel. In addition, the overall stress of the external structure of the piezoelectric generator was less than 14 MPa, so it met the design requirements.

The resonance of the structure is destructive to the proposed piezoelectric generator. Therefore, in order to avoid the resonance of the piezoelectric generator in the working process, the excitation frequency should be far away from the resonance frequency. Therefore, modal characteristics of piezoelectric generator were analyzed in this paper. Table 3 shows the natural frequencies of the piezoelectric generator, and Figure 8 presents the modal shapes of the piezoelectric generator.

Table 3. Natural frequencies of the piezoelectric generator.

Order	First Order	Second Order	Third Order	Fourth Order
Frequencies (Hz)	2107.8	2353.9	2905.1	3456.1

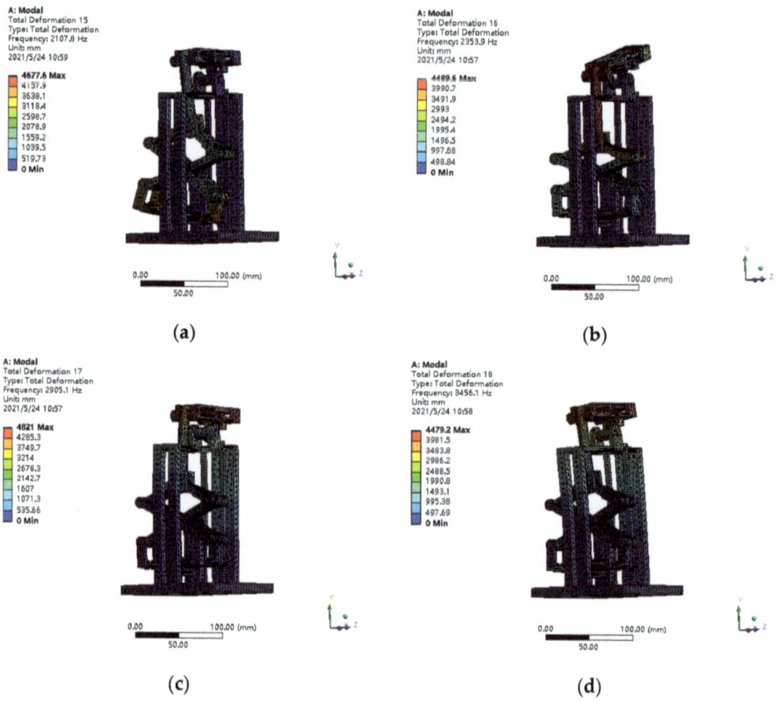

Figure 8. Modal shapes of piezoelectric generator. (**a**) First-order modal shape; (**b**) second-order modal shape; (**c**) third-order modal shape; (**d**) fourth-order modal shape.

According to Figure 8, the first-order natural frequency of the piezoelectric is 2107.8 Hz, and the main vibration mode is the bending vibration of the elastic steel plate. At the same time, with the increase in the frequency order, the main mode shape changes to the displacement amplification mechanism.

Since the piezoelectric generator designed in this paper mainly works in the low-frequency range less than 100 Hz, it does not generate resonance during the working process.

5. Experimental Analysis

In order to test the performance of the piezoelectric generator, an experimental platform was manufactured, as shown in Figure 9. The generator was manufactured using 3D printing technology. The vibration excitation of the generator was realized by the reciprocating motion of a DC motor, and a speed controller was used to adjust the excitation frequency. An adjustable resistor varying in the range of 0–10 MΩ was utilized as load resistor. The output voltage of the developed generator was collected by an NI USB-6002 data acquisition (DAQ) card. The upper computer installed with LabVIEW acquisition software was used to analyze the test data.

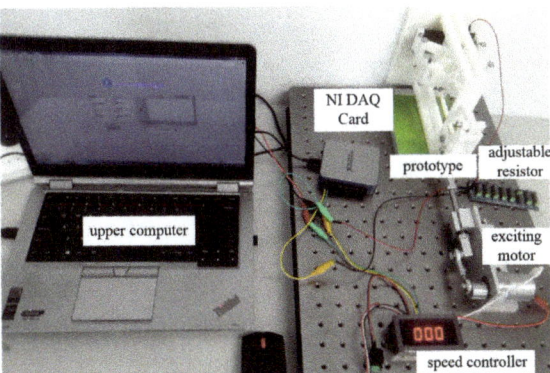

Figure 9. Experimental test system of the proposed piezoelectric generator.

Firstly, the voltage generated by the piezoelectric generator under instantaneous excitation was evaluated, as illustrated in Figure 10a,b. Here, through changing the load resistance, the instantaneous response law of the output voltage was obtained. Figure 10c shows the instantaneous maximum voltage and maximum power of the generator vary with load resistance. Results show that:

(1) Under instantaneous vibration excitation, the generator generated transient voltage response, and the response time lasted 0.2 s–0.3 s. Under different load resistances, the voltage response curves of the piezoelectric generator were similar, and the response amplitudes were different. When the load resistance increased, the maximum voltage response amplitude also increased.

(2) As the load resistance R increased, the maximum output voltage increased. For the output power, the maximum instantaneous power occurred at $R = 35$ KΩ and was 2.44 mW.

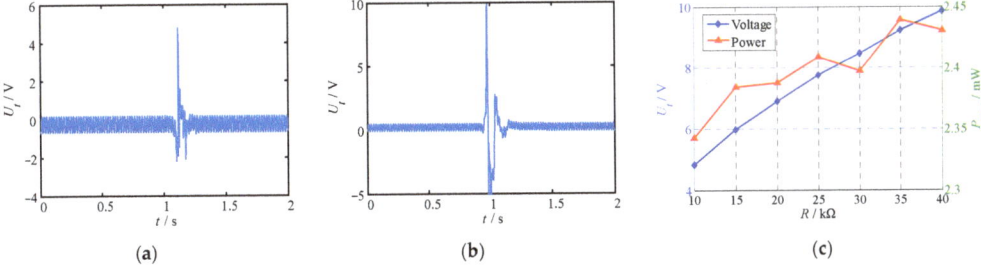

Figure 10. Instantaneous voltage and power of the proposed piezoelectric generator. (**a**) Instantaneous voltage at $R = 10$ KΩ; (**b**) instantaneous voltage at $R = 40$ KΩ; (**c**) instantaneous power.

The voltage response of the piezoelectric generator under continuous excitation was also tested. Using a laser displacement sensor with a range of ±5 mm, the vibration amplitude of the piezoelectric MFC was measured, and the test result is shown in Figure 11a. Under continuous vibration excitation, the piezoelectric MFC generated a continuous voltage signal. The images in Figure 11b,c show the voltage-response test results of the piezoelectric generator under continuous excitation with different frequencies and different load resistances. According to Figure 11, it can be noted that:

(1) Under the excitation of a continuous vibration signal, the displacement of the piezoelectric MFC changed regularly and continuously in the range of 0–2 mm. The average displacement was 1.23 mm.

(2) At different exciting frequencies and load resistances, the voltage response of the piezoelectric generator was similar. The difference was that the maximum output voltage increased with the increase in excitation frequency and load resistance.
(3) When the excitation frequency and load resistance were constant, the output voltage of the generator varied regularly with time. The output voltage waveform approximated sinusoidal law.

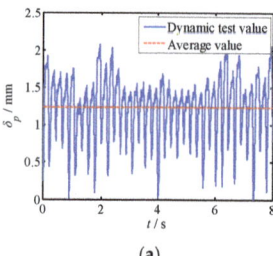
(a)

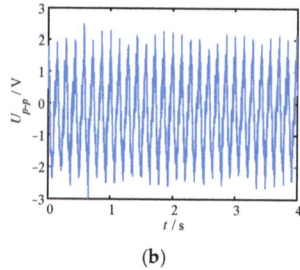
(b)

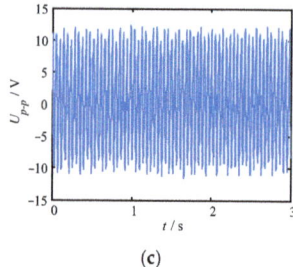
(c)

Figure 11. MFC displacement and continuous voltage response of the piezoelectric generator. (**a**) MFC displacement; (**b**) continuous voltage at R = 10 kΩ, f = 10 Hz; (**c**) continuous voltage at R = 40 kΩ, f = 25 Hz.

The variation in voltage with load resistance at different frequencies was tested experimentally. From the relationship between voltage and power, the variation of power with load resistance was obtained. The images in Figure 12a,b show the output voltage variations with the load resistance. The images in Figure 12c,d show the output power variations with the load resistance. Additionally, the electric generating test results under instantaneous excitation and continuous excitation were compared with the respective simulation results. The comparison results are shown in Figure 12e,f. From Figure 12, it can be observed:

(1) With the increase in load resistance, the peak voltage and effective value first increased, and then tended to be stable or even slightly decrease. The maximum peak voltage and effective voltage were 12.22 V and 8.64 V, respectively, and occurred at the load resistance of 40 kΩ and the excitation frequency of 25 Hz. When the load resistance was constant, the peak voltage and effective voltage increased with the increase in excitation frequency. When the load resistance was 40 kΩ, the voltage change was more obvious.

(2) When the excitation frequency was less than 20 Hz, the maximum power and average power tended to be stable with the change in load resistance. When the excitation frequency was greater than 20 Hz, the output power was not stable. The power decreased with the increase in load resistance. The maximum power and average power of the piezoelectric generator were 4.82 mW and 2.41 mW, respectively, occurring at the load resistance of 20 kΩ and the excitation frequency of 25 Hz.

(3) The instantaneous voltage of the generator was very close to the simulated voltage. The maximum error between the instantaneous voltage corresponding to different load resistor and the simulation voltage was 15.7%, which occurred when R was 10 kΩ. The instantaneous voltage and simulation voltage had the same variation rule, and the instantaneous voltage was less than the continuous voltage.

(4) As the load resistance changed, the maximum simulation power values remained unchanged. The reason was that the output power was not affected by the load resistance based on Equation (12). The instantaneous power fluctuated within the range of 2.34 mW–2.44 mW. The continuous power increased first and then decreased, and its power range was 3.25 mW–4.82 mW. When the load resistance R was 20 kΩ and 25 kΩ, the continuous power was close to the average power, and the errors were 5.24% and 2.84%, respectively. There was a significant difference between

instantaneous power and continuous power. For instantaneous power, which was generated only for a very short time, due to the lack of continuous excitation, its power was relatively small. For continuous power, the maximum output power was close to the calculated value. When the load resistance was small, the energy was not fully released and the power was small, while when the load resistance was large, the stability of the output power was reduced and the power was slightly reduced. Hence, the instantaneous power and continuous power in Figure 12f were inconsistent with the calculated results.

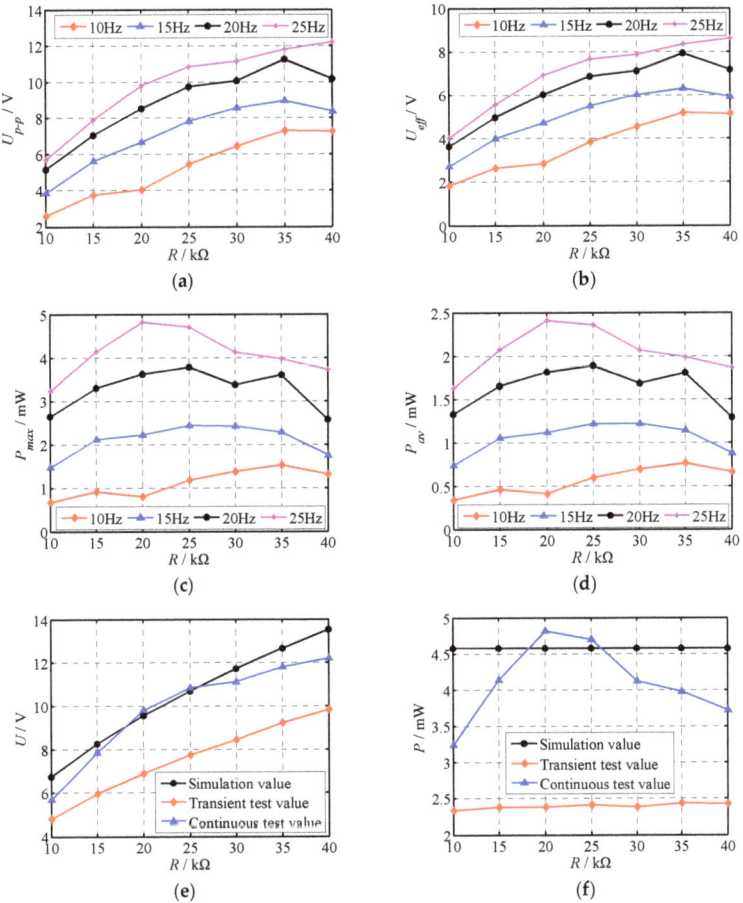

Figure 12. The output voltage and power vary with the load resistance. (**a**) Maximum voltage; (**b**) effective voltage; (**c**) maximum power; (**d**) average power; (**e**) comparison of experimental and simulated voltages; (**f**) Comparison of experimental and simulated powers.

Therefore, the output power of the piezoelectric generator under continuous excitation is closer to the simulated power, especially under certain loads. When the load resistance is 20 kΩ, the generator shows better performance.

6. Conclusions

In this paper, a low-frequency vibration-type piezoelectric generator based on the PZT-PI composite film was proposed. The theoretical model and working principle of the piezoelectric generator were presented and discussed. In addition, a home-made experimental platform was built to test the output characteristics of the piezoelectric generator. Results show that:

(1) The transient excitation voltage of the proposed piezoelectric generator increased with the increase in load resistance, while the continuous excitation voltage increased first and then tended to be stable or slightly decreased.
(2) The maximum continuous power produced by the piezoelectric generator was about 4.82 mW.
(3) The simulation results agreed well with the experimental results on continuous excitation voltage and power with load resistances of 20 kΩ and 25 kΩ.

Author Contributions: Conceptualization, J.W.; investigation, Y.T.; validation, Z.Z.; writing—original draft, C.L.; writing-review & editing, J.S. All authors have read and agreed to the published version of the manuscript.

Funding: This research was funded by the National Key Research and Development Program of China (No.2018YFC0309100) and the Postgraduate Research & Practice Innovation Program of Jiangsu Province (No.KYCX21_3445).

Institutional Review Board Statement: Not applicable.

Informed Consent Statement: Not applicable.

Data Availability Statement: Not applicable.

Conflicts of Interest: The authors declare no conflict of interest.

References

1. Shakeel, M.; Rehman, K.; Ahmad, S.; Amin, M.; Iqbal, N.; Khan, A. A low-cost printed organic thermoelectric generator for low-temperature energy harvesting. *Renew. Energy* **2021**, *167*, 853–860. [CrossRef]
2. Mishra, S.; Unnikrishnan, L.; Nayak, S.K.; Mohanty, S. Advances in piezoelectric polymer composites for energy harvesting applications: A systematic review. *Macromol. Mater. Eng.* **2019**, *304*, 1800463. [CrossRef]
3. Teng, H.C.; Kok, B.C.; Uttraphan, C.; Yee, M.H. A review on energy harvesting potential from living plants: Future energy resource. *Int. J. Renew. Energy Res.* **2018**, *8*, 2598–2614.
4. Viet, N.; Xie, X.; Liew, K.; Banthia, N.; Wang, Q. Energy harvesting from ocean waves by a floating energy harvester. *Energy* **2016**, *112*, 1219–1226. [CrossRef]
5. Sari, I.; Balkan, T.; Kulah, H. An electromagnetic micro power generator for wideband environmental vibrations. *Sens. Actuators A Phys.* **2008**, *145*, 405–413. [CrossRef]
6. El-Rayes, K.; Gabran, S.; Abdel-Rahman, E.; Melek, W. Variable-Flux Biaxial Vibration Energy Harvester. *IEEE Sens. J.* **2018**, *18*, 3218–3227. [CrossRef]
7. Hadas, Z.; Vetiska, V.; Vetiska, J.; Krejsa, J. Analysis and efficiency measurement of electromagnetic vibration energy harvesting system. *Microsyst. Technol.-Micro-Nanosyst.-Inf. Storage Process. Syst.* **2016**, *22*, 1767–1779. [CrossRef]
8. Wang, X.; John, S.; Watkins, S.; Yu, X.; Xiao, H.; Liang, X.; Wei, H. Similarity and duality of electromagnetic and piezoelectric vibration energy harvesters. *Mech. Syst. Signal Process.* **2014**, *52–53*, 672–684.
9. Liu, C.C.; Jing, X.J. Vibration energy harvesting with a nonlinear structure. *Nonlinear Dyn.* **2016**, *84*, 2079–2098. [CrossRef]
10. Triplett, A.; Quinn, D.D. The effect of non-linear piezoelectric coupling on vibration-based energy harvesting. *J. Intell. Mater. Syst. Struct.* **2009**, *20*, 1959–1967. [CrossRef]
11. Iqbal, M.; Nauman, M.M.; Khan, F.U.; Abas, P.E.; Cheok, Q.; Iqbal, A.; Aissa, B. Vibration-based piezoelectric, electromagnetic, and hybrid energy harvesters for microsystems applications: A contributed review. *Int. J. Energy Res.* **2021**, *45*, 65–102.
12. Zhang, Y.; Wang, T.; Luo, A.; Hu, Y.; Li, X.; Wang, F. Micro electrostatic energy harvester with both broad bandwidth and high normalized power density. *Appl. Energy* **2018**, *212*, 362–371. [CrossRef]
13. Shao, J.; Zhou, L.; Chen, Y.; Liu, X.; Ji, M. Model-based dielectric constant estimation of polymeric nanocomposite. *Polymers* **2022**, *14*, 1121. [CrossRef]
14. Moradian, S.; Akhkandi, P.; Huang, J.; Gong, X.; Abdolvand, R. A Battery-Less Wireless Respiratory Sensor Using Micro-Machined Thin-Film Piezoelectric Resonators. *Micromachines* **2021**, *12*, 363. [CrossRef] [PubMed]
15. Titsch, C.; Li, Q.; Kimme, S.; Drossel, W.G. Proof of Principle of a Rotating Actuator Based on Magnetostrictive Material with Simultaneous Vibration Amplitude. *Actuators* **2020**, *9*, 81. [CrossRef]
16. Zhang, B.; Li, H.; Zhou, S.; Liang, J.; Gao, J.; Yurchenko, D. Modeling and analysis of a three-degree-of-freedom piezoelectric vibration energy harvester for broadening bandwidth. *Mech. Syst. Signal Process.* **2022**, *176*, 109169. [CrossRef]
17. Liu, M.; Qian, F.; Mi, J.; Zuo, L. Dynamic interaction of energy-harvesting backpack and the human body to improve walking comfort. *Mech. Syst. Signal Process.* **2022**, *174*, 109101. [CrossRef]

18. Wang, L.; Wu, Z.T.; Liu, S.; Wang, Q.; Sun, J.; Zhang, Y.; Qin, G.; Lu, D.; Yang, P.; Zhao, L.; et al. Uniform Stress Distribution of Bimorph by Arc Mechanical Stopper for Maximum Piezoelectric Vibration Energy Harvesting. *Energies* **2022**, *15*, 3268. [CrossRef]
19. Morel, A.; Brenes, A.; Gibus, D.; Lefeuvre, E.; Gasnier, P.; Pillonnet, G.; Badel, A. A comparative study of electrical interfaces for tunable piezoelectric vibration energy harvesting. *Smart Mater. Struct.* **2022**, *98*, 107209.
20. Rojas, E.F.; Faroughi, S.; Abdelkefi, A.; Park, Y.H. Investigations on the performance of piezoelectric-flexoelectric energy harvesters. *Appl. Energy* **2021**, *288*, 116611. [CrossRef]
21. Sharma, M.; Chauhan, A.; Vaish, R. Energy harvesting using piezoelectric cementitious composites for water cleaning applications. *Mater. Res. Bull.* **2021**, *137*, 111205. [CrossRef]
22. Pan, J.; Qin, W.; Yang, Y.; Yang, Y. A collision impact based energy harvester using piezoelectric polyline beams with electret coupling. *J. Phys. D Appl. Phys.* **2021**, *54*, 225502. [CrossRef]
23. Liu, M.; Xia, H.; Liu, G.Q. Experimental and numerical study of underwater piezoelectric generator based on Vortex-induced Vibration. *Eng. Res. Express* **2021**, *3*, 045056.
24. Zhou, L.; Liu, Y.; Ma, L.; Wu, Y. Piezoelectric generator with nonlinear structure. *Proc. Inst. Mech. Eng. Part L J. Mater. Des. Appl.* **2021**. [CrossRef]
25. Zou, H.; Shi, Q.; Cai, H.; Liu, J. The multi-harmonic excitation characteristic of airflow piezoelectric generator. *J. Vibroeng.* **2021**, *23*, 1219–1229. [CrossRef]
26. Ha, Y.; Jeong, S.; Cheo, S.; Lee, B.; Kim, M.; Park, T. Generating characteristics of an eye-shaped piezoelectric generator. *Ceram. Int.* **2015**, *41*, S691–S694. [CrossRef]
27. Ramadan, K.S.; Sameoto, D.; Evoy, S. A review of piezoelectric polymers as functional materials for electromechanical transducers. *Smart Mater. Struct.* **2014**, *23*, 033001. [CrossRef]
28. Baur, C.; Apo, D.J.; Maurya, D.; Priya, S.; Voit, W. Advances in piezoelectric polymer composites for vibrational energy harvesting. In *Polymer Composites for Energy Harvesting, Conversion, and Storage*; American Chemical Society: Washington, DC, USA, 2014; pp. 1–27.
29. Covaci, C.; Gontean, A. Piezoelectric energy harvesting solutions: A review. *Sensors* **2020**, *20*, 3512. [CrossRef]
30. Shao, J.; Liao, X.; Ji, M.; Liu, X. A Modeling Study of the Dielectric Property of Polymeric Nanocomposites Based on the Developed Rayleigh Model. *ACS Appl. Polym. Mater.* **2021**, *3*, 6338–6344. [CrossRef]
31. Lin, J.; Malakooti, M.H.; Sodano, H.A. Thermally stable poly (vinylidene fluoride) for high-performance printable piezoelectric devices. *ACS Appl. Mater. Interfaces* **2020**, *12*, 21871–21882. [CrossRef]
32. Gong, S.; Zhang, B.; Zhang, J.; Wang, Z.L.; Ren, K. Biocompatible poly (lactic acid)-based hybrid piezoelectric and electret nanogenerator for electronic skin applications. *Adv. Funct. Mater.* **2020**, *30*, 1908724. [CrossRef]
33. Yazdani, A.; Manesh, H.D.; Zebarjad, S.M. Piezoelectric properties and damping behavior of highly loaded PZT/polyurethane particulate composites. *Ceram. Int.* **2021**, *7*, 126. [CrossRef]
34. Sappati, K.K.; Bhadra, S. Piezoelectric polymer and paper substrates: A review. *Sensors* **2018**, *18*, 3605. [CrossRef] [PubMed]
35. Dagdeviren, C.; Yang, B.D.; Su, Y.; Tran, P.L.; Joe, P.; Anderson, E.; Xia, J.; Doraiswamy, V.; Dehdashti, B.; Feng, X.; et al. Conformal piezoelectric energy harvesting and storage from motions of the heart, lung, and diaphragm. *Proc. Natl. Acad. Sci. USA* **2014**, *111*, 1927–1932. [CrossRef] [PubMed]
36. Deng, J.; Guasch, O.; Zheng, L.; Song, T.; Cao, Y. Semi-analytical model of an acoustic black hole piezoelectric bimorph cantilever for energy harvesting. *J. Sound Vib.* **2021**, *494*, 115790. [CrossRef]
37. Derakhshani, M.; Momenzadeh, N.; Berfield, T.A. Analytical and experimental study of a clamped-clamped, bistable buckled beam low-frequency PVDF vibration energy harvester. *J. Sound Vib.* **2021**, *497*, 115937. [CrossRef]
38. Kim, J.H.; Kim, B.; Kim, S.W.; Kang, H.W.; Park, M.-C.; Park, D.H.; Ju, B.K.; Choi, W.K. High-performance coaxial piezoelectric energy generator (C-PEG) yarn of Cu/PVDF-TrFE/PDMS/Nylon/Ag. *Nanotechnology* **2021**, *32*, 145401. [CrossRef]
39. Ma, Y.; Wang, J.; Li, C.; Fu, X. A Micro-Power Generator Based on Two Piezoelectric MFC Films. *Crystals* **2021**, *11*, 861. [CrossRef]
40. Bai, F.X.; Zhang, M.J.; Sun, J.Z.; Sun, J.Z.; Dong, W.J. Research and design of wearable cylindrical shell piezoelectric energy harvester. *Pie-Zoeletrics Acoustoopics* **2021**, *43*, 39–44.
41. Wilkie, W.K.; Bryant, R.G.; Fox, R.L.; Hellbaum, R.F.; High, G.W.; Jalink, A., Jr.; Little, B.D.; Mirick, P.M. Method of Fabricating a Piezoelectric Composite Apparatus. US Patent No. 6629341, 7 October 2003.

Communication

Refractive Index and Temperature Sensing Performance of Microfiber Modified by UV Glue Distributed Nanoparticles

Hongtao Dang [1], Yan Zhang [1], Yukun Qiao [1] and Jin Li [1,2,3,*]

1. Shaanxi Engineering Research Center of Controllable Neutron Source, School of Electronic Information, Xijing University, Xi'an 710123, China; skydht@163.com (H.D.); zy353615@163.com (Y.Z.); qiaoyukun6530@163.com (Y.Q.)
2. College of Information Science and Engineering, Northeastern University, Shenyang 110819, China
3. Hebei Key Laboratory of Micro-Nano Precision Optical Sensing and Measurement Technology, Qinhuangdao 066004, China
* Correspondence: lijin@ise.neu.edu.cn

Abstract: Dielectric materials with high refractive index have been widely studied to develop novel photonic devices for modulating optical signals. In this paper, the microfibers were modified by silicon nanoparticles (NPs) and silver NPs mixed in UV glue with ultra-low refractive index, respectively, whose corresponding optical and sensing properties have been studied and compared. The influence from either the morphological parameters of microfiber or the concentration of NPs on the refractive index sensing performance of microfiber has been investigated. The refractive index sensitivities for the microfiber tapers elaborated with silver NPs and silicon NPs were experimentally demonstrated to be 1382.3 nm/RIU and 1769.7 nm/RIU, respectively. Furthermore, the proposed microfiber was encapsulated in one cut of capillary to develop a miniature temperature probe, whose sensitivity was determined as 2.08 nm/°C, ranging from 28 °C to 43 °C.

Keywords: silicon nanoparticles; surface plasmon resonance; microfiber sensor; refractive index sensing; temperature sensing

1. Introduction

Gold and silver nanoparticles (NPs) can confine the light field in the sub-wavelength scale to overcome the optical diffraction limit due to the localized surface plasma resonance (LSPR) excited at their surface [1]. LSPR based optics modulation has been widely involved in astrophysics, life science, optics, photonics, materials and other fields [2,3]. Metal or metal oxide NPs can be doped in polymer gel [4] and elaborated on the surface or inside of the optical fibers to develop compact sensing probes [5,6]. A highly sensitive performance has been experimentally demonstrated in many works reported in recent years [7–10], in which noble metal or metal oxide nano-films or NPs were elaborated on the different fiber structures (microfiber taper, D-shaped, U-shaped interface) to excite long range surface plasmon resonance (LRSPR) or LSPR effect [11]. These fiber sensors have been used for detection of analytes such as hormones, contaminants, etc. [12]. In addition to traditional noble metal NPs, other metal oxides (SnO_2, WO_3, etc.) also exhibit excellent sensing performance [13,14]. Various NPs have been doped into silica or polymer optical fibers for determination of bio-molecular or chemical composition [15–17], where internal doping and surface coating have become the most typical modification methods. Compared with crystal or dielectric materials, the scattering loss of metal particles is extremely serious. During the LSRP excitation process, the transition of free electrons among different energy levels will increase the surrounding temperature, resulting in the Joule heating effect [18,19]. Although it has been proven to be a merit in photo-thermal cancer therapy, photo-thermal imaging and photo-thermal bio-sensing [20], the local heating is extremely unfavorable for photonics devices and sensors with high requirements for environmental stability, such as in vivo detection [21], where the local heater will deform or even erase the nanostructure, thereby affecting the optical modulation performance of the photonic device [22]. More issues can be found in other works, for example, it may change the refractive index of the sensing material or structure due to the thermo-optical effect [23], and it can volatilize the liquid and cause thermal damage to in vivo samples [24]. In addition, LSPR devices also suffer due to poor compatibility with the semiconductor's oxide. To overcome the optical loss and self-heating problem, dielectric materials with high refractive index have been introduced, including silicon, germanium, gallium phosphide, etc. The magneto-optics and direction-scattering characteristics for all-dielectric NPs or structures have been verified [25–27]. Near field optical manipulation has been demonstrated based on dielectric NPs with high refractive index, which has proved a promising prospect in solving the scattering loss and self-heating problem in photonics devices.

Mie scattering mainly depends on dielectric constant and size for lossless, non-magnetic NPs. The maximum scattering efficiency of subwavelength NPs is only related to the resonance frequency and is irrelevant to the material [28]. Therefore, the LSPR-like effect, similar to that of metal NPs, can also be realized for dielectric NPs with a high refractive index. The scattering efficiency will be significantly enhanced due to the higher refractive index [29,30]. It has been revealed that all-dielectric nanostructures have similar characteristics to plasmonic devices, including enhanced scattering, high-frequency magnetic fields, and negative refractive index [31]. By precisely designing the structures and optimizing their parameters, all-dielectric nanostructures display an even better performance [32,33]. Furthermore, dielectric NPs (or structures) can play a similar role to plasmonic devices, being self-assembled (or processed) to act as photonic crystals and meta-surfaces on 2D substrate or optical fibers [34,35].

In recent years, most of the related research on the nonlinear optical properties and sensing applications of dielectric nanostructures have been focused on the meta-surface structure on planar substrates. The controllable modification and sensing application of all dielectric nanoparticles with high refractive index are rarely reported. In this work, silicon or silver NPs have been elaborated on the surface of microfibers and their corresponding refractive index sensing performance experimentally characterized. The influence of NPs' concentration on the sensing performance has been verified to optimize the sensitivity. Based on the refractive sensing mechanism, a temperature probe was experimentally demonstrated as the Si NPs' elaborated microfiber probe.

2. Materials and Methods

The microfiber was fabricated by a two-steps point-fixed heating-drawing technique, from the common single-mode fiber (SMF-28, Corning Inc., Corning, NY, USA) with refractive index of 1.4682 and 1.4628 for the corresponding core and cladding, respectively. The preparation process of the microfiber taper is divided into two main steps, as shown in Figure 1: pre-stretching by a fiber splicer, and point-fixed stretching by a fiber tapering machine.

The single-mode fiber taper is made via a manual mode using a fiber splicing machine (Furukawa, S178A, Furukawa Electric Co., Ltd., Tokyo, Japan), where the discharging time and moving process of the fiber holders can be flexibly adjusted. In detail, the ordinary single-mode optical fiber was fixed in the fiber splicing machine. After selecting the corresponding optical fiber matching mode, the working mode was changed to manual operation. The pulse value of two fixed motors Z_L and Z_R was set as ~3400. After repeating the arc discharges twice, the bi-conical microfiber with a waist diameter of ~40 μm could be obtained; the same single-mode fiber taper was scan-heated and further stretched by a fiber stretching machine (Idealphotonics, IPCS-5000, Idealphotonics Inc., Vancouver, BC, Canada) to obtain microfiber with a diameter of less than 5 μm and a length of 5–8 mm.

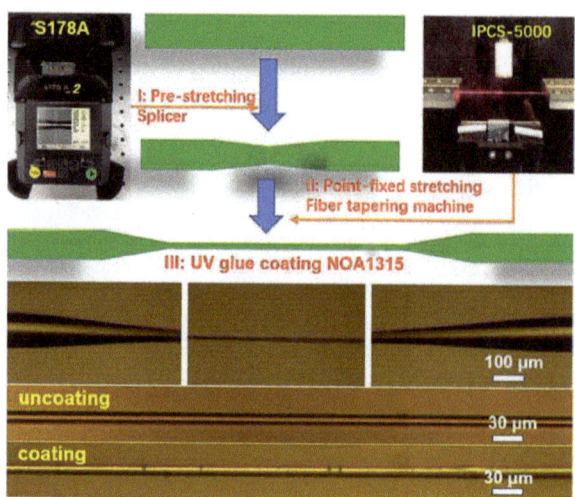

Figure 1. Schematic diagram of fabrication process of NPs-coated microfiber using two-steps point-fixed stretching method.

A hydrogen generator was used to provide the hydrogen-oxygen flame with temperature up to 1500 °C. During the heating-stretching process, the torch scans in a fixed region, where the desired fiber length and scanning range can be set via the operating board. The target parameters (fiber length) are easily obtained by the equal volume calculation. Meanwhile, the scanning range should be limited to less than the length of the thin waist region to maintain the uniform diameter of the long-tapered microfiber. In order to effectively reduce the optical loss and improve the binding efficiency of the light field, UV glue with ultra-low refractive index (NOA1315, refractive index is 1.315, Norland Products Inc., Jamesburg, NJ, USA) was introduced to disperse the silver or silicon NPs and uniformly elaborate them on the outer surface of the microfiber region. NOA1315 has good fluidity due to its ultra-low viscosity of only 15 cps. The NPs were added and the mixture was dispersed by magnetic stirring for 10 min to distribute the NPs uniformly. The UV glue-NPs mixture was injected into a syringe and dripped onto the surface of the microfibers, whose surface morphology can be compared according to the microscope pictures in Figure 1.

The transmission spectra of clean microfiber tapers were compared with those of microfibers elaborated by NPs, as shown in Figure 2a,b, respectively.

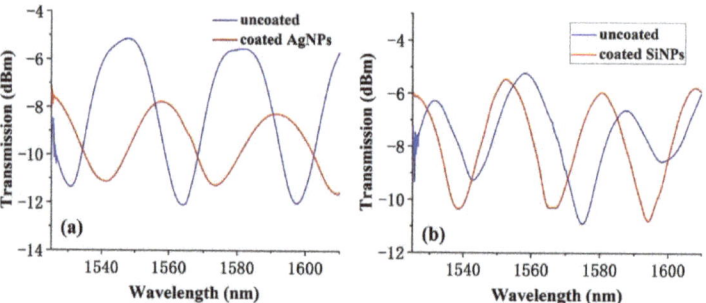

Figure 2. Spectra comparison of a clean microfiber taper with an elaborated microfiber taper by (**a**) silver and (**b**) silicon NPs.

The significant influence on the spectra is revealed, due to the introduction of the UV/silver or UV/silicon NPs composite film. The interference spectrum still has a good waveform in the later experiment. To accurately control the sensing length, too long or too thin microfiber was excluded due to poor fabrication repeatability. In this work, the microfiber length was strictly controlled at ~6 mm to obtain a stable spectrum with the relative-fixed free range (FSR) of ~40 nm. During the first step, the manual mode of the fiber splicer was used to fabricate a short taper with waist diameter of ~50 μm, as the inserted image shows. This short taper was further stretched using the flame scanning method by a multifunctional fiber drawing machine. Then, the flame torch was, respectively, fixed on point 1 and 2 to obtain microfiber double-tapers with thinner diameter of ~10 μm.

In Mie's theory, the extinction spectra Q_{ext} of the spherical NPs can be presented [36] as

$$Q_{ext} = \frac{18\pi \varepsilon_m^{3/2} V}{\lambda} \cdot \frac{\varepsilon_2(\lambda)}{[\varepsilon_1(\lambda) + 2\varepsilon_m]^2 + \varepsilon_2^2(\lambda)} \quad (1)$$

Here, ε_m, is the dielectric constant of the surrounding environment; ε_1 and ε_2 refer to the real and imaginary part of the complex dielectric constant, respectively; λ is the working wavelength; and V is the volume of spherical NPs. In periodic metal nanostructures, the excitation spectra depend on the structure size [37]. However, in our proposed microfiber sensor, it mainly relies on the multimode interference (MMI) effect to detect the change in external refractive index [38], where the wavelength shift of the MMI spectrum ($\Delta\lambda$) depends on the change of the equivalent refractive index (n_{eff}) of the surrounding medium environment, which can be expressed as [39]

$$\Delta \lambda = p \frac{n_{eff} W_{eff}^2}{L} \quad (2)$$

where p-th is the image of the input field; L is the MMI length along the microfiber; W_{eff} is the effective optical diameter of microfiber, and depends on the physical diameter W, the refractive index of core n_r and the cladding n_c of the microfiber. This value can be calculated by [40]

$$W_{eff} = W + \frac{1}{2} \left(\frac{\lambda_0}{\pi} \right) \left(n_r^2 - n_c^2 \right)^{-1/2} \left[\left(\frac{n_c}{n_r} \right)^2 + 1 \right] \quad (3)$$

Here, λ_0 is the working wavelength in free-space. Because the diameter of the microfiber is very small, its core and cladding have been integrated during the fiber stretching process. The evanescent field will propagate along the microfiber, and the refractive index of the whole microfiber will be usually equivalent to the same value. In this case, it assumes that $W_{eff} \approx W$ [38]. The second term of Equation (3) will become significant with the decrease in W; therefore, the sensitivity of the MMI based microfiber can be improved by reducing the diameter of the microfiber and extending the sensing region.

3. Results and Discussion

3.1. Refractive Index Sensing Performance Comparison

The refractive index sensing performance of a UV-glue-coated-microfiber long-taper with the length of ~6 mm has first been demonstrated. In Figure 3a, the transmission spectra red-shifted when the refractive index changed from 1.3313 to 1.3422. The corresponding sensitivity was determined to be 1093.26 nm/RIU for the linear fitting results, shown in Figure 3b. It should be noted that the introduction of the UV glue layer with low refractive index resulted in an interference spectrum with high quality.

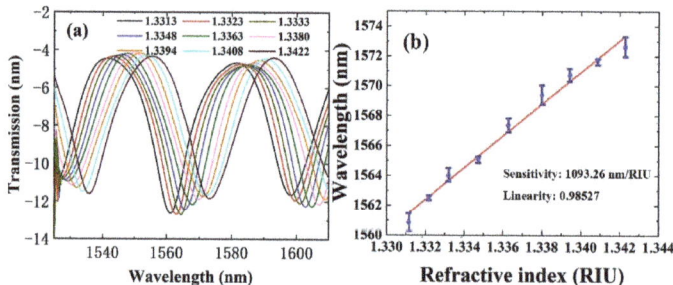

Figure 3. Refractive index sensing performance of a UV-glue-coated-microfiber. (**a**) Spectra change as a function of refractive index change from 1.3313 to 1.3422 and (**b**) corresponding sensing characteristics curve.

Furthermore, the UV-glue-coated-microfiber can sense the refractive index change for its environment. The influence of NPs' concentration on the optical and sensing performance of the function film coated microfiber should be studied. Too low concentration will support a weaker LSPR effect or LSPR-like effect for either silver or silicon NPs, which will also exert little impact on the sensing performance. By contrary, too high concentration will introduce a serious loss signal due to the light scattering on the microfiber surface. The concentration range was finally chosen in the range of 0.01 mol/L–0.5 mol/L in this experiment. The output spectrum of the silver NPs' (0.01 mol/L) elaborated microfiber is shown in Figure 4a. The sensitivity is determined to be 1382.3 nm/RIU after linear fitting (Figure 4b) of the experimental data for the wavelength shift. During the experiment, different refractive index environments are provided by the slowly cooling distilled water from 40 °C to room temperature, so as to reduce the influence of experimental environment fluctuation on the transmission spectrum of microfiber.

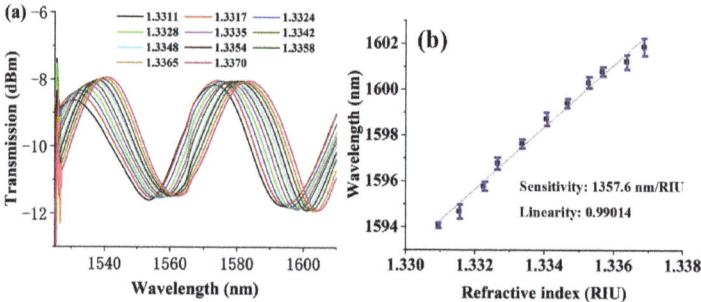

Figure 4. (**a**) Interference spectra and (**b**) responding curve as a function of refractive index for microfiber taper elaborated by silver NPs with concentration of 0.1 mol/mL.

The refractive index values were measured by an Abe refractometer (WYA-2S, Shanghai Yidian physical optical instrument Co., Ltd., Shanghai, China); meanwhile, the corresponding spectra were recorded by a spectrum analyzer. Further experimental results show that, with the increase of concentration, as compared in Table 1, this sensitivity cannot be further enhanced. At low concentrations (0.01 mol/L, 0.05 mol/L, 0.1 mol/L), the sensitivity is ~1360 nm/RIU with a good linearity of $R2 = $ ~0.99. When the concentration is 0.5 mol/L, the sensitivity has the highest value of up to 1842.3 nm/RIU, $R2 = 0.99167$. Figure 5a shows the output spectrum of the microfiber elaborated by silicon NPs with a concentration of 0.01 mol/L. The experimental fitting line reveals a refractive index sensitivity of 1769.7 nm/RIU in Figure 5b. Further experimental results indicate that, as the concentration increases, the sensitivity was reduced significantly.

Table 1. Sensing performance comparison for 6 mm length of microfiber elaborated by silver and silicon NPs with different concentrations.

Materials	NPs Concentration	Sensitivity	Linearity
Silver	0.01 mol/L	1382.3 nm/RIU	0.99291
	0.05 mol/L	1364.8 nm/RIU	0.98768
	0.1 mol/L	1357.6 nm/RIU	0.99014
	0.5 mol/L	1842.3 nm/RIU	0.99167
Silicon	0.01 mol/L	1769.7 nm/RIU	0.98992
	0.05 mol/L	1471.8 nm/RIU	0.98968
	0.1 mol/L	1468.1 nm/RIU	0.99534
	0.5 mol/L	1321.8 nm/RIU	0.99252

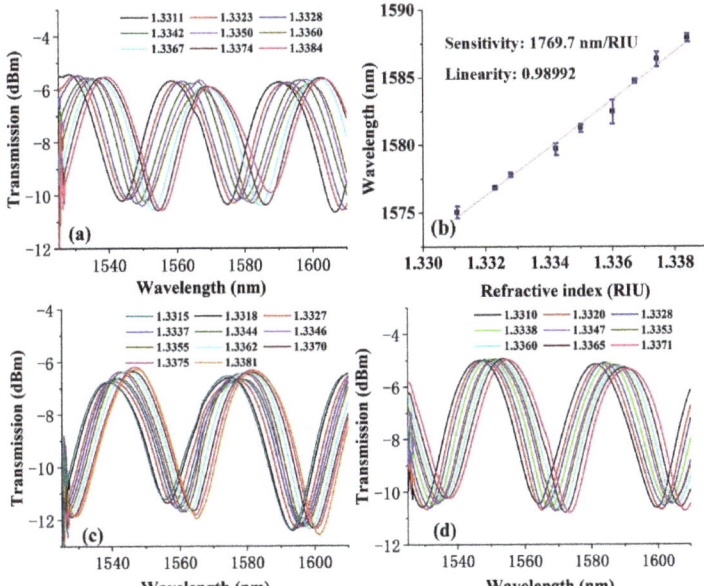

Figure 5. (a) Interference spectra as a function of refractive index and (b) corresponding fitting curve for microfiber taper elaborated by silicon NPs with a concentration of 0.01 mol/mL. (c,d) Interference spectra for other two probes with similar morphological parameters and coating layers.

It should be noticed that the preparing process of the proposed microfiber RI sensor is controllable either for the fabrication of microfiber (the parameters of fiber splicer and fiber tapering machine will be fixed), or for the UV-glue coating process (the mass of NPs and volume of UV-glue can be determined). The spectra response of two more probes (with a diameter of ~10 μm, length of ~6 mm and similar silicon NPs concentration of 0.01 mol/mL) are included in Figure 5c,d. Their spectral waveforms show little difference, where the differentiated positions of the peaks and dips are affected by the scale difference (diameter and length) of the microfiber, which is less than one quarter of the optical wavelength. The wave spectra and corresponding response to refractive index change are similar to each other. More than five rounds of experiments to prepare microfiber sensors of the same size and similar coating film, the RI sensitivity difference was limited to less than 100 nm/RIU. For the highest concentration of 0.5 mol/L, the corresponding sensitivity was 1321.8 nm/RIU. The decreasing sensitivity may have contributed to the evanescent field scattering by the high dose of silicon NPs. However, in general, the introduction of the silicon NPs significantly improved the sensitivity. Each sensor probe will be calibrated separately before it is used. During the detection process for refractive index, the specific value of refractive index is calculated by the relative wavelength shift of special peaks or dips. Therefore, the little differences in either spectrum or sensitivity for different probes will not exert a significant effect on their refractive index sensing performance.

Compared with the previous experimental data, it can be concluded that the refractive index sensitivity of a clean microfiber without coating layer is 1093.3 nm/RIU. The comparison of the sensing performance of the microfibers elaborated by silver and silicon NPs with different concentrations is shown in Table 1.

The spectra of microfiber elaborated by silver and silicon NPs with the same concentration of 0.5 mol/L can be seen in Figure 6. Although the microfiber is destroyed during the measurement process, and the silicon NP modified microfiber had only six spectra corresponding to different refractive indexes, the monotonic and significant refractive index response curve can be obtained, with corresponding sensitivity of 1321.8 nm/RIU.

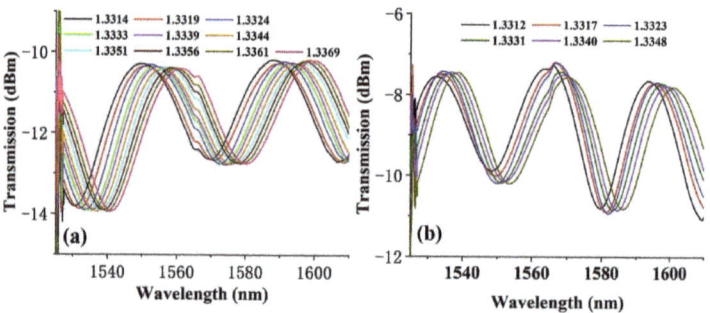

Figure 6. Interference spectra as a function of refractive index of the microfiber elaborated with (a) silver and (b) silicon NPs with same concentration of 0.5 mol/L.

It was found that the microfiber elaborated by silver NPs has the highest sensitivity of 1842.3 nm/RIU; for silicon NPs, the highest sensitivity is up to 1769.7 nm/RIU. It is revealed that, after being elaborated by the NP layer, the refractive index sensitivity of microfiber was significantly improved. NPs have a significant effect on optimizing the sensing performance. Furthermore, the dielectric NPs have a similar LSPR effect, similar to that of metal NPs. The refractive index sensing curve has also been verified with a high linearity coefficient R^2. The sensitivity of the proposed microfiber sensors has been compared with those of other metal NPs-based fiber sensors in Table 2.

Table 2. Comparison of sensitivity for reported works and proposed sensor.

Sensitive Materials	Sensitivity	Refs.
Bi-layered Au NPs	49.63 a.u./RIU	[1]
Ag NPs based works	349.1–8600 nm/RIU	[2]
Au NPs based works	900–5140 nm/RIU	[2]
Au film (40 nm)	~2459–20,863 nm/RIU #	[4]
Au NPs + ZnO NPs	~6 nm/μm	[5]
Au film + TiO$_2$ layer	30,000 nm/RIU #	[7]
Ag NPs	1842.3 nm/RIU	This work
Si NPs	1769.7 nm/RIU	

Simulation results.

The sensitivity can be theoretically increased to 10^4 levels by the compositing of other function materials with the metal NPs. The experimentally demonstrated sensitivity for the Ag NPs and Au NPs ranges from 349.1–8600 nm/RIU to 900–5140 nm/RIU, respectively, referring to the earlier review (Ref. [2]). The sensitivity in this work is comparable with that reported in the references. In this work, the silicon NPs were elaborated on the microfiber to experimentally demonstrate its possible LSPR-like effect, similar to that of metal NPs.

This sensitivity can be further improved by reducing the diameter of the microfiber or introducing other function materials.

3.2. Temperature Sensing Properties Analysis

A high-sensitivity temperature probe was proposed by encapsulating the silicon-NPs-elaborated-microfiber sensor with high refractive index sensitivity in one cut of silica capillary filled with glycerin. Both ends of the capillary were sealed by photosensitive glue. The refractive index of glycerin is inversely proportional to the temperature, which supplies the refractive index changing environment. Although the refractive index of glycerol is greater than that of silica micro fiber, a certain amount of optical signal is still limited near the interface of the low refractive index UV coating layer and silica microfiber in the form of evanescent field. The spectra as a function of temperature curve are shown in Figure 7 during 28 °C–43 °C.

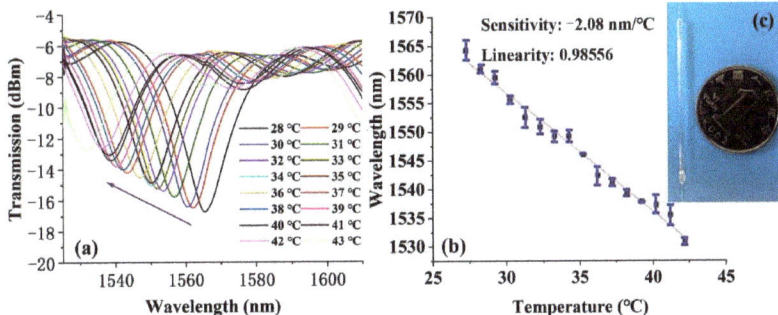

Figure 7. (a) Interference spectra as a function of environment temperature and (b) temperature sensing characteristic curve for microfiber taper elaborated by silicon NPs during 28 °C–43 °C. (c) Picture of microfiber packaged in glycerin by a capillary.

The photo of the microfiber encapsulated in the silica capillary is illustrated in Figure 7c. Its actual size can be determined by comparing with the 1 Yuan RMB coin. In the temperature range of 28 °C–43 °C, the wavelength blue-shifted by 2.08 nm for every 1 °C of temperature decrease, resulting in a temperature sensitivity of 2.08 nm/°C. High sensitivity also means a narrow working range of between 1520 nm–1610 nm. Although the sensitivity of the proposed temperature probe is verified in a limited range (28 °C–43 °C), normally it may work in a larger temperature range by demodulating the phase change of the spectral wavelength. For this temperature probe, the temperature sensing performance is mainly limited by the thermal stability of glycerol. Relevant studies show that the water in glycerol aqueous solution will evaporate rapidly when the temperature is higher than 120 °C [41], and the oxidation temperature of glycerol is about 150 °C. Therefore, the probe may work normally at a temperature less than 100 °C. Therefore, it will be suitable for high-precision monitoring of temperature fluctuations in special areas desiring a high sensitivity, rather than a wide working range.

4. Conclusions

In this paper, silver NPs and silicon NPs were uniformly dispersed in low-refractive-index UV glue and coated on the surface of a microfiber to prepare refractive index probes with high sensitivity. The corresponding refractive index sensitivities of 1842.3 nm/RIU and 1769.7 nm/RIU have been experimentally obtained, respectively, for silver NPs' and silicon NPs' elaborated microfiber with a diameter of ~10 μm and length of ~6 mm. The silicon NPs' modified microfiber with high refractive index sensitivity has also been used for determining environmental temperature with a sensitivity of 2.08 nm/°C. The sensing probe is easily obtained from ordinary single-mode optical fiber, and can be easily connected into the conventional fiber network for exploring high-performance sensors and photonic devices.

Author Contributions: H.D.: Data curation and Formal analysis; Y.Z. and Y.Q.: Investigation, Methodology, Writing-original draft; J.L.: Supervision, Writing-review & editing, Funding acquisition and Project administration. All authors have read and agreed to the published version of the manuscript.

Funding: This work was supported by Hebei Natural Science Foundation (F2020501040).

Institutional Review Board Statement: Not applicable.

Informed Consent Statement: Not applicable.

Data Availability Statement: Not applicable.

Conflicts of Interest: The authors declare no conflict of interest.

References

1. Kim, H.-M.; Park, J.-H.; Lee, S.-K. Fabrication and Measurement of Fiber Optic Localized Surface Plasmon Resonance Sensor Based on Hybrid Structure of Dielectric Thin Film and Bi-Layered Gold Nanoparticles. *IEEE Trans. Instrum. Meas.* **2021**, *70*, 1–8. [CrossRef]
2. Li, J.; Wang, H.; Li, Z.; Su, Z.; Zhu, Y. Preparation and Application of Metal Nanoparticals Elaborated Fiber Sensors. *Sensors* **2020**, *20*, 5155. [CrossRef]
3. Chen, H.; Liu, S.; Zi, J.; Lin, Z. Fano Resonance-Induced Negative Optical Scattering Force on Plasmonic Nanoparticles. *ACS Nano* **2015**, *9*, 1926–1935. [CrossRef]
4. Zhao, S.; Shen, Y.; Yan, X.; Zhou, P.; Yin, Y.; Lu, R.; Han, C.; Cui, B.; Wei, D. Complex-surfactant-assisted hydrothermal synthesis of one-dimensional ZnO nanorods for high-performance ethanol gas sensor. *Sens. Actuators B Chem.* **2019**, *286*, 501–511. [CrossRef]
5. Li, J.; Chen, G.; Meng, F. A Fiber-Optic Formic Acid Gas Sensor Based on Molybdenum Disulfide Nanosheets and Chitosan Works at Room Temperature. *Opt. Laser Technol.* **2022**, *150*, 107975. [CrossRef]
6. Chen, G.; Li, J.; Meng, F. Formic Acid Gas Sensor Based on Coreless Optical Fiber Coated by Molybdenum Disulfide Nanosheet. *J. Alloy. Compd.* **2022**, *896*, 163063. [CrossRef]
7. Gangwar, R.K.; Amorim, V.A.; Marques, P.V.S. High Performance Titanium Oxide Coated D-Shaped Optical Fiber Plasmonic Sensor. *IEEE Sens. J.* **2019**, *19*, 9244–9248. [CrossRef]
8. Bi, W.; Wu, Y.; Chen, C.; Zhou, D.; Song, Z.; Li, D.; Chen, G.; Dai, Q.; Zhu, Y.; Song, H. Dye Sensitization and Local Surface Plasmon Resonance-Enhanced Upconversion Luminescence for Efficient Perovskite Solar Cells. *ACS Appl. Mater. Interfaces* **2020**, *12*, 24737–24746. [CrossRef]
9. Mola, G.T.; Mthethwa, M.C.; Hamed, M.S.; Adedeji, M.A.; Mbuyise, X.G.; Kumar, A.; Sharma, G.; Zang, Y. Local surface plasmon resonance assisted energy harvesting in thin film organic solar cells. *J. Alloy. Compd.* **2021**, *856*, 158172. [CrossRef]
10. Hu, J.; Zhao, J.; Zhu, H.; Chen, Q.; Hu, X.; Koh, K.; Chen, H. AuNPs network structures as a plasmonic matrix for ultrasensitive immunoassay based on surface plasmon resonance spectroscopy. *Sens. Actuators B Chem.* **2021**, *340*, 129948. [CrossRef]
11. Wang, Y.; Zhu, G.; Li, M.; Singh, R.; Marques, C.; Min, R.; Kaushik, B.K.; Zhang, B.Y.; Jha, R.; Kumar, S. Water Pollutants p-Cresol Detection Based on Au-ZnO Nanoparticles Modified Tapered Optical Fiber. *IEEE Trans Nanobioscience* **2021**, *20*, 377–384. [CrossRef]
12. Soares, M.S.; Vidal, M.; Santos, N.F.; Costa, F.M.; Marques, C.; Pereira, S.O.; Leitão, C. Immunosensing Based on Optical Fiber Technology: Recent Advances. *Biosensors* **2021**, *11*, 305. [CrossRef]
13. Meng, F.; Zheng, H.; Chang, Y.; Zhao, Y.; Li, M.; Wang, C.; Sun, Y.; Liu, J. One-Step Synthesis of Au/SnO2/RGO Nanocomposites and Their VOC Sensing Properties. *IEEE Trans. Nanotechnol.* **2018**, *17*, 212–219. [CrossRef]
14. Zhao, S.; Shen, Y.; Zhou, P.; Zhong, X.; Han, C.; Zhao, Q.; Wei, D. Design of Au@WO3 core−shell structured nanospheres for ppb-level NO2 sensing. *Sens. Actuators B Chem.* **2019**, *282*, 917–926. [CrossRef]
15. Turlier, J.; Fourmont, J.; Bidault, X.; Blanc, W.; Chaussedent, S. In situ formation of rare-earth-doped nanoparticles in a silica matrix from Molecular Dynamics simulations. *Ceram. Int.* **2020**, *46*, 26264–26272. [CrossRef]
16. Sypabekova, M.; Aitkulov, A.; Blanc, W.; Tosi, D. Reflector-less nanoparticles doped optical fiber biosensor for the detection of proteins: Case thrombin. *Biosens. Bioelectron.* **2020**, *165*, 112365. [CrossRef]
17. Sharma, A.K.; Pandey, A.K.; Kaur, B. A Review of advancements (2007–2017) in plasmonics-based optical fiber sensors. *Opt. Fiber Technol.* **2018**, *43*, 20–34. [CrossRef]
18. Baffou, G.; Quidant, R. Thermo-plasmonics: Using metallic nanostructures as nano-sources of heat. *Laser Photon-Rev.* **2013**, *7*, 171–187. [CrossRef]
19. Lu, Y.R.; Nikrityuk, P.A. Steam methane reforming driven by the Joule heating. *Chem. Eng. Sci.* **2022**, *251*, 117446. [CrossRef]
20. Lal, S.; Clare, S.E.; Halas, N.J. Nanoshell-Enabled Photothermal Cancer Therapy: Impending Clinical Impact. *Acc. Chem. Res.* **2008**, *41*, 1842–1851. [CrossRef]

21. Boyer, D.; Tamarat, P.; Maali, A.; Lounis, B.; Orrit, M. Photothermal Imaging of Nanometer-Sized Metal Particles Among Scatterers. *Science* **2002**, *297*, 1160–1163. [CrossRef] [PubMed]
22. Faruk, O.; Ahmed, A.; Jalil, M.A.; Islam, M.T.; Shamim, A.M.; Adak, B.; Hossain, M.; Mukhopadhyay, S. Functional textiles and composite based wearable thermal devices for Joule heating: Progress and perspectives. *Appl. Mater. Today* **2021**, *23*, 101025. [CrossRef]
23. Deng, B.; Luong, D.X.; Wang, Z.; Kittrell, C.; McHugh, E.A.; Tour, J.M. Urban mining by flash Joule heating. *Nat. Commun.* **2021**, *12*, 1–8. [CrossRef] [PubMed]
24. Xuan, X. Review of nonlinear electrokinetic flows in insulator-based dielectrophoresis: From induced charge to Joule heating effects. *Electrophoresis* **2022**, *43*, 167–189. [CrossRef] [PubMed]
25. Kuznetsov, A.I.; Miroshnichenko, A.E.; Brongersma, M.L.; Kivshar, Y.S.; Luk'Yanchuk, B. Optically resonant dielectric nanostructures. *Science* **2016**, *354*, aag2472. [CrossRef]
26. Zograf, G.P.; Petrov, M.I.; Makarov, S.V.; Kivshar, Y.S. All-dielectric thermonanophotonics. *Adv. Opt. Photon-* **2021**, *13*, 643–702. [CrossRef]
27. Yan, J.; Liu, X.; Ma, C.; Huang, Y.; Yang, G. All-dielectric materials and related nanophotonic applications. *Mater. Sci. Eng. R: Rep.* **2020**, *141*, 100563. [CrossRef]
28. Barreda, Á.I.; Saleh, H.; Litman, A.; González, F.; Geffrin, J.M.; Moreno, F. On the scattering directionality of a dielectric particle dimer of high refractive index. *Sci. Rep.* **2018**, *8*, 7976. [CrossRef]
29. Terekhov, P.D.; Baryshnikova, K.V.; Greenberg, Y.; Fu, Y.H.; Evlyukhin, A.B.; Shalin, A.S.; Karabchevsky, A. Enhanced absorption in all-dielectric metasurfaces due to magnetic dipole excitation. *Sci. Rep.* **2019**, *9*, 1–9. [CrossRef]
30. Frizyuk, K.; Volkovskaya, I.; Smirnova, D.; Poddubny, A.; Petrov, M. Second-harmonic generation in Mie-resonant dielectric nanoparticles made of noncentrosymmetric materials. *Phys. Rev. B* **2019**, *99*, 075425. [CrossRef]
31. Sain, B.; Meier, C.; Zentgraf, T. Nonlinear optics in all-dielectric nanoantennas and metasurfaces: A review. *Adv. Photon-* **2019**, *1*, 024002. [CrossRef]
32. Genevet, P.; Capasso, F.; Aieta, F.; Khorasaninejad, M.; Devlin, R. Recent advances in planar optics: From plasmonic to dielectric metasurfaces. *Optica* **2017**, *4*, 139–152. [CrossRef]
33. Castellanos, G.W.; Murai, S.; Raziman, T.V.; Wang, S.; Ramezani, M.; Curto, A.G.; Gómez Rivas, J. Exciton-Polaritons with Magnetic and Electric Character in All-Dielectric Metasurfaces. *ACS Photonics* **2020**, *7*, 1226–1234. [CrossRef]
34. Li, J.; Yan, H.; Dang, H.; Meng, F. Structure design and application of hollow core microstructured optical fiber gas sensor: A review. *Opt. Laser Technol.* **2021**, *135*, 106658. [CrossRef]
35. Li, Y.; Xin, H.; Zhang, Y.; Li, B. Optical Fiber Technologies for Nanomanipulation and Biodetection: A Review. *J. Light. Technol.* **2021**, *39*, 251–262. [CrossRef]
36. Zhou, J.; Wang, Y.; Zhang, L.; Li, X. Plasmonic biosensing based on non-noble-metal materials. *Chin. Chem. Lett.* **2018**, *29*, 54–60. [CrossRef]
37. Butt, M.; Khonina, S.; Kazanskiy, N. Plasmonics: A Necessity in the Field of Sensing-A Review (Invited). *Fiber Integr. Opt.* **2021**, *40*, 14–47. [CrossRef]
38. Guzmán-Sepúlveda, J.R.; Guzmán-Cabrera, R.; Castillo-Guzmán, A.A. Optical Sensing Using Fiber-Optic Multimode Interference Devices: A Review of Nonconventional Sensing Schemes. *Sensors* **2021**, *21*, 1862. [CrossRef]
39. Soldano, L.; Pennings, E. Optical multi-mode interference devices based on self-imaging: Principles and applications. *J. Light. Technol.* **1995**, *13*, 615–627. [CrossRef]
40. Okamoto, K. *Fundamentals of Optical Waveguides*; Academic Press: London, UK, 2006.
41. Castelló, M.; Dweck, J.; Aranda, D.A.G. Thermal stability and water content determination of glycerol by thermogravimetry. *J. Therm. Anal.* **2009**, *97*, 627–630. [CrossRef]

MDPI AG
Grosspeteranlage 5
4052 Basel
Switzerland
Tel.: +41 61 683 77 34

Polymers Editorial Office
E-mail: polymers@mdpi.com
www.mdpi.com/journal/polymers

Disclaimer/Publisher's Note: The title and front matter of this reprint are at the discretion of the . The publisher is not responsible for their content or any associated concerns. The statements, opinions and data contained in all individual articles are solely those of the individual Editor and contributors and not of MDPI. MDPI disclaims responsibility for any injury to people or property resulting from any ideas, methods, instructions or products referred to in the content.

www.ingramcontent.com/pod-product-compliance
Lightning Source LLC
LaVergne TN
LVHW070715100526
838202LV00013B/1100